Industrial Engineering: Design, Simulation and Optimization

Industrial Engineering: Design, Simulation and Optimization

Editor: Justin Riggs

NY RESEARCH
P R E S S

New York

Published by NY Research Press
118-35 Queens Blvd., Suite 400,
Forest Hills, NY 11375, USA
www.nyresearchpress.com

Industrial Engineering: Design, Simulation and Optimization
Edited by Justin Riggs

International Standard Book Number: 978-1-63238-654-0 (Hardback)

Cataloging-in-Publication Data

Industrial engineering : design, simulation and optimization / edited by Justin Riggs.
    p. cm.
Includes bibliographical references and index.
ISBN 978-1-63238-654-0
1. Industrial engineering. 2. Engineering design. 3. Simulation methods.
4. Mathematical optimization. I. Riggs, Justin.
T56.24 .I53 2019
658--dc23

Contents

Permissions

List of Contributors

Index

Preface

Industrial engineering is a branch of engineering that is aimed at optimizing a complex set of processes, systems and organizations by eliminating waste in terms of human work hours, time and energy. It strives to improve quality and productivity. Some of the areas of application of the principles of this field are in flow process charting, process mapping, strategizing for operational logistics, developing algorithms, streamlining operations, etc. This field has evolved to include motion time system, computer simulation, mathematical optimization, etc. This book contains some path-breaking studies in the field of industrial engineering. It provides significant information of this discipline to help develop a good understanding of its design, simulation and optimization. With valuable contributions from experts working in this area of study, this book targets students and professionals alike.

The information contained in this book is the result of intensive hard work done by researchers in this field. All due efforts have been made to make this book serve as a complete guiding source for students and researchers. The topics in this book have been comprehensively explained to help readers understand the growing trends in the field.

I would like to thank the entire group of writers who made sincere efforts in this book and my family who supported me in my efforts of working on this book. I take this opportunity to thank all those who have been a guiding force throughout my life.

Editor

Optimization of process parameters through GRA, TOPSIS and RSA models

Suresh Nipanikar[a]*, Vikas Sargade[b] and Ramesh Guttedar[c]

[a]*Research Scholar, Department of Mechanical Engineering, Dr. Babasaheb Ambedkar Technological University, Lonere-402103, Maharashtra, India*
[b]*Professor, Department of Mechanical Engineering, Dr. Babasaheb Ambedkar Technological University, Lonere-402103, Maharashtra, India*
[c]*PG Student, Department of Mechanical Engineering, Dr. Babasaheb Ambedkar Technological University, Lonere-402103, Maharashtra, India*

CHRONICLE	ABSTRACT
	This article investigates the effect of cutting parameters on the surface roughness and flank wear during machining of titanium alloy Ti-6Al-4V ELI(Extra Low Interstitial) in minimum quantity lubrication environment by using PVD TiAlN insert. Full factorial design of experiment was used for the machining 2 factors 3 levels and 2 factors 2 levels. Turning parameters studied were cutting speed (50, 65, 80 m/min), feed (0.08, 0.15, 0.2 mm/rev) and depth of cut 0.5 mm constant. The results show that 44.61 % contribution of feed and 43.57 % contribution of cutting speed on surface roughness also 53.16 % contribution of cutting tool and 26.47 % contribution of cutting speed on tool flank wear. Grey relational analysis and TOPSIS method suggest the optimum combinations of machining parameters as cutting speed: 50 m/min, feed: 0.8 mm/rev., cutting tool: PVD TiAlN, cutting fluid: Palm oil.
Keywords: *Ti6Al4V ELI* *Surface roughness* *Flank wear* *PVD TiAlN* *MQL*	

Nomenclature

Vc	Cutting speed, m/min	ANOVA	Analysis of variance
f	(Feed (mm/rev	MQL	Minimum quantity lubrication
ap	depth of cut, mm	SS	Sequential sum of square
Ct	Cutting tool insert	MS	Adjusted mean squares
Ra	Surface roughness (µm)	VB	Flank wear, mm.

1. Introduction

Ti-6Al-4V ELI alloy (Extra Low Interstitial) is a higher purity grade of Ti-6Al-4V alloy. This grade has low oxygen, iron and carbon. It has biomedical applications such as joint replacements, bone fixation

* Corresponding author
E-mail: sureshnipanikar15@gmail.com (S. Nipanikar)

devices, surgical clips, cryogenic vessels because of its good fatigue strength and low modulus and is the preferred grade for marine and aerospace applications. Surface roughness affects the performance of mechanical components and their production costs because it influences on different factors, such as geometrical tolerances, ease of handling, friction, electrical and thermal conductivity, etc. Workpiece and tool insert material properties and machining conditions influence on surface roughness. The functions of cutting fluids are cooling, lubrication and assistance in chip flow. Therefore, the effect of fluid abandonment is highly mechanical and thermal effect on the cutting tool insert and the machined surface which increases tool wear and surface roughness.

Escamilla et al. (2013) observed that despite the extended use of titanium alloy in numerous fields, it posses assorted machining problems and considered a difficult to cut material. Khanna and Davim (2015) found that majority of heat developed gets transmitted to the cutting tool in the machining of titanium alloys due to its low thermal conductivity, hence making a prominent heat concentration on the leading edge of tool, which prompts to hasty tool failure. Wu and Guo (2014) explored that high cutting temperature accelerates the tool wear, which may result in short tool life. It likewise tends to weld on cutting tool while machining, which prompts to crack and untimely breakdown of tools. Islam et al. (2013) explored that the enhancement of machinability of titanium along with its alloys depends on a vast degree on the viability of the efficacy of cooling and lubrication method. Sharma et al. (2015) found that heat developed amid machining is not uprooted and is one of the main causes of the reduction in tool life and surface finish. MQL shows important results in reducing the machining cost, cutting fluid quantity as well as surface roughness produced after machining. Supreme task of MQL is the carriage of chips out of the contact zones to avoid contact between hot chips and the produced surface. The use of MQL during turning was analyzed by many researchers. Some good results were obtained with this technique. Liu et al. (2013) observed that the wear execution of different coated tool inserts in high speed dry and MQL turning of Ti6Al4V titanium alloy, MQL was found to be superior and feed rate was the principle variable influencing on cutting forces and surface roughness. Revankar et al. (2014) observed that the surface roughness is minimum in MQL environment as compared to dry and wet condition.

Sargade et al. (2016) observed that the feed was the most dominant factor for surface roughness having 97.34% contribution during turning the Ti6Al4V ELI by using PVD TiAlN insert in dry environment. Shetty et al. (2014) reported that the impact of lubrication was highest physically as well as statistical influence on surface roughness of about 95.1% when turning Ti6Al4V by implementing PCBN tool under dry and near dry condition. Ramana et al. (2012) found that machining performance under MQL environment shows better results as compared to dry and flooded conditions in reduction of surface roughness. Ali et al. (2011) observed that MQL provides the proper lubrication that minimizes the friction resulting in retention of tool sharpness for a longer period. Retention of cutting edge sharpness due to reduction of cutting zone temperature seemed to be the main reason behind reduction of cutting forces by the MQL application of MQL jet in machining medium carbon steel. Dimensional accuracy and surface finish has been substantially improved mainly due to reduction of wear and damage at the tool tip due to application of MQL. Attanasio et al. (2006) found that lubricating the flank surface of a tip by the MQL technique reduces the tool wear and increases the tool life. Khan et al. (2009) observed that the significant contribution of MQL jet in reducing the flank wear and that was remarkable improvement in tool life also MQL reduces deep grooving which is very detrimental and may cause premature and catastrophic failure of the cutting tools. Surface finish was also improved due to reduction in wear and damage at the tool tip by the application of MQL. Dhar et al. (2007) reported that the MQL machining could be better than that of dry because MQL reduces machining temperature and improves the chip tool interaction and maintains sharp cutting edge. Xu et al. (2012) observed that machining performance and tool life were improved due to machining of Ti6Al4V in MQL environment.

It is evident from the literature review and to the best perception of the author, the application of MQL in machining provides very rewarding results. It was also found that no systematic study has been conducted to analysis the machining of Ti6Al4V ELI.

Performance of Titanium alloy Ti6Al4V ELI is investigated by utilizing three optimization methods i.e. Grey relational analysis, TOPSIS and Response surface analysis approaches. Consequently, the key goal of this study is the parameter optimization of the turning Titanium alloy Ti6Al4V ELI in MQL environment by using PVD TiAlN coated insert and uncoated insert for surface roughness and tool wear. Experimental observations are analyzed by using Grey relational analysis, TOPSIS and Response surface analysis. Henceforth, the use of the above mentioned optimization methods for machining of Ti6Al4V ELI in the present work is quite innovative.

Therefore, the main purpose of this study is to explore the effects of machining conditions on surface roughness and tool flank wear in turning of Ti6Al4V ELI in minimum quantity lubrication environment and compare the performance with palm oil and coconut oil at various machining parameters with coated and uncoated inserts.

2. Design of experiment

There are various ways in which design of experiments may be designed and it always depends on the number of factors and levels of each factors.

Full factorial design of experiment: A full factorial design of experiment contains of two or more than two factors, each with distinct probable values or levels and experiments are performed for all probable combinations of these levels across all such factors. This experiment allows us to study the outcome of each factor on the output variable, as well as the effects of interactions between factors on the response variable. Full factorial DOE was designed in the presented work by considering two machining parameters such as cutting speed and feed with three levels and two parameters such as cutting tool and cutting fluid with two levels of operations for every factor and the response variables are surface roughness and tool wear.

2.1 Grey relational analysis

The objective of grey relational analysis is to convert the multi objective optimization problem into a single objective problem. This methodology gives the rank of the experiment based on grey relational grade. The highest grey relational grade identifies the optimum cutting condition combination.

Step1: Calculation of Signal to noise ratio for surface roughness and flank wear considering "smaller is better" type of signal to noise ratio.

$$\text{Signal to noise ratio} = -10log \frac{1}{n} \left(\sum_{i=1}^{n} yi^2 \right) \tag{1}$$

where, n is the number of observations and y is the observed data.

Step 2: Distribute the data evenly and convert the data into acceptable range for further analysis. For calculating the normalized value of kth performance characteristic of ith experiment is defined as follows,

$$xi(k) = \frac{max \, x_i^0(k) - x_i^0(k)}{max \, x_i^0(k) - min \, x_i^0(k)} \tag{2}$$

where, $xi(k)$ is the normalized value.

Step 3: The aim of the grey relational coefficient is to express the relationship between the best and actual normalized experimental results. The grey relational coefficient is calculated as,

$$\xi_i(k) = \frac{\Delta_{min} + \varsigma\Delta_{max}}{\Delta_{0i}(k) + \varsigma\Delta_{max}},$$
(3)

where, $\Delta_{0i}(k) = \| x_0(k) - x_i(k) \|$ is the difference of absolute value between $x_0(k)$ and $x_i(k)$; ς = distinguishing coefficient, Δ_{min} is the smallest value of Δ_{0i} and Δ_{max} is the largest value of Δ_{0i}

Step 4: The grey relational grade is determined by averaging the grey relational coefficients corresponding to each performance characteristics.

$$\gamma_i = \frac{1}{n}\sum_{k=1}^{n}\mathcal{E}i(k),$$
(4)

where, γ_i is the grey relational grade for the ith experiment and k is the number of performance characteristics.

Step 5: Determination of the optimal set is the final step. Maximum value of grey relational grade indicates the optimum set.

2.2 Techniques for order preferences by similarity to ideal solution (TOPSIS)

According to Wang et al. (2016) TOPSIS is one of the well-known classical multiple criteria decision making methods, which was originally developed by Hwang and Yoon in 1981, with further development by Chen and Hwang in 1992. The TOPSIS method introduces two reference points; a positive ideal solution and negative ideal solution. The positive ideal solution is the one that maximizes the profit criteria and minimizes the cost criteria, whereas the negative ideal solution maximizes the cost criteria and minimizes the profit criteria. TOPSIS determines the best alternative by minimizing the distance to the ideal solution and by maximizing the distance to the negative ideal solution. TOPSIS method has been applied for converting the multi response into single response. Following steps followed for the TOPSIS in the present article are given below.

Step 1: By using the following equation normalized the decision matrix

$$r_{ij} = \frac{aij}{\sqrt{\sum_{i=1}^{m} a_{ij}^2}}$$
(5)

where, i=1 ….. m and j= 1 ….n

a_{ij} represents the actual value of the ith value of jth experimental run and γ_{ij} represents the corresponding normalized value.

Step 2: Weight for each output is calculated

Step 3: The weighted normalized decision matrix is calculated by multiplying the normalized decision matrix by its associated weights.

$$V_{ij} = Wi \; x \; r_{ij}$$
(6)

where, i=1, …, m and j=1, …, n.

Step 4: Positive ideal solution (PIS) and negative ideal solution (NIS) are determined as follows:

$V^+ = (V_1^+, V_2^+, V_3^+, \dots V_n^+)$ maximum values

$V^- = (V_1^-, V_2^-, V_3^-, \dots V_n^-)$ minimum values

Step 5: The separation of each alternative from positive ideal solution and negative ideal solution is calculated as

$$S_i^+ = \sqrt{\sum_{j=1}^{M}(V_{ij} - V_j^+)^2} \tag{7}$$

$$S_i^- = \sqrt{\sum_{j=1}^{M}(V_{ij} - V_j^-)^2} \tag{8}$$

where i=1, 2, ...N

Step 6: The closeness coefficient is calculated as

$$CC_i = \frac{S_i^-}{S_i^+ + S_i^-} \tag{9}$$

2.3 Response surface methodology

It is a collection of mathematical and statistical techniques for empirical modeling. By careful design of experiments, the objective is to optimize a response variable which is influenced by several independent variables (input variables). Generally a second order model is developed in response surface methodology. The initial step in RSM is to determine an appropriate approximation for the functional relationship between the response factor *y* and a set of independent variables as follows,

$$y = \beta_0 + \sum_{i=1}^{k} \beta_i x_i + \sum_{i=1}^{k} \beta_{ii} x_i^2 + \sum_i \sum_j \beta_{iix_j} + \varepsilon, \tag{10}$$

where, *y* is the estimated response, β_0 is constant, β_i, β_{ii} and β_{ij} represent the coefficient of linear, quadratic and cross product terms respectively and *n* is the number of process parameters. The β coefficients, used in the above model, can be calculated by means of least square method. The quadratic model is normally used when the response function is unknown or non-linear.

3. Experimental procedures

3.1 Workpiece material

The workpiece material used during the turning process was in the form of a cylinder bar of alpha-beta titanium alloy Ti-6Al-4V ELI. The composition of the Ti-6Al-4V ELI (in wt. %) are given in Table 1.

Table 1
Chemical composition of Ti6Al4V ELI

Composition	C	Si	Fe	Al	N	V	S	O	H	Ti
Wt %	0.08	0.03	0.22	6.1	0.006	3.8	0.003	0.12	0.003	Balance

The workpiece has a microstructure, which consisted of elongated alpha phase surrounded by fine, dark etching of beta matrix. This material offers high strength and depth hardenability (32 HRC). Fig. 1 shows the microstructure of Ti6Al4V ELI. The microstructure shows acicular alpha and aged beta. Alpha platelets at the prior beta grain boundaries. HF + HNO_3+H_2O etchant was used. Fig. 2 shows the photographic view of Kistler 3-D dynamometer.

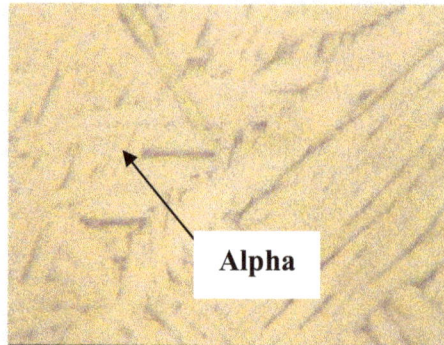

Fig. 1. Microstructure of Ti6Al4V ELI

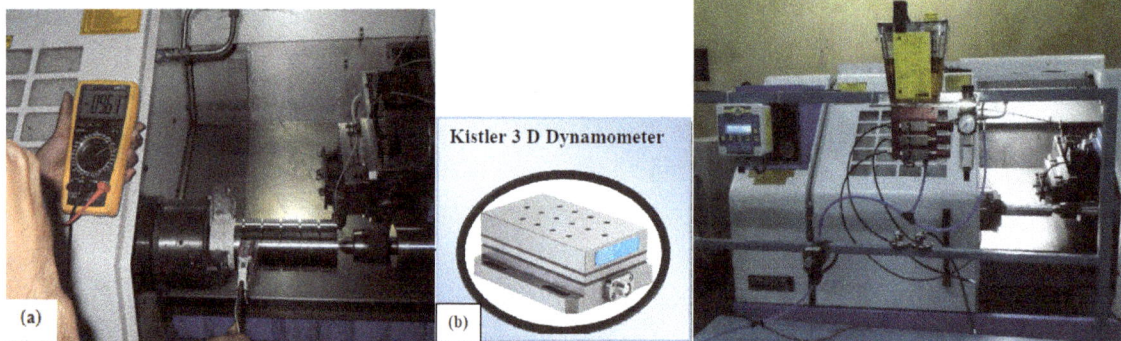

Fig. 2. (a) Photographic view of (a) experimental setup (b) Kistler 3-D Dynamometer unit (c) MQL setup

3.2   Cutting tool material

A cutting tool insert with ISO designation CNMG 120408-QM-1105 PVD TiAlN was used for the turning experiments. Fig. 3 also shows the photographic view of TiAlN insert, surface morphology and EDAX profile of PVD TiAlN insert. The Vickers hardness is 3100 HV.

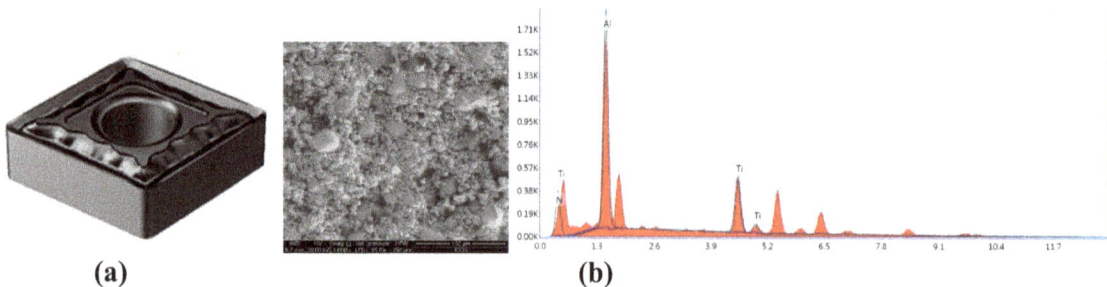

(a)   **(b)**

Fig. 3. Photographic view of (a) PVD TiAlN insert (b) Surface morphology and EDAX profile of PVD TiAlN coating

3.3 Machining tests

All the machining experiments were conducted on ACE CNC LATHE JOBBER XL, with FANUC Oi Mate- TC as a controller. During the experiments, the combinations of the machining process parameter values were designed by using L36 mixed orthogonal array design of experiment. The cutting speeds were set at 50, 65 and 80 m/min, while the feed were 0.08, 0.15 and 0.2 mm/rev. The depth of cut was 0.5 mm is constant during the machining process. The machining experiments were carried out in MQL environment. The cutting conditions are shown in Table 2.

Table 2
Cutting condition for experimental works

Machine tool	ACE CNC LATHE JOBBER XL, FANUC Oi Mate- TC as a controller
Workpiece Material	Titanium alloy, Ti-6Al-4V ELI
Cutting tool (insert)	Cutting insert : Uncoated Carbide insert, ISO CNMG 120408-QM-1105 Sandvik make PVD TiAlN insert,
Tool holder	PCLNL 2525 M12
Machining parameters	Cutting speed (*Vc*): 50, 65 and 80 m/min Feed (*f*): 0.08, 0.15 and 0.2 mm/rev Depth of cut (*d*): 0.5mm
Machining Environment	MQL
Cutting Fluid	Coconut oil (Viscosity: 80 cP), Palm oil (Viscosity:130 cP)
Cutting fluid supply	For MQL cooling: air:6 bar, flow rate 54 ml/hr (through external nozzle)

Turning parameters and their levels are shown in Table 3

Table 3
Turning Parameters and their levels

Machining Parameters	Level 1	Level 2	Level 3
Cutting speed, Vc (m/min)	50	65	80
Feed, f (mm/rev)	0.08	0.15	0.2
Cutting tool, Ct	Coated (PVD TiAlN)	Uncoated	--
Cutting fluid	Palm oil	Coconut oil	--

Experimental Design layout is shown in Table 4.

Table 4
Experimental Design layout

Expt. No.	Cutting Environment	Cutting Tool Insert	Cutting speed (m/min)	Feed (mm/rev)	Expt. No.	Cutting Environment	Cutting Tool Insert	Cutting speed (m/min)	Feed (mm/rev)
1	MQL (Palm Oil)	Coated	50	0.08	19	MQL (Coconut Oil)	Coated	50	0.08
2			50	0.15	20			50	0.15
3			50	0.2	21			50	0.2
4			65	0.08	22			65	0.08
5			65	0.15	23			65	0.15
6			65	0.2	24			65	0.2
7			80	0.08	25			80	0.08
8			80	0.15	26			80	0.15
9			80	0.2	27			80	0.2
10	MQL (Palm Oil)	Uncoated	50	0.08	28	MQL (Coconut Oil)	Uncoated	50	0.08
11			50	0.15	29			50	0.15
12			50	0.2	30			50	0.2
13			65	0.08	31			65	0.08
14			65	0.15	32			65	0.15
15			65	0.2	33			65	0.2
16			80	0.08	34			80	0.08
17			80	0.15	35			80	0.15
18			80	0.2	36			80	0.2

4. Results and discussions

4.1 Grey Relational Analysis

Initially, analysis and evaluation of single performance characteristic was performed. Then, multiple performance analysis were conducted by grey relational theory. From single performance analysis point of view the effect of machining parameters like cutting speed, feed rate, cutting tool insert and cutting fluid on the surface roughness and flank wear during turning of Ti6Al4V ELI in MQL environment was analyzed using response graphs which were drawn by using response table with the average values. The response table for surface roughness and tool flank wear is shown in Table 5.

Table 5
Observed response values, S/N ratio of responses, normalized values, grey relational coefficient and grey relational grade

Expt. No.	Responses		S/N Ratio		Normalized data		Grey relational Coefficient		Grey relational grade
	Ra	VB	Ra	VB	Ra	VB	Ra	VB	
1	0.413	20.7	7.68	-26.32	0.97	1.00	0.94	1.00	**0.97**
2	0.827	30.2	1.65	-29.60	0.78	0.89	0.70	0.82	0.76
3	0.99	43.02	0.09	-32.67	0.71	0.73	0.63	0.65	0.64
4	0.343	25.02	9.29	-27.97	1.00	0.95	1.00	0.91	0.95
5	0.79	32.2	2.05	-30.16	0.80	0.86	0.72	0.79	0.75
6	0.98	45.04	0.18	-33.07	0.72	0.71	0.64	0.63	0.64
7	0.383	34.07	8.34	-30.65	0.98	0.84	0.97	0.76	0.86
8	0.77	52.02	2.27	-34.32	0.81	0.63	0.72	0.57	0.65
9	0.97	70.27	0.26	-36.94	0.72	0.41	0.64	0.46	0.55
10	1.017	45.09	-0.15	-33.08	0.70	0.71	0.63	0.63	0.63
11	1.77	53.9	-4.96	-34.63	0.37	0.61	0.44	0.56	0.50
12	2.463	69.97	-7.83	-36.90	0.06	0.41	0.35	0.46	0.40
13	0.857	76.27	1.34	-37.65	0.77	0.34	0.69	0.43	0.56
14	1.867	68.9	-5.42	-36.76	0.32	0.43	0.42	0.47	0.45
15	2.31	72.98	-7.27	-37.26	0.13	0.38	0.36	0.45	0.40
16	0.9	78.09	0.92	-37.85	0.75	0.32	0.67	0.42	0.55
17	1.733	98.27	-4.78	-39.85	0.38	0.08	0.45	0.35	0.40
18	2.593	102	-8.28	-40.17	0.00	0.03	0.33	0.34	0.34
19	0.46	21	6.74	-26.44	0.95	1.00	0.91	0.99	0.95
20	0.84	34.06	1.51	-30.64	0.78	0.84	0.69	0.76	0.73
21	1.07	45.02	-0.59	-33.07	0.68	0.71	0.61	0.63	0.62
22	0.457	27.09	6.80	-28.66	0.95	0.92	0.91	0.87	0.89
23	0.83	34.7	1.62	-30.81	0.78	0.83	0.70	0.75	0.72
24	1.03	47.09	-0.26	-33.46	0.69	0.69	0.62	0.61	0.62
25	0.437	36.97	7.19	-31.36	0.96	0.81	0.92	0.72	0.82
26	0.8	56.9	1.94	-35.10	0.80	0.57	0.71	0.54	0.62
27	0.99	72.09	0.09	-37.16	0.71	0.39	0.63	0.45	0.54
28	0.81	46.51	1.83	-33.35	0.79	0.69	0.71	0.62	0.66
29	1.44	55.9	-3.17	-34.95	0.51	0.58	0.51	0.54	0.53
30	2.42	70.02	-7.68	-36.90	0.08	0.41	0.35	0.46	0.41
31	0.6	76.97	4.44	-37.73	0.89	0.33	0.81	0.43	0.62
32	1.553	69.9	-3.82	-36.89	0.46	0.42	0.48	0.46	0.47
33	2.327	74.97	-7.34	-37.50	0.12	0.36	0.36	0.44	0.40
34	0.73	79.09	2.73	-37.96	0.83	0.31	0.74	0.42	0.58
35	1.557	98.44	-3.85	-39.86	0.46	0.08	0.48	0.35	0.42
36	2.58	104.9	-8.23	-40.42	0.01	0.00	0.33	0.33	0.33

Table 5 shows higher grey relational grade for experiment number 1 i.e. optimum set of machining parameters. Optimal cutting condition is as follows.

- *Vc*: 50 m/min   *f*: 0.08 mm/rev.,   *Ct*: PVD TiAlN   *cutting fluid*: Palm oil.

Mean table for grey relational grade is shown in Table 6. Effect of process parameters on grey relational grade is shown in Fig. 5. In the grey relational analysis, to obtain better performance a greater grey relational grade is required. From Table 6 and Fig. 5, the optimum machining parameter combination is determined as *Vc*: 50 m/min, *f*: 0.08 mm/rev., *Ct*: PVD TiAlN, cutting fluid: Palm oil for simultaneously achieving minimum surface roughness and minimum flank wear.

Fig. 4(a) indicates the grey relational grade of various levels of machining process parameters. It shows that the first level of each machining parameter indicates the highest grey relational grade. Feed was the dominant factor influencing the surface roughness and flank wear, simultaneously. Fig. 4(b) shows the main effect plots for grey relational grade.

Table 6

Mean table for Grey relational grade

Level	Cutting speed	Feed	Cutting fluid	Tool
1	**0.65**	**0.75**	**0.61**	**0.73**
2	0.62	0.58	0.60	0.48
3	0.56	0.49	--	--
Delta	0.09	0.26	0.01	0.25
Rank	3	1	4	2

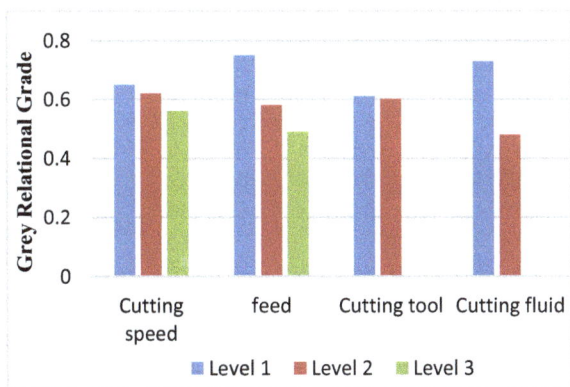

(a)

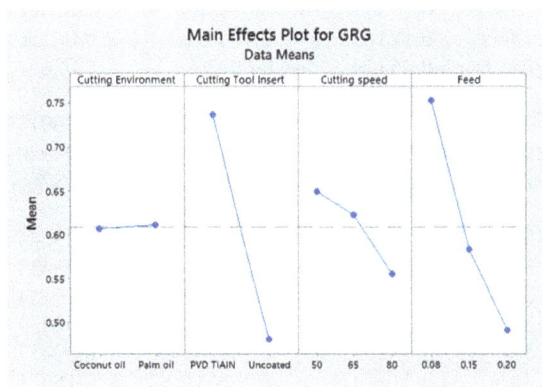

(b)

Fig. 4. (a) Effect of process parameters on grey relational grade and (b) Main effect plot for GRG

4.2 TOPSIS

The two response parameters such as surface roughness Ra and tool flank wear VB are normalized. In this article the same priority is given to both the responses i.e. surface roughness and flank wear weight are taken as 0.5 (i.e. W_{Ra}= 0.5 and W_{VB} = 0.5). With the proper weight criteria the relative normalized weight matrix has been calculated. The weight criteria are multiplied to get the normalized weighted matrix using Eq. (6). The ideal and the negative ideal solutions are calculated from the normalized weighted matrix table. The separation measures of each criterion from the ideal and negative ideal solutions were calculated with Eqs. (7-8). Finally, the relative closeness coefficient (CC_i) value for each combination of factors of turning process is calculated using Eq. (9). It was understood that the experiment number 1 is the best experiment. Table 7 shows the normalized, weighted normalized data, separation measures and closeness coefficient.

Table 7
Normalized, weighted normalized data, Separation measures and Closeness coefficient values

Exp. No.	Normalized data		Weighted normalized data		Separation measures		Closeness coefficient
	Ra	VB	Ra	VB	S^+	S^-	CCi
1	0.05017	0.05558	0.02509	0.02779	0.00425	0.17410	**0.97616**
2	0.10046	0.08108	0.05023	0.04054	0.03205	0.14684	0.82086
3	0.12027	0.11550	0.06013	0.05775	0.04942	0.12799	0.72144
4	0.04167	0.06717	0.02083	0.03359	0.00580	0.17371	0.96769
5	0.09597	0.08645	0.04798	0.04323	0.03123	0.14669	0.82446
6	0.11905	0.12093	0.05953	0.06046	0.05064	0.12671	0.71446
7	0.04653	0.09147	0.02326	0.04574	0.01811	0.16450	0.90082
8	0.09354	0.13967	0.04677	0.06983	0.04940	0.13153	0.72696
9	0.11784	0.18866	0.05892	0.09433	0.07667	0.10899	0.58704
10	0.12355	0.12106	0.06177	0.06053	0.05242	0.12494	0.70444
11	0.21502	0.14471	0.10751	0.07236	0.09746	0.08477	0.46518
12	0.29921	0.18786	0.14960	0.09393	0.14476	0.04755	0.24726
13	0.10411	0.20477	0.05205	0.10239	0.08087	0.11223	0.58121
14	0.22680	0.18499	0.11340	0.09249	0.11294	0.06542	0.36679
15	0.28062	0.19594	0.14031	0.09797	0.13856	0.04617	0.24992
16	0.10933	0.20966	0.05467	0.10483	0.08414	0.10895	0.56423
17	0.21053	0.26384	0.10526	0.13192	0.13406	0.05299	0.28329
18	0.31500	0.27385	0.15750	0.13693	0.17490	0.00389	0.02177
19	0.05588	0.05638	0.02794	0.02819	0.00712	0.17167	0.96019
20	0.10204	0.09145	0.05102	0.04572	0.03511	0.14276	0.80259
21	0.12998	0.12087	0.06499	0.06044	0.05492	0.12255	0.69056
22	0.05552	0.07273	0.02776	0.03637	0.01102	0.16656	0.93792
23	0.10083	0.09316	0.05041	0.04658	0.03505	0.14265	0.80277
24	0.12512	0.12643	0.06256	0.06321	0.05474	0.12262	0.69137
25	0.05309	0.09926	0.02654	0.04963	0.02258	0.15958	0.87607
26	0.09718	0.15277	0.04859	0.07638	0.05596	0.12654	0.69335
27	0.12027	0.19355	0.06013	0.09678	0.07940	0.10687	0.57374
28	0.09840	0.12487	0.04920	0.06244	0.04478	0.13369	0.74910
29	0.17493	0.15008	0.08747	0.07504	0.08169	0.09608	0.54049
30	0.29398	0.18799	0.14699	0.09400	0.14248	0.04799	0.25196
31	0.07289	0.20665	0.03644	0.10333	0.07713	0.12673	0.62164
32	0.18866	0.18767	0.09433	0.09384	0.09881	0.07873	0.44344
33	0.28268	0.20128	0.14134	0.10064	0.14082	0.04331	0.23520
34	0.08868	0.21234	0.04434	0.10617	0.08183	0.11834	0.59120
35	0.18915	0.26430	0.09457	0.13215	0.12778	0.06352	0.33205
36	0.31342	0.28164	0.15671	0.14082	0.17674	0.00079	0.00445

From Table 7 it was understood that the experiment number 1 is the best experiment among the 36 experiment because this experiment shows the maximum closeness coefficient considering both responses and experiment number 36 shows the poor performance because it shows the lowest closeness coefficient among the 36 experiments. From Table 7, the optimum machining parameter combination determined as *Vc*: 50 m/min, *f*: 0.08 mm/rev., *Ct*: PVD TiAlN, cutting fluid: Palm oil for simultaneously achieving minimum surface roughness and minimum flank wear.

4.3 Response Surface Analysis

4.3.1 Surface roughness, Ra (µm)

Surface finish is an important index of machinability as the performance and service life of the machined part are often affected by its surface roughness, nature and extent of residual stresses and presence of surface or subsurface micro cracks, particularly when that part is to be used under dynamic loading.

Fig. 7 and Fig. 8 show the effect of the feed rate at various cutting speeds on the surface roughness. It is realized that surface roughness increases with an increase in the feed rate. The surface roughness is observed to be minimum at high cutting speed with low feed rate. Increasing feed rate, leading to vibration and generating more heat and consequently contributing to a higher surface roughness. Surface roughness decreases abruptly with the increase in cutting speed for a given value of feed rate. This is owing to the fact that, as cutting speed increases, the temperature increases at the cutting zone which leads to the tempering of material and thus reduces the surface roughness. Application of the cutting fluid at cutting zone decreases the coefficient of friction at the interface of the tool-chip over the rake face which gets better surface finish. The analysis of variance (ANOVA) was applied to study the effect of the machining process parameters on the surface roughness. Table 8 gives the statistical model summary of linear and quadratic model for Ra. It reveals that the quadratic model is the best model. So, for further investigation quadratic model was used.

Table 8

Model summary statistics for Ra

Source	Standard deviation	R^2	Adj. R^2
Linear model	0.160356	88.46 %	86.08 %
Quadratic model	0.150838	98.79 %	98.15 % suggested

Table 9 gives the response surface regression of surface roughness. The value 'p' i.e. probability of obtaining a result equal to or more extreme than what was actually observed. If 'p' value for the model is less than 0.05 which shows that the model terms are important. In order to understand the turning process, the experimental results were used to develop the second order model using response surface methodology (RSM). MINITAB 17 was used for the statistical analysis of experimental results. The proposed quadratic model was developed from the functional relationship using RSM method for following conditions separately. When, cutting tool: PVD TiAlN, cutting fluid: Coconut oil

$$Ra = 1.321 - 0.0311\ Vc + 0.68\ f + 0.000199\ Vc*Vc + 9.47\ f*f + 0.0304\ Vc*f \tag{11}$$

When, cutting tool: Uncoated, cutting fluid: Coconut oil

$$Ra = 0737- 0.0288\ Vc + 9.20\ f + 0.000199\ Vc*Vc + 9.47\ f*f + 0.0304\ Vc*f \tag{12}$$

When, cutting tool: PVD TiAlN, cutting fluid: Palm oil

$$Ra = 1.430 - 0.0321\ Vc + 0.04\ f + 0.000199\ Vc*Vc + 9.47\ f*f + 0.0304\ Vc*f \tag{13}$$

When, cutting tool: Uncoated, cutting fluid: Coconut oil

$$Ra = 1.061 - 0.0298\ Vc + 8.56\ f + 0.000199\ Vc*Vc + 9.47\ f*f + 0.0304\ Vc*f \tag{14}$$

Table 9

Response surface design for Surface Roughness (Ra)

Source	DF	Adj SS	Adj MS	F	P
Vc	1	0.0005	0.00053	0.06	0.807
f	1	7.3882	7.38816	841.70	**0.000**
Ct	1	6.7602	6.76023	770.16	**0.000**
Cutting fluid	1	0.0325	0.03246	3.70	0.067
$Vc \times Vc$	1	0.0161	0.01605	1.83	0.189
$f \times f$	1	0.0087	0.00870	0.99	0.330
$Vc \times f$	1	0.0121	0.01207	1.37	0.253
$Vc \times Ct$	1	0.0075	0.00746	0.85	0.366
$Vc \times$ Cutting fluid	1	0.0014	0.00143	0.16	0.691
$f \times Ct$	1	1.5812	1.58123	180.14	**0.000**
$f \times$ Cutting fluid	1	0.0089	0.00894	1.02	0.324
$Ct \times$ Cutting fluid	1	0.1047	0.10465	11.92	**0.002**
Error	23	0.2019	0.00878		
Total	35	16.6230			

Table 10

Analysis of variance for Ra

Source	DF	Adj SS	Adj MS	F-Value	P-Value	% contribution
Vc	2	0.0163	0.00815	0.12	0.885	0.09
f	2	7.4163	3.70816	56.08	**0.000**	44.61
Ct	1	7.2424	7.24238	109.52	**0.000**	43.57
Cutting fluid	1	0.0303	0.03033	0.46	0.504	0.18
Error	29	1.9177	0.06613			11.55
Total	**35**	16.6230				100

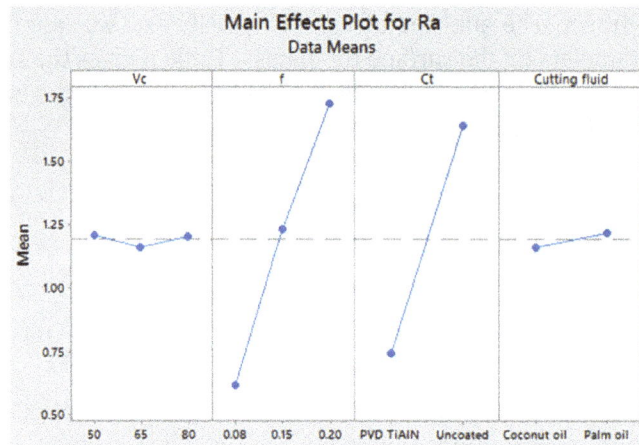

Fig. 6. Main effect plot for Flank wear, Ra (µm)

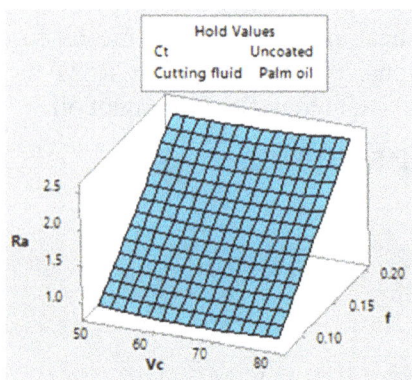

(a)

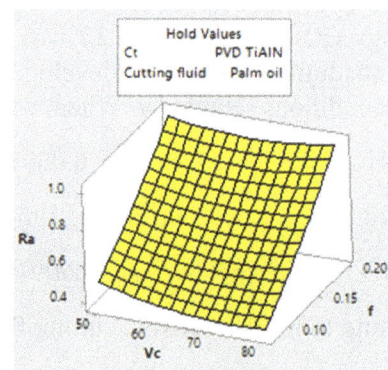

(b)

Fig. 7. Surface plot of Ra in MQL enivironment (Palm oil) (a) Unoated (b) PVD TiAlN insert

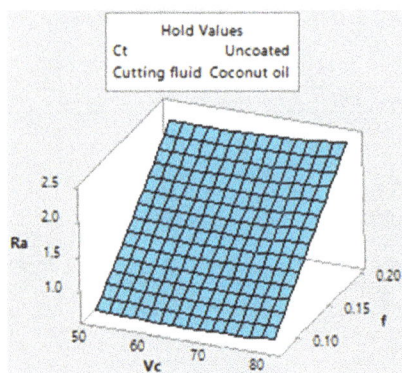

(a)

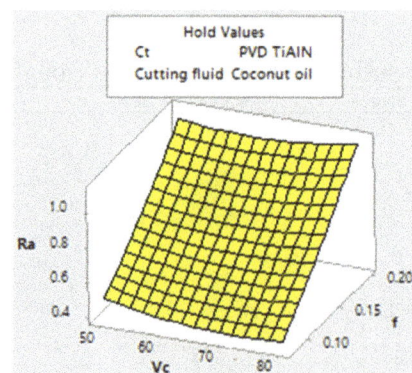

(b)

Fig. 8. Surface plot of Ra in MQL enivironment (Coconut oil) (a) Unoated (b) PVD TiAlN insert

The acceptability of the model has been investigated by the examination of residuals. The residuals, which are the variance between the observed output and the predicted output are examined using normal probability plots of the residuals and the plots of the residuals versus the predicted response. If the model is aceptable, the points on the normal probability plots of the residuals should form a straight line. The plots of the residual versus the predicted response should be structureless, they should contain no noticeable pattern. The normal probability plots of the residuals and the plots of the residuals versus the predicted response for the surfac roughness values are shown in Fig. 9. It reveales that the residuals generally fall on a straight line indicating that the errors are distributed normally.

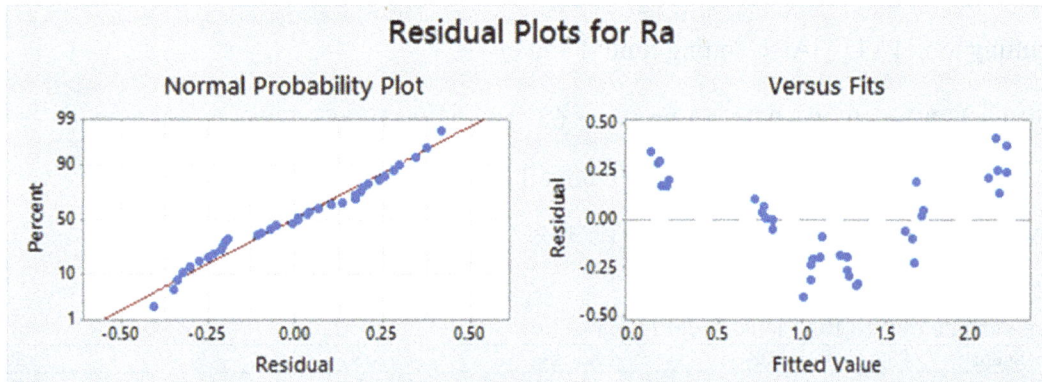

Fig. 9. Normal probability plot of residual and Plot of residual vs. fitted surface roughness values

4.3.2 Flank wear, VB (μm)

The cutting tool inserts in turning generally fail by gradual wear by abrasion, diffusion, adhesion, galvanic action and chemical erosion etc. depending on the workpiece and tool material and machining conditions. Tool wear initially starts with a relatively faster rate because of attrition and micro chipping at the sharp cutting edges. Cutting tools often fail prematurely, randomly and catastrophically by mechanical breakage and plastic deformation under adverse machining conditions caused by intensive pressure, temperature, dynamic loading at the tool tips if the tool material lacks strength, fracture toughness and hot hardness.

Fig. 11 and Fig. 12 show that flank wear increases with an increase in the cutting speed and feed rate. The flank wear is found to be minimum at low cutting speed with low feed rate. Increasing feed rate, leading to vibration and generating more heat and consequently contributing to a high flank wear. Flank wear increases abruptly with the increase in cutting speed. This is owing to the fact that, as cutting speed increases, the temperature increases at the cutting zone because Ti6Al4V ELI having low thermal conductivity (6.7 W/m-K). Due to the high temperature at the cutting zone, the tool loose its strength and thus plastic deformation took place. Application of the cutting fluid at cutting zone reduces the temperature through heat convection. As a result it leads to reduction in tool wear.

The analysis of variance (ANOVA) was applied to study the effect of the machining process parameters on the flank wear. Table 11 gives the statistical model summary of linear and quadratic model for VB. It reveals that the quadratic model is the best model. So, for further investigation quadratic model was used.

Table 11
Model summary statistics for VB

Source	Standard deviation	R^2	Adj. R^2	
Linear model	0.160356	93.04 %	91.60 %	
Quadratic model	0.150838	95.95 %	93.84 %	suggested

Table 12 gives the response surface regression of flank wear. The proposed quadratic model was developed from the functional relationship using RSM method for following conditions separately.

When, cutting tool: PVD TiAlN, cutting fluid: Coconut oil

$$VB = 80.5 - 2.36\ Vc - 93\ f + 0.02153\ Vc \times Vc + 657\ f \times f + 1.99\ Vc \times f \qquad (15)$$

When, cutting tool: Uncoated, cutting fluid: Coconut oil

$$VB = 93.9 - 1.85\ Vc - 182\ f + 0.02153\ Vc \times Vc + 657\ f \times f + 1.99\ Vc \times f \qquad (16)$$

When, cutting tool: PVD TiAlN, cutting fluid: Palm oil

$$VB = 80.1 - 2.38\ Vc - 97\ f + 0.02153\ Vc \times Vc + 657\ f \times f + 1.99\ Vc \times f \qquad (17)$$

When, cutting tool: Uncoated, cutting fluid: Palm oil

$$VB = 94.7 - 1.88\ Vc - 186\ f + 0.02153\ Vc \times Vc + 657\ f \times f + 1.99\ Vc \times f \qquad (18)$$

Table 12

Response surface design for Flank Wear (VB)

Source	DF	Adj SS	Adj MS	F	P
Vc	1	4946.2	4946.2	142.41	**0.000**
f	1	2614.6	2614.6	75.28	**0.000**
Ct	1	10631.1	10631.1	306.09	**0.000**
Cutting fluid	1	30.8	30.8	0.89	0.356
$Vc \times Vc$	1	187.7	187.7	5.40	**0.029**
$f \times f$	1	41.9	41.9	1.21	0.283
$Vc \times f$	1	51.8	51.8	1.49	0.234
$Vc \times Ct$	1	345.6	345.6	9.95	**0.004**
$Vc \times$ Cutting fluid	1	0.7	0.7	0.02	0.890
$f \times Ct$	1	173.4	173.4	4.99	**0.035**
$f \times$Cutting fluid	1	0.4	0.4	0.01	0.920
$Ct \times$ Cutting fluid	1	3.5	3.5	0.10	0.755
Error	23	798.8	34.7		
Total	35	19744.5			

Table 13

Analysis of variance for VB

Source	DF	Adj SS	Adj MS	F-Value	P-Value	% contribution
Vc	2	5225.5	2612.8	55.14	**0.000**	26.47
f	2	2617.1	1308.6	27.62	**0.000**	13.25
Ct	1	10496.3	10496.3	221.51	**0.000**	53.16
Cutting fluid	1	31.4	31.4	0.66	0.422	0.16
Error	29	1374.2	47.4			6.96
Total	**35**	19744.5				100

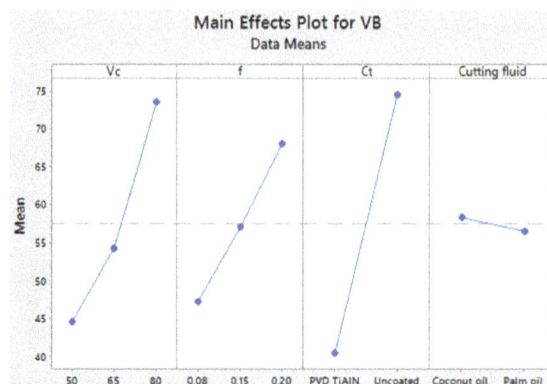

Fig. 10. Main effect plot for Flank wear, VB (μm)

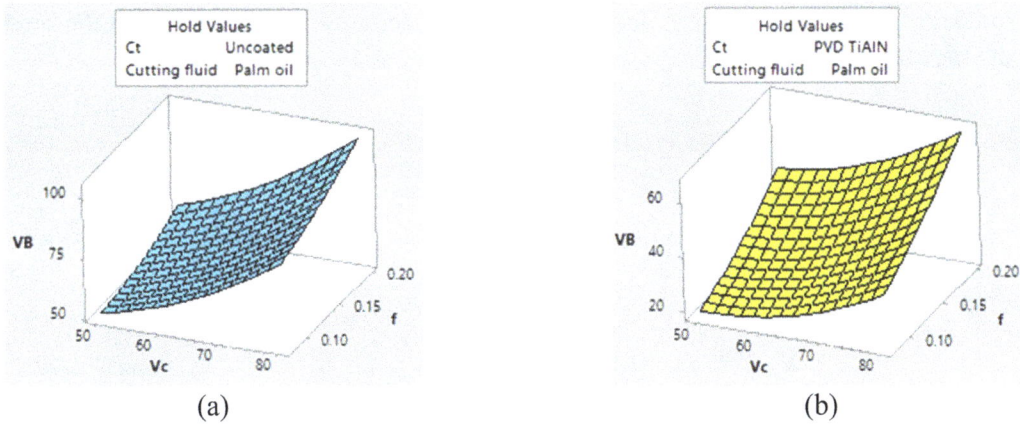

(a)   (b)

Fig. 11. Surface plot of VB in MQL enivironment (Palm oil) (a) Unoated (b) PVD TiAlN insert

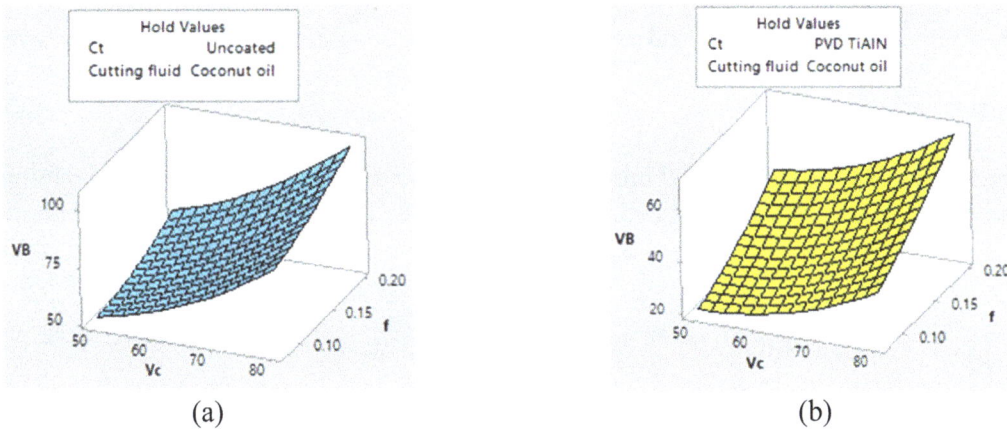

(a)   (b)

Fig. 12. Surface plot of VB in MQL enivironment (Coconut oil) (a) Unoated (b) PVD TiAlN insert

The normal probability plots of the residuals and the plots of the residuals versus the predicted response for the flank wear values are shown in Fig. 13. It reveales that the residuals usually fall on a straight line indicating that the errors are distributed normally.

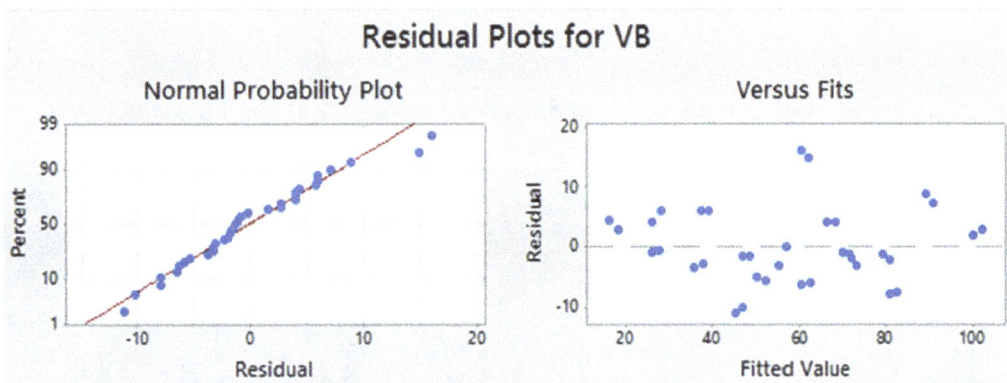

Fig. 13. Normal probability plot of residual and Plot of residual vs. fitted flank wear values

Fig. 14 shows comparison of the second order model with the experimental value of surface roughness and flank wear. Correlations of experimental results with the second order model are 0.99 and 0.98 for surface roughness and flank wear, respectively and correlations of experimental results with linear model are 0.94 and 0.96 for surface roughness and flank wear, respectively. So it is clear that the second order model is the most reliable.

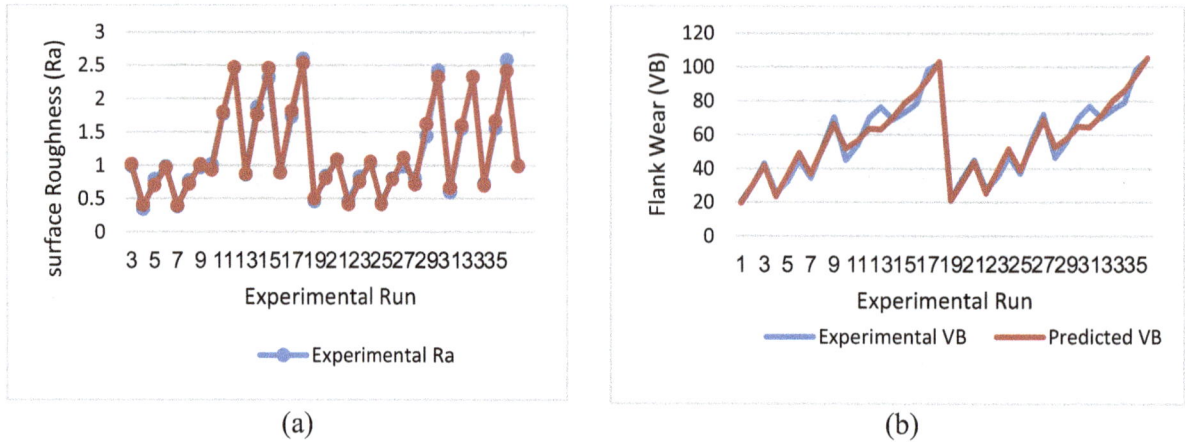

(a)   (b)

Fig. 14. Comparison of measured and predicted (a) surface roughness, *Ra (μm)* (b) Flank wear, *VB (μm)*

4.4 Optimization of Response

The aim of experiments related to machining is to achieve the desired surface finish and minimum flank wear of cutting tool insert with the optimal cutting parameters. To attain this end, the RSM optimization seems to be a useful technique. The goal is to minimize surface roughness (*Ra*) and minimize flank wear (*VB*).

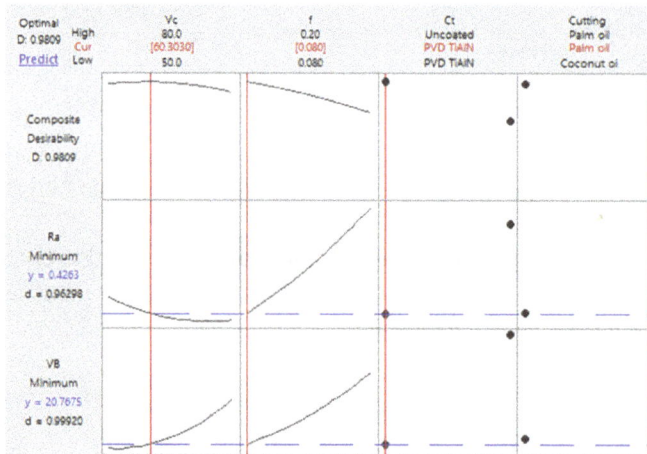

Fig. 15. Response Optimization for Surface roughness (Ra) and flank wear (VB)

Table 14

Confirmation test

Sr. No.	Optimum Condition				Experimental Ra (μm)	RSM Predicted Ra, (μm)	Experimental VB (μm)	RSM Predicted VB, (μm)
	Vc (m/min)	*f* (mm/rev)	Cutting tool insert	Cutting fluid				
1	60.3030	0.08	PVD TiAlN	Palm oil	0.37	0.43	23.7	20.92

PVD TiAlN tool wear pattern is observed under scanning electron microscope at constant cutting speed of 80 m/min and at 0.08, 0.15 and 0.2 mm/rev feed under dry and minimum quantity lubrication environment.

Cutting condition	Dry	MQL
Vc: 80 m/min, f: 0.08 mm/rev, ap:0.5 mm		
Vc: 80 m/min, f: 0.15 mm/rev, ap: 0.5 mm		
Vc: 80 m/min, f: 0.2 mm/rev, ap: 0.5 mm		

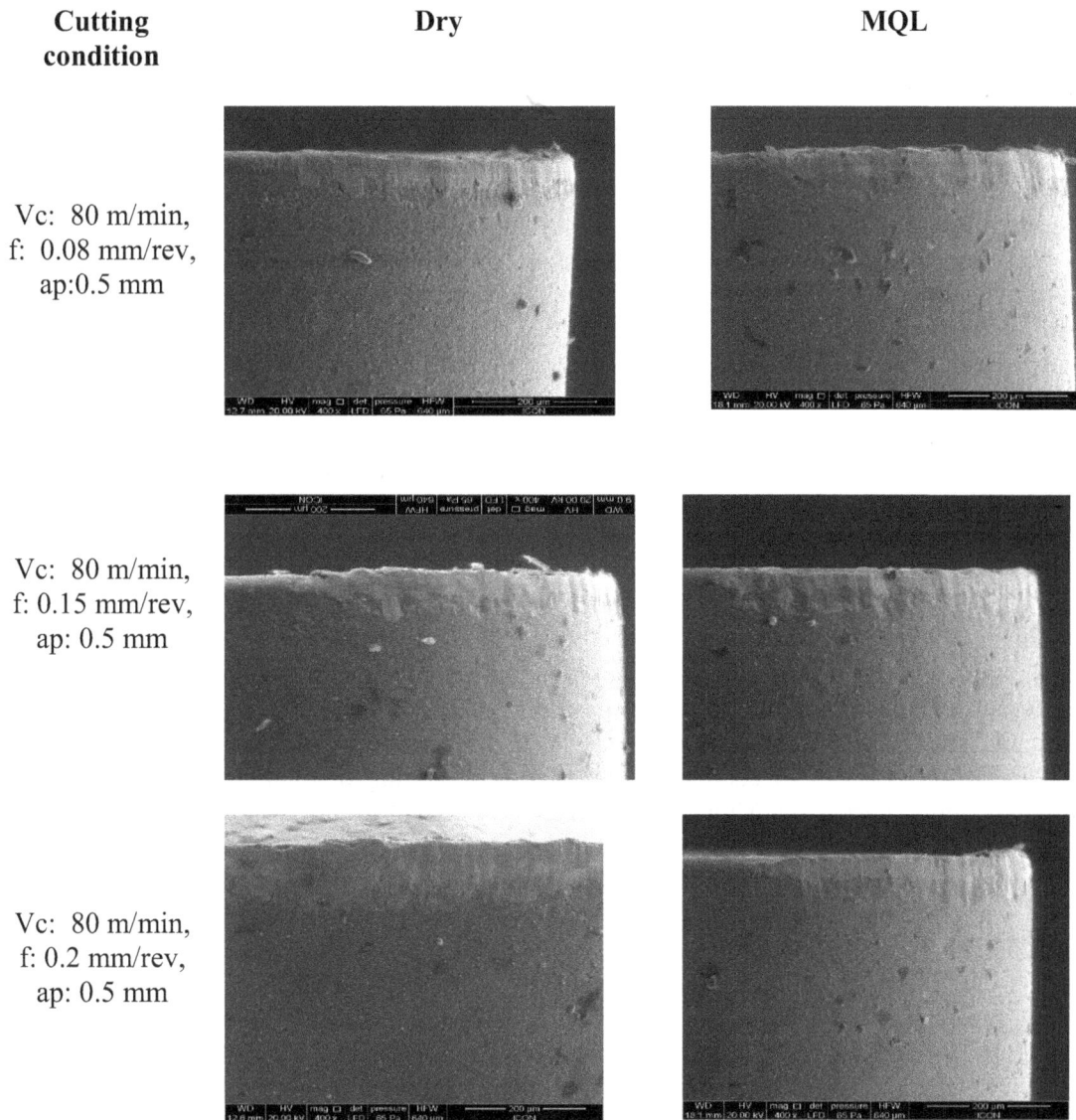

Fig. 16. SEM images of tool wear pattern for PVD TiAlN coated insert

5. Conclusions

The following can be concluded from the results obtained when turning of titanium alloy Ti-6Al-4V ELI under MQL environment using PVD TiAlN cutting tool:

- Feed rate has the highest influence on the surface roughness and accounts for 44.61% contribution in the total variability of the model followed by cutting tool insert 43.57%.
- Cutting tool has the highest influence on the flank wear and accounts for 53.16 % contribution in the total variability of the model followed by cutting speed insert 26.47%.
- From the Grey relational analysis and TOPSIS method, the optimum combination of process parameters are considered as Vc: 50m/min, f: 0.08 mm/rev., Ct: PVD TiAlN, Cutting fluid: Palm oil
- The optimum cutting process parameters were obtained through RSM as
Vc: 60.3030 m/min, f: 0.08 mm/rev., Ct: PVD TiAlN and cutting fluid: Palm oil
- Developed second order model has high square values of the regression coefficients which indicated high correlation with variances in the predictor values.
- The developed second order model seems to be satisfactory.

References

Ali, S. M., Dhar, N. R., & Dey, S. K. (2011). Effect of minimum quantity lubrication (MQL) on cutting performance in turning medium carbon steel by uncoated carbide insert at different speed-feed combinations. *Advances in Production Engineering & Management, 6*(3).

Attanasio, A., Gelfi, M., Giardini, C., & Remino, C. (2006). Minimal quantity lubrication in turning: effect on tool wear. *Wear, 260*(3), 333-338.

Dhar, N. R., Islam, S., & Kamruzzaman, M. (2007). Effect of minimum quantity lubrication (MQL) on tool wear, surface roughness and dimensional deviation in turning AISI-4340 steel. *Gazi University Journal of Science,20*(2), 23-32.

Escamilla-Salazar, I. G., Torres-Treviño, L. M., González-Ortíz, B., & Zambrano, P. C. (2013). Machining optimization using swarm intelligence in titanium (6Al 4V) alloy. *The International Journal of Advanced Manufacturing Technology, 67*(1-4), 535-544.

Islam, M. N., Anggono, J. M., Pramanik, A., & Boswell, B. (2013). Effect of cooling methods on dimensional accuracy and surface finish of a turned titanium part. *The International Journal of Advanced Manufacturing Technology, 69*(9-12), 2711-2722.

Khan, M. M. A., Mithu, M. A. H., & Dhar, N. R. (2009). Effects of minimum quantity lubrication on turning AISI 9310 alloy steel using vegetable oil-based cutting fluid. *Journal of materials processing Technology, 209*(15), 5573-5583.

Khanna, N., & Davim, J. P. (2015). Design-of-experiments application in machining titanium alloys for aerospace structural components. *Measurement, 61*, 280-290.

Liu, Z., An, Q., Xu, J., Chen, M., & Han, S. (2013). Wear performance of (nc-AlTiN)/(a-Si 3 N 4) coating and (nc-AlCrN)/(a-Si 3 N 4) coating in high-speed machining of titanium alloys under dry and minimum quantity lubrication (MQL) conditions. *Wear, 305*(1), 249-259.

Ramana, M. V., Vishnu, A. V., Rao, G. K. M., & Rao, D. H. (2012). Experimental Investigations, Optimization Of Process Parameters And Mathematical Modeling In Turning Of Titanium Alloy Under Different Lubricant Conditions. *Journal of Engineering (IOSRJEN) www. iosrjen. org ISSN, 2250*, 3021.

Revankar, G. D., Shetty, R., Rao, S. S., & Gaitonde, V. N. (2014). Analysis of surface roughness and hardness in titanium alloy machining with polycrystalline diamond tool under different lubricating modes. *Materials Research*, (AHEAD), 1010-1022.

Sargade, V., Nipanikar, S., & Meshram, S. (2016). Analysis of surface roughness and cutting force during turning of Ti6Al4V ELI in dry environment. *International Journal of Industrial Engineering Computations, 7*(2), 257-266.

Sharma, V. S., Singh, G., & Sørby, K. (2015). A review on minimum quantity lubrication for machining processes. *Materials and Manufacturing Processes, 30*(8), 935-953.

Shetty, R., Jose, T. K., Revankar, G. D., Rao, S. S., & Shetty, D. S. (2014). Surface Roughness Analysis during Turning of Ti-6Al-4V under Near Dry Machining using Statistical Tool. *International Journal of Current Engineering and Technology, 4*(3), 2061-2067.

Wang, P., Zhu, Z., & Wang, Y. (2016). A novel hybrid MCDM model combining the SAW, TOPSIS and GRA methods based on experimental design. *Information Sciences, 345*, 27-45.

Wu, H., & Guo, L. (2014). Machinability of titanium alloy TC21 under orthogonal turning process. *Materials and Manufacturing Processes, 29*(11-12), 1441-1445.

Xu, J. Y., Liu, Z. Q., An, Q. L., & Chen, M. (2012, April). Wear Mechanism of High-Speed Turning Ti-6Al-4V with TiAlN and AlTiN Coated Tools in Dry and MQL Conditions. In *Advanced Materials Research* (Vol. 497, pp. 30-34)

2

Pricing and lot sizing optimization in a two-echelon supply chain with a constrained Logit demand function

Yeison Díaz-Mateus[a], Bibiana Forero[a], Héctor López-Ospina[b*] and Gabriel Zambrano-Rey[a]

[a]*Industrial Engineering Department, Pontificia Universidad Javeriana, Cra 7 #40-62, Ed. Jose Gabriel Maldonado P.3, Bogotá, Colombia*
[b]*Department of Industrial Engineering, Universidad del Norte, Km. 5 Vía Puerto Colombia, Barranquilla, Colombia*

CHRONICLE	ABSTRACT
Keywords: *Constrained multinomial logit* *Pricing* *Lotsizing* *Supply chain optimization* *PSO*	Decision making in supply chains is influenced by demand variations, and hence sales, purchase orders and inventory levels are therefore concerned. This paper presents a non-linear optimization model for a two-echelon supply chain, for a unique product. In addition, the model includes the consumers' maximum willingness to pay, taking socioeconomic differences into account. To do so, the constrained multinomial logit for discrete choices is used to estimate demand levels. Then, a metaheuristic approach based on particle swarm optimization is proposed to determine the optimal product sales price and inventory coordination variables. To validate the proposed model, a supply chain of a technological product was chosen and three scenarios are analyzed: discounts, demand segmentation and demand overestimation. Results are analyzed on the basis of profits, lotsizing and inventory turnover and market share. It can be concluded that the maximum willingness to pay must be taken into consideration, otherwise fictitious profits may mislead decision making, and although the market share would seem to improve, overall profits are not in fact necessarily better.

1. Introduction

While supply chain models focus on minimizing logistic costs, marketing models focus on maximizing revenues by adjusting products' price and by trying to follow closer demand changes. Obviously, a joint approach allows to determine in a more global way the expected profits of organizations. Consequently, models that incorporate the concepts of inventory theory and optimal price have been developed, where demand depends on the price and other factors (e.g., sales, seasonality, and product life-cycle management) for a planning horizon. For example, Kim and Lee (1998) studied the optimal selling price and inventory levels, with sensitive demand to price sales. Since then, there have been some related models that combine production and prices (Deng & Yano 2006; Ardjmand et al., 2016), requirements planning and price (Geunes et al., 2006; Chen & Chen 2015), inventory levels and supplier price (van den Heuvel & Wagelmans 2006; Taleizadeh & Noori-daryan 2016). In terms of marketing, certain

* Corresponding author
E-mail: hhlopez@uninorte.edu.co (H. López-Ospina)

models have introduced specific conditions such as discounts (Berger & Bechwati 2001), payment due dates (Ghoreishi, et al., 2014), and segmentation of demand (Ghoniem & Maddah 2015). On the other hand, Shavandi et al. (2012) present a constrained multi-product pricing and inventory model in which perishable products are put into three categories: substitute, complementary and independent; to solve the model genetic algorithm is developed. However, it is important to note that in most of these studies, the demand is modeled with linear and price elasticity functions, which do not necessarily capture all key aspects in the customers purchase behavior. Then, this paper proposes a model that incorporates the changes in demand and its influence on the price adjustment and on the logistic costs associated with inventory levels, in a coordinated problem between a supplier and a vendor. Also, the model considers that there are different socioeconomic groups with different product valuation, hence with different maximum willingness to pay. For the proposed model, the I-JPLMSP model (Yaghin et al., 2014) (*Integrated join pricing and lotsizing model with sales promotions*) is taken as reference because it integrates the level of inventories with the optimal price in a two echelon supply chain, and also it includes discount and sales policies to meet demand. This logistic model assumes a single vendor and supplier, and a unique non-perishable product without seasonality, with an infinite production rate. However, the I-JPLMSP model is built upon a non-linear function of demand based on a multinomial logit distribution (MNL) (Márquez-Díaz et al., 2011). Thus, demand is only evaluated for a single group of customers with the same rating for the product attributes. The I-JPLMSP model aims is to optimize the multi-echelon profits between the vendor and the supplier.

The main contribution of this paper is to extend the I-JPLMSP model to take into account the following conditions. First, customer segmentation to analyze different socioeconomic groups, with different product valuations, for obtaining a more appropriate demand estimation. Second, price constraints are introduced by using the constrained multinomial logit (CMNL) (Martínez et al., 2009) instead of the unconstrained multinomial logit. The main advantage of using the CMNL over traditional MNL is the possibility to include constraints on the product's attributes through penalty functions (Pérez et al., 2016). In particular, the CMNL allows to include a constraint associated with the maximum willingness to pay by customers that directly affects the estimation of demand and optimal price. Finally, to validate the model, a case study in the Colombian market was used, where socio-economic segmentation is quite present and has an impact on demand and product price estimation.

The rest of the paper is organized as follows. Section 2 explains the CMNL for demand estimation. Then, Section 3 starts by presenting the problem statement, and then the formulation of demand based on the CMNL is explained, followed by the formulation of the logistic multi-echelon model, and last by the Model resolution procedure.

2. The constrained multinomial logit for demand estimation

In this research, the discrete choice model known as constrained multinomial logit (CMNL) is used (Martínez et al., 2009) to include constraints on the product selling price, taking into account the consumers' maximum willingness to pay. The **CMNL** assumes that the perceived utility by an agent, i.e., a consumer, who belongs to the socio-economic group (h) associated with the discrete product (x) denoted by V_{hx} is split into a compensatory part (a in Eq. (1)) and another non-compensatory part (b in Eq. (1)) which indicates the feasibility of that alternative to h,

$$U_{hx} = \underset{(a)}{V_{hx}} + \underset{(b)}{\frac{1}{\mu}Ln(\varphi_{hx})} + \xi_{hx} \tag{1}$$

where $Ln(\varphi_{hx})$ is a boundary or penalty function imposed by group h to the attributes of the discrete product x. The stated penalty, with a logarithm function, allows constraints to be subtly broken by the decision maker (Martínez et al., 2009). The random component ξ_{hx} represents and reflects the inability of the analyst to model all the attributes and changes in preferences and behaviors of individuals, measurement and modeling errors, lack of accurate information, among others. If such inaccuracies are Gumbel distributed with scale parameter μ, then the probability of a consumer, who belongs to group (h), to purchase product I can be represented as in equation (2).

$$P_{hx} = \frac{\varphi_{hx}e^{\mu V_{hx}}}{1 + \sum_j \varphi_{hj}e^{\mu V_{hj}}}. \qquad \forall \, h \in H \qquad (2)$$

This probability is known as the constrained multinomial logit model (CMNL) (Martínez et al., 2009). There are some interesting and novel applications that are used on modeling demand in a discrete choices context, in areas such as the mode of transport choice (Castro, et al., 2013), the location of schools and their capacity (Castillo-López & López-Ospina 2015; Martinez, et al., 2011), the optimal price and packaging (Pérez et al., 2016), the subway route choice (Herrera, 2014), place of residence and housing choice (Martínez & Donoso 2010; López-Ospina, et al., 2016; López-Ospina, et al., 2017), food choice (Ding et al., 2012), parking management (Caicedo, et al., 2016) among others. These applications require constrained variables in different contexts, which imply high non-linearity in demand estimation, which also involves high non-linearity when attributes are decision variables within optimization models, such as the selling price. The following sections describe the detailed logistic problem and the non-linear formulation proposed.

3. The logistic non-linear optimization problem

3.1 Problem statement

The problem modeled in this article is focused on a two-echelon supply chain with a single vendor and a single supplier, trading a single non-perishable product under the assumption of coordination of cycle times for both supplier and vendor. To formulate the problem the following parameters and variables are taken into account:

Parameters

$H = set\ of\ socio-economic\ groups, h \in H$
$I_v = holding\ inventory\ cost\ for\ the\ vendor$
$K_v = ordering\ cost\ for\ vendor$
$C_v = purshase\ cost for\ the\ vendor$
$I_p = holding\ inventory\ cost\ for\ the\ supplier$
$A_p = ordering\ cost\ for\ the\ supplier$
$C_p = purshase\ cost for\ the\ supplier$
$D_h(P) = demand\ of\ each\ group\ h\ depending\ on\ the\ price\ P\ of\ the\ product$
$Q = economic\ order\ quantity\ (EOQ)$

Variables

$m = number\ of\ vendor's\ orders\ within\ one\ supplier's\ cycle. Coordination\ constant$
$T_v = duration\ of\ vendor's\ inventory\ (annual)$
$P = product\ selling\ price$
$T_p = duration\ of\ supplier's\ inventory\ (annual)$

Assumptions

- The maximum willingness to pay, that each group h assigns to the product, is known
- The inventory replenishment time is negligible
- Planning time is infinite
- The supplier delivery rate is infinite
- There is a coordination of inventory cycle times between the vendor and the supplier

3.2. Formulation of demand based on the CMNL

From the point of view of microeconomic modeling, discrete choice analysis on a product is based on the principle of utility maximization where the price is directly related to observable characteristics of the product. In addition, it is assumed that each individual makes the decision based on the perceived utility of the product, good or service. Hence, this situation can be modeled by random utility models (*RUM*) initially developed by Block and Marschak (1960). In 1975, Mc Fadden (1975) makes an econometric extension of this theory by considering that a population of individuals do the same choice on a set of alternatives, i.e., that the population can be split up having as a reference common socio-economic factors in a group of individuals which conditions their choices. Each group h within the population is called cluster (Martínez et al., 2009). A particular case of these models is the multinomial logit, used by Yaghin et al. (2014) to estimate demand D, under the assumption of a single socio-economic group ($H=1$), as in equation (3).

$$D(P) = \alpha \frac{e^{-bP}}{1 + e^{-bP}}, \qquad \alpha, b > 0 \tag{3}$$

In Eq. (3) the utility function is given by $V = -bP$, assuming a single product valuation attribute, i.e., the price P, and its coefficient of variation $-b$, that shows the variation per price unit changed in the utility function. The utility function is negative because for any consumer, an increase in the product price decreases its perceived utility. In addition, the market size is represented by α, which allows to obtain a deterministic demand since the demand for that product, at price P, can be obtained by multiplying the number of customers by the purchase probability.

This paper proposes a new model for demand estimation, but using the constrained multinomial logit (CMNL), and introducing the following aspects that make demand estimation more realistic: 1. multiple socio-economic groups with similar characteristics to consumers from the same group in order to analyze and describe logistic and demand impacts. It is important to clarify that an aggregate demand will be obtained, that is, the sum of demands for all clusters. 2. The utility function U_h associated with the product depends on the product selling price P and on a set of characteristics $A_k, k = 1,2,...,K$ that defines product attractiveness, which are assessed for each sub-group h, as suggested in equation (4).

$$U_h = \underbrace{-b_h P}_{Price\ valuation} + \underbrace{\sum_{k}^{K} \beta_{hk} A_k}_{Product\ attractiveness} + \underbrace{\varepsilon_h}_{Modeling\ errors} \qquad \forall\, h \in H \tag{4}$$

In Eq. (4), β_{hk} are the coefficients associated with each characteristic k of the product, within characteristics A_k assessed by group h. Therefore, if it is assumed that the error is independent and identically distributed (IID) Gumbel with scale parameter μ. Then, if U_h is introduced in Eq. (3) replacing V, the demand for each group or cluster h for the product is defined as:

$$D_h(P) = \alpha_h \frac{e^{-b_h P + \sum_k^K \beta_{hk} A_k}}{1 + e^{-b_h P + \sum_k^K \beta_{hk} A_k}}, \qquad \forall\, h \in H \tag{5}$$

Hence, the aggregate demand can be calculated as:

$$AggD = \sum_h D_h(P) \tag{6}$$

Based on Eq. (5), the probability of not purchasing for each group h is defined as:

$$(\text{probability of not purchasing})_h = \frac{1}{1 + e^{-b_h P + \sum_k^K \beta_{hk} A_k}}, \qquad \forall h \in H \tag{7}$$

A third aspect that is integrated to this model is a penalty associated with the maximum willingness to pay for the product, for each group h; making demand estimation more realistic. Thus, demand modeling changes from the classic multinomial logit model to the discrete choice model known as CMNL (Eq. 8).

$$D_h(P) = \alpha_h \frac{\varphi_h^U(P)\varphi^L(P)e^{V_h}}{1 + \varphi_h^U(P)\varphi^L(P)e^{V_h}}. \qquad \forall h \in H \tag{8}$$

Therefore, the perceived utility presented in Eq. (4) becomes Eq. (9) to introduce the maximum willingness to pay for the product, for each group h, denoted as $\frac{1}{\mu}\ln(\varphi_h^U(P))$ and an exogenous lower bound for the product price $\frac{1}{\mu}ln(\varphi_h^L(P))$. These two constraints are defined by a binomial logit model (Castro et al., 2012) (Eq. (11) and Eq. 12), so that each subgroup h can delimit the product choice based on price. The CMNL allows to smoothly integrate those constrains.

$$U_h = V_h + \frac{1}{\mu}\ln(\varphi_h^U(P)) + \frac{1}{\mu}ln\left(\varphi_h^L(P)\right) + \varepsilon_h \quad \forall h \in H \tag{9}$$

$$V_h = -b_h P + \sum_{k=1}^K \beta_{hk} A_k \tag{10}$$

$$\varphi_h^U = \frac{1}{1 + Exp\,(w(P - a_h + \rho))} \tag{11}$$

$$\varphi^L = \frac{1}{1 + Exp\,(w(\gamma - P + \rho))}. \tag{12}$$

In Eqs. (11-12), γ and a_h are the lower and upper bounds for the product's price, respectively, w is the scale parameter of the binomial logit, and ρ is a parameter defined by Eq. (13), that includes η as the value associated with the population's proportion that overrides the associated constraint. It is important to note that if the parameter η is set to 0 or 1, then the given expression for ρ is undefined, because the binominal logit functions can only predict deterministic choices (probabilities equal to zero or one) when the variables tend to infinity (Martínez et al., 2009).

$$\rho = \frac{1}{w}\ln(\frac{1-\eta}{\eta}). \tag{13}$$

Once the new demand function is defined with the inclusion of the new conditions, the following section shows how this demand is used in a logistic model.

3.3. Formulation of the logistic multi-echelon model

This section provides an optimal policy for a two-echelon supply chain. The main idea is to optimize, simultaneously, the selling price, the size of the purchase order to the supplier and the number of purchase orders per cycle, in order to maximize the profit. The utility function F, for this particular case, is described in Eq. (14), where the subscripts v y p define the vendor and the supplier, respectively.

$$F = (Sales_v - Costs_v) + (Sales_p - Costs_p).$$

(14)

For this model, it is assumed that the vendor is a retailer that purchases the product and resale it. For this reason its total annual profit function, depending on the estimated aggregated demand, is described as follows:

$$F_v = \sum_h D_h(P) P - \left\{0,5 I_v T_v \sum_h D_h(P) + \frac{K_v}{T_v} + \sum_h D_h(P) C_v\right\}.$$

(15)

For the supplier, it is assumed that it is a wholesaler who purchases a product to market it though its distribution channels and make a profit. Thus, an EOQ (economic order quantity) model is assumed resulting in a profit function similar to the vendor's, but with the following differences: (1) supplier profit is calculated based on the vendor purchase price; (2) a formulation (Eq. (16)) was used to coordinate inventories based on reorder time of the supplier, to avoid inventory shortages in the model

$$T_p = m T_v.$$

(16)

Then, the joint utility function under the assumption of coordination between the two echelons of the supply chain is described by Eq. (17).

$$F(P, m, T_v, T_p) = \sum_h D_h(P) (P + C_v) - \left\{0,5(m-1)T_v I_p \sum_h D_h(P) + \frac{K_p}{mT_v} + \sum_h D_h(P) C_p\right\} -$$
$$\left\{0,5 I_v T_v \sum_h D_h(P) + \frac{K_v}{T_v} + \sum_h D_h(P) C_v\right\}.$$

(17)

3.4. Model resolution procedure

The proposed logistic model implies a high non-linearity in estimating demand, where there is a direct relationship, and non-linear, between the restricted demand and the logistic model choices. Consequently, it is not possible to obtain a solution analytically or using an exact method, and therefore heuristics and meta-heuristics must be used. Given that PSO (*particle swarm optimization*) has shown the effectiveness in highly non-linear optimization problems (Moghadam & Seyedhosseini, 2010; Hashemi, et al., 2010; Zahara & Hu, 2008; Xu et al., 2013; Jafari et al., 2013; Karimi-Nasab, et al., 2015; Bai et al., 2016; Kumar, et al., 2016, Guedria, 2016; among others), in this work a PSO was used to solve the logistic model. The Eq. (19) and Eq. (18) show the velocity and position of each particle i for each dimension d at iteration t, where ω is the particle inertia, C_1 and C_2 are acceleration constants, $best_{id}$ is the best position the particle has reached to the current iteration, and $best_{gd}$ is the position of the best particle of the whole swarm, considering a fully connected swarm (Kennedy & Mendes 2002). Fig. 1 shows the resolution procedure pseudo-code.

$$Vel_{id}(t) = \omega * Vel_{id}(t-1) + C_1 * rnd * \left(Best_{id} - pos_{id}(t-1)\right) +$$
$$C_2 * rnd * \left(Best_{gd} - pos_{id}(t-1)\right),$$

(18)

$$pos_{id}(t) = pos_{id}(t-1) + Vel_{id}(t).$$

(19)

The particle coding scheme has three dimensions based on three decision variables: the selling price of the product (P), the number of shipments within the single supplier cycle (m), and finally, the duration of the inventory for the vendor (T_v). The fourth decision variable (T_v) is not integrated into the coding scheme because of its dependent relationship to (T_v), according to equation (16).

```
Define particle's dimensions
Define the feasible space for each dimension
Set swarm size, c₁, c₂, ω
Do
{
         Initialize randomly each particle's position (feasible space)
         Initialize randomly each particle's velocity (feasible space)
         Calculate each particle's fitness (calculate demand())
         Bestᵢ=current particle's position
} for all particles in the swarm
Bestg=best particle
Do
{
      Do
      {
      Calculate each particle's new velocity (feasible space)
      Calculate each particle's new position (feasible space)
      Calculate each particle's fitness(calculate demand())
      Bestᵢ=current particle's position
      } for all particles in the swarm
      Bestg=best particle
} While the max number of iterations is not attained
```

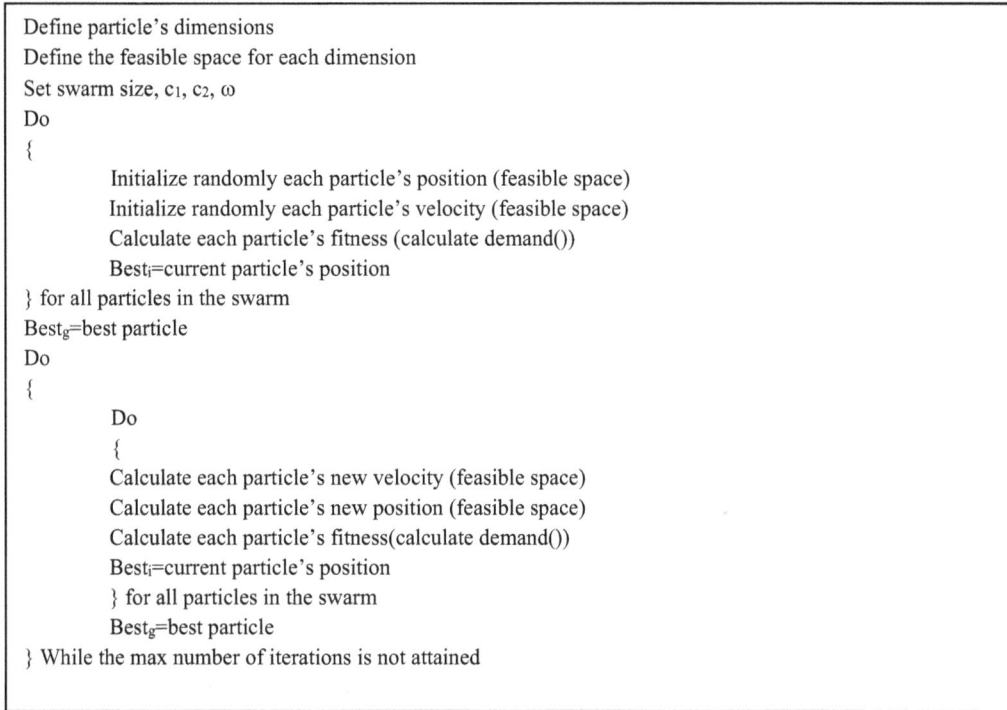

Fig 1. Resolution procedure based on PSO

Eq. (17) is used as the PSO objective function, and for each variable the following feasibility constraints are true as it is explained below.

Feasibility constraint for m

To analyze this constraint, it is assumed that variable m is continuous, thus the first derivate of Eq. (17) regarding m is defined as:

$$\frac{\partial F}{\partial m} = \sum_h D_h(P)\,(P + C_v) - \left\{ 0,5(m-1)T_v I_p \sum_h D_h(P) + \frac{K_p}{mT_v} + \sum_h D_h(P)\,C_p \right\} \tag{20}$$

$$- \left\{ 0,5 I_v T_v \sum_h D_h(P) + \frac{K_v}{T_v} + \sum_h D_h(P)\,C_v \right\},$$

then:

$$\frac{\partial F}{\partial m} = \left[\frac{K_p}{m^2 T_v} - K_v \right] - 0,5 T_v I_p \sum_h D_h(P). \tag{21}$$

Therefore, a critical point of the function is found, obtaining:

$$m = \frac{\sqrt{2}\,\sqrt{K_p}}{\sqrt{T_v[T_v I_p \sum_h D_h(P) + 2K_v]}}. \tag{22}$$

Analyzing the second derivative regarding m:

$$\frac{\partial^2 F}{\partial m^2} = -\frac{2K_p}{m^3 T_v}. \tag{23}$$

Since the second derivative is negative, then the critical point is a relative maximum. Through Eq. (22), it can be stated that m is a non-decreasing function of P. Thus, the maximum number of shipments can be set as

- $m_{max} \rightarrow T_{v\,min} \neq 0 \rightarrow 0{,}1$
- $P \rightarrow P_{max}, i.e \sum_h D_h(P) \rightarrow 0$

$$m_{max} = \left\lceil \sqrt{\frac{K_p}{0{,}1 * A_v}} \right\rceil . \tag{24}$$

Thus it can be concluded that the maximum value of the profit function is associated with an integer value m in the interval $m = [1, m_{max}]$.

Feasibility constraint for T_v

T_v assumes a lower and an upper level of duration, which depends on market dynamics and product obsolescence. As a result, T_v strongly depends on the chosen market and product that are evaluated on the proposed model, considering that $T_v \neq 0$.

Feasibility constraint for P

Within the logistic model a purchase price C_v was established for the vendor, which is defined as the lower bound of the price P, and the upper bound of the price is given by the maximum willingness to pay among all socio-economic groups h. Taken into account that the CMNL allows the price not to strictly respect these constraints, by a soft penalty (as shown in Fig. 2), hard upper and lower bounds were established for the PSO, up to half of the purchase price C_v (lower bound, $C_v/2$) and to double of the maximum willingness to pay among all socio-economic groups (upper bound, $2*max\{a_h\}$).

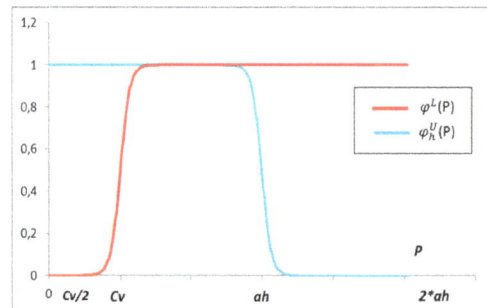

Fig 2. Penalty functions of constrained Logit

4. Implementation and results

To validate the proposed model the digital television market was chosen, particularly in the Colombian context, based on the previous study of González and Serna (2013). From this work, the 32-inch LED television was taken as the reference because of its variance in demand. Only in 2013, 1'700.000 screens were sold in the country, and in 2014 the sales of these products were around 2 million units. As far as the sizes, 32-inch models are consolidated as the preferred size by Colombians with more than 52% of sales, and LED technology accounts for 90% of sales (Tiempo, 2016a; Tiempo, 2016b). In the following sections, the socio-economic groups, costs, and logit parameters are defined, and the implementation procedure and results analysis are reported.

4.1. Socio-economic groups and attributes definition

In order to obtain the characteristics of people that directly affect the maximum willingness to pay, a survey with the following study variables was designed: age, gender, zone, neighborhood and price. The zone and neighborhood variables are allowed to define three socio-economic groups, i.e., lower, middle, upper social classes or stratification; according to the demographic characterization of Bogotá (2003).

From the obtained results, an analysis of variance was performed from which it was concluded, with a significance level of 5%, that the single studied variable that affects the maximum willingness to pay is the socio-economic group to which the individual belongs. Additionally, to define the upper and lower bound constraints, the maximum willingness to pay for each socio-economic group and the market size (193436 inhabitants) based on the target population defined by surveys were obtained. Such values are presented in Table 1.

Table 1

Socio-economic groups

Socio-economic groups	Maximum willingness to pay (COP$) 1 dollar is equivalent to 3000 COP$	Market size (inhabitants)
Lower	894.615	50 294
Middle	1.190.604	115 288
Upper	1.958.333	27855

From these results, the price sensitivity coefficient is also defined as shown in Table 2.

Table 2

Price sensitivity coefficient

Socio-economic groups	Price sensitivity coefficient
Lower	0.725
Middle	0.5375
Upper	0.225

Finally, through direct observation of the market, the minimum selling price of 32-inches LED televisions is set to COP$ 620.999.

4.2. Costs definition

Because the costs associated in Eq. (17) are strategic values of enterprises and therefore is not readily available information, additional parameters were included so such values could be estimated. Then, the parameter PV was included and it refers to the average current selling price in the market, COP$ 927.104. As well, the parameter PG was included to represent the expected profit percentage. Three reference values of 30%, 40 and 50% were fixed for PG. The parameter PV is equal to both supplier and vendor, as the logistic model aims to balance the perceived profits for the two echelons in the chain. Consequently, to set the values of C_v and C_p the relationship between the vendor and supplier was defined as shown in Fig. 3.

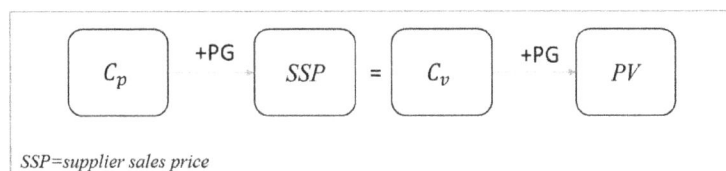

SSP=supplier sales price

Fig 3. Relationship between vendor and supplier costs

Similarly, the costs of holding inventory and ordering were defined under the established relationships in Table 3.

Table 3

Holding inventory costs and ordering costs for vendor and supplier

	Cost	Relation
Vendor	Holding	0,15 times
	Ordering	41,67 times
Supplier	Holding	0,08 times
	Ordering	125 times

4.3. Parameters for the CMNL

As proposed by Castro et al. (2013), the penalty is defined as the product of both, the upper and lower constraints. Taking into account that both constraints and their thresholds are independent, the results of constraints by varying the parameters (w) and (η) in Eqs. (11-13) were analyzed. Then, it was evident that the penalties were too strict when η took values lower than 0.9, and after this value, variations were not significant. In addition, it was noted that the more w increased, there were not large variations. For this reason, it was established that an intermediate penalty would be used, for which $w = \mathbf{0.5}$ was fixed.

4.4. Procedure implementation

The implementation of the proposed model was run on R Statistics software. To implement the solution, and based on some control instances, the parameters in Table 4 were used, which allowed convergence of the model.

Table 4
Values for PSO parameters

Parameters	Values
Par (Components of each individual)	Empty array of length 3
Fn (Adaptability function)	Joint gains function equation (17)
Lower (Lower constraints)	Array containing lower bounds for each variable
Upper(Upper constraints)	Array containing upper bounds for each variable
Maxit (Number of iterations)	1500
S (Population size)	60
W (Decreasing inertia)	Minimum value: 0,02 Maximum value: 2

4.5. Results analysis

Given the general description of the optimization problem, the proposed model was evaluated against three scenarios. The first scenario analyzes the impact of discounts by means of sensitivity analysis. The second scenario compares the model without including the price constraints in the logit demand function. More, to obtain an approach that depicts the influence of customer behavior in demand estimation, a third scenario including socio-economic segmented demand was assessed. In order to assess the model accuracy and its behavior in the proposed scenarios, the base model was considered regardless of the product attractiveness, i.e., assessment of the attributes other than price in the utility function. Hence, the analysis of the base model and product attractiveness are presented at first, followed by the scenarios.

Analysis of the base model

Results obtained from the base model are reported in Table 5. When performing a comparative analysis among the three expected profit percentages, it can be noted that the biggest profit obtained was when *PG*=50%. Profit is 1.39 and 1.15 times above profits obtained with *PG*=30% and *PG*=40%, respectively. This behavior occurs because when the expected profit for the company increases, the objective function is affected in the same way due to their direct relationship. Regarding the demand, it is evident that the lower the selling price, the more increase in demand by the following relation: for every two percentage points that the price decreases, demand increases about 2.7%. As for the inventory level, it can be observed that for an expected profit of 50%, 7 orders per year are made to meet its annual demand. These orders will be made every 53 days using an economic order quantity of 6243 units. In contrast, the cycle time for the supplier inventory will be 106 days within which the vendor will make 2 orders. For the expected profits of 30% and 40%, it is perceived to have the same number of orders per year compared to the value of 50%. However, it is possible to observe that the total costs increase as the expected profit percentage decreases. Therefore, the model with expected profit of 50% has the lowest total costs, being 47.4% and 19.6% lower than those models with expected profits set to 30% and 40%, respectively. This phenomenon is associated to holding inventory costs, since the more the duration of inventory increases for the vendor and supplier, the lower the expected profit percentage. Finally, when comparing the average price obtained by direct observation of the market ($ 927,104) to the three studied values, the model that gets closest to this number is the model with *PG*=50%. For this reason, an expected profit of 50% was used for the following analysis.

Table 5

Results of base model

Parameters	PG=30%	PG=40%	PG=50%
Objective Function (COP$)	24,171,500,000	29,250,350,000	33,720,110,000
Price (COP $)	1,079,779	1,050,611	1,030,514
Number of shipments within one supplier's cycle	2	2	2
Duration of vendor's inventory (days)	55	54	53
Number of shipments per year	7	7	7
Duration of supplier's inventory (days)	110	107	106
Q	6023	6159	6243
Demand	40081	41903	43050
Total costs for vendor (COP $)	6,484,300,000	23,723,450,000	20,305,930,000
Total costs for supplier (COP $)	18,634,300,000	14,359,020,000	10,293,500,000

To observe the behavior of PSO throughout iterations, the convergence curve for the objective function for the case of an expected profit set to 50% is shown in Fig. 4.

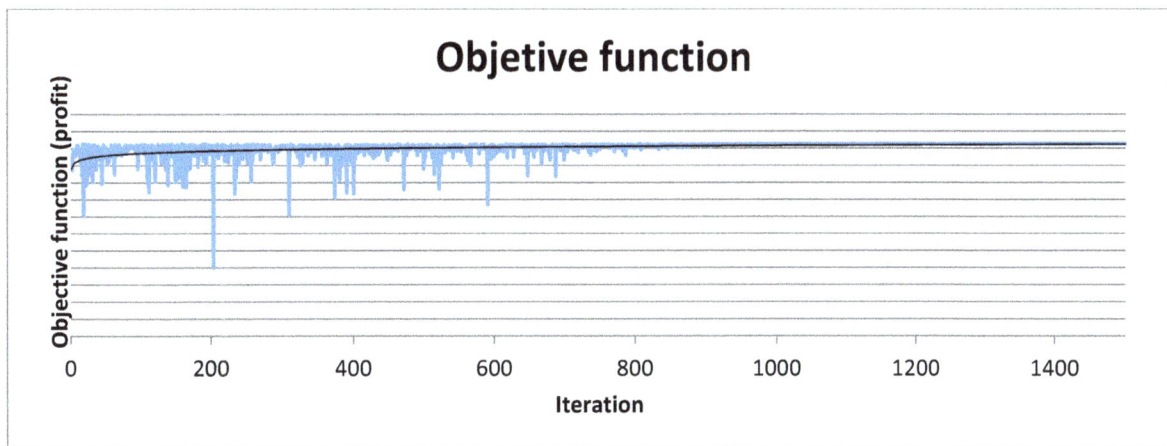

Fig 4. Convergence chart for base model

Analysis of product attractiveness

To analyze product attractiveness, a sensibility analysis study on two factors was conducted: socio-economic group and attractiveness, each of them with three levels (lower, middle, upper), being the price the response variable.

Table 6

Sensibility analysis

Analysis of variance					
Origin of variations	Sums of squares (COP$)	Degrees of freedom	Average of squares	F	P-value
Social stratum (E)	6,484,778,577,204	2	8.24E+12	1.49E+06	0.00%
Attraction (A)	9,750,856,924	2	4.88E+09	8.83E+02	0.00%
E&A	2,409,729,336	4	6.02E+08	1.09E+02	0.00%
Error	1,590,787,406	288	5.52E+06		
Total	16,498,529,950,870	296	5.57E+10		
		$\alpha = 5\%$			

All possible values of attractiveness were evaluated within the intervals defined above and 33 results were obtained for each possible combination between the levels of social stratification and product attractiveness. Once these data obtained, an analysis of variance was performed as shown in Table 6. With a significance level of 5% it was evident that social stratification, product attractiveness and the combination of both factors significantly affect the optimal price, and therefore, they also affect significantly the entire logistic model. Based on the adjusted coefficient of determination, it was found that 99.9% of the observed price variation is explained by the model. Given the above results, it can be noted that product attractiveness becomes relevant in this model for both price definition and inventory management. Nevertheless, it was stated that attractiveness would be equal to 0 in the following analysis, i.e., the utility function depends only on the price variable, similar to the model developed on the model I-JPLMSP designed and analyzed by Yaghin et al. (2014).

4.5.1. Scenario 1: Discounts analysis

In order to analyze the logistic impact by incorporating the sales discounts (r), r is defined based on the expected profit percentage (PG). To avoid losses, the following relationships were established among the selling price found in the market (including discounts) and the purchase cost of vendor:

$$C_v \le PV * (1 - r) \tag{25}$$
$$C_v \le \frac{C_v}{(1 - PG)} (1 - r) \tag{26}$$

Thus:

$$r \le PG \tag{27}$$

Then, the utility function of CMNL model was adjusted as shown in equations (28)-(30)

$$U_h = -b_h P (1 - r) \quad \forall h \in N \tag{28}$$
$$D_h(P) = \alpha_h \frac{e^{-b_h P(1-r)}}{1 + e^{-b_h P(1-r)}} \quad \forall h \in N \tag{29}$$

$$F(P, m, T_v, T_p,) = \sum_h D_h(P) (P(1-r) + C_v) - \left\{ 0,5(m-1)T_v I_p \sum_h D_h(P) + \frac{A_p}{mT_v} + \sum_h D_h(P) C_p \right\}$$
$$- \left\{ 0,5 I_v T_v \sum_h D_h(P) + \frac{A_v}{T_v} + \sum_h D_h(P) C_v \right\} \tag{30}$$

Finally, based on an expected profit of 50% (upper bound) and on direct market observation, it was established that the most common discount on the selling price is 20%, as reported in Table 7. From these results it is noted that by incorporating the discount, all the associated parameters are affected, especially the demand and profits. With a 20% discount, demand increased 4.29% with respect to base model, but the total profit decreased by 36.6%.

Table 7
Results for 20 % discount

Parameters	PG=50%
Objective Function (COP$)	24,690,610,000
Price(COP$)	1,030,514
Number of shipments within one supplier's cycle (m)	2
Duration of vendor's inventory (days)	53
Number of shipments per year	7
Duration of supplier's inventory (days)	106
Q	6510
Demand	44896
Total costs for vendor (COP$)	20,305,930,000
Total costs for supplier (COP$)	10,293,500,000
Price with 20% discount (COP$)	824,411

4.5.2. Scenario 2: Logit model

The objective of this scenario was to analyze the impact of constraints and hence, these were omitted in the calculation of demand with the classical MNL. The results of this scenario are shown in Table 8. From those, it can be observed that by not considering the constraints in the model, demand was overestimated at 26.8% compared to the base model. In turn, this overestimation means that profit increases 4 times above the base model. Additionally, it is evident that price tended to the upper bound imposed, so that for a product such as 32-inch LED television, a high price that is 4,2 times the one found in the market was held.

Table 8

Results of Logit model

Parameters	PG=50%
Objective Function (COP$)	200,486,000,000
Price (COP$)	$ 3,916,667
Number of shipments within one supplier's cycle (m)	2
Duration of vendor's inventory (days)	47
Number of shipments per year	8
Duration of supplier's inventory (days)	94
Q	7031
Demand	54611
Total costs for vendor (COP$)	25,709,260,000
Total costs for supplier (COP$)	3,012,920,000

4.5.3. Scenario 3: Segmented demand

In order to analyze the influence and logistic impacts of each socio-economic group defined within the demand, a model was run for each group, taking into account a product attractiveness equal to 0, an expected profit equal to 50% and an estimated demand through the CMNL. The results obtained are presented in Table 9. Meanwhile, Table 10 shows the contribution of each socio-economic group to the aggregate demand of the base model. From these results it can be noted that, as for the base model, the greatest demand is associated to middle-stratum, followed by the lower-stratum and finally by the upper-stratum. This demand behavior is explained due to the distribution of the surveyed population, since 26% corresponds to lower-social stratum, 59.6% to middle-stratum, and 14.4% to upper-stratum. Furthermore, it is noted that demand for each group varies in different proportions in relation to the selling price, and therefore, by taking the price of the base model as a reference, it was evident that price decreased by 7% in lower-stratum, and demand increased by 12%. For middle-social stratum, price increased 1% and consequently demand decreased 1.4%. For the latter group, the upper-social stratum, the price increased by 26% and demand decreased by 7%.

Table 9

Results for aggregate demand

Parameters	*Lower*	*Middle*	*Upper*
Objective Function (COP$)	$ 912,741,000	$ 22,901,800,000	$ 1,457,199,000
Price (COP$)	$ 962,381	$ 1,042,319	$ 1,386,959
Number of shipments within one supplier's cycle	2	2	2
Duration of vendor's inventory (days)	93	65	297
Number of shipments per year	4	6	1.2
Duration of supplier's inventory (days)	185	129	595
Q	3571	5117	823
Demand	14089	28928	1010
Total costs for vendor (COP$)	$,731,220,000	$ 13,696,550,000	$ 694,587,900
Total costs for supplier (COP$)	$,446,009,000	$ 6,963,481,000	$ 372,309,900

Regarding the duration of supplier's inventory, it can be observed that an order is made every 297 days for the upper-social stratum in order to meet their demand, being its inventory cycle of 595 days. This long duration may significantly affect the devaluation of inventory, meaning that as time passes the monetary unit will lose their commercial value because of technology advances, which significantly impact the televisions market, with high risk of technology obsolescence.

Finally it is noted that the target population is located at middle-social stratum since this stratum contributes 65.7% of total demand. For this reason, and to facilitate comparison among different the proposed scenarios, the impact of this scenario was assessed only by using the middle-stratum.

Table 10

Influence of groups on aggregate demand

Sub-groups	Lower	Middle	Upper
PG= 50%	12623	29341	1086

5. Conclusions

This paper has presented a logistic optimization, non-linear, multi-echelon model that integrates the selling price of a product and the level of inventories, including a constraint about the maximum willingness to pay of customers, taking into account different socio-economic groups. To estimate the demand, the discrete choice multinomial restricted logit (CMNL) model was used in order to include soft constraints on the purchase price. Moreover, the optimal purchase price, as well as the coordination inventory variables, were obtained through a metaheuristic based on particle swarm optimization (PSO).

To validate the proposed model, the population was segmented into lower, middle and upper classes depending on their income, according to the Colombian social stratification. In assessing the behavior of each socio-economic group faced to the choice of the same product, the choice is conditioned by the maximum willingness to pay. In contrast, from the vendor's perspective it is important to consider the minimum selling price. Therefore, there is a need to use the constrained multinomial logit model for modeling demand, in order to include the aforementioned constraints.

Through the application example, the behavior of the proposed aggregate demand model could be analyzed against three scenarios. The first scenario focused on the base model with a sensitivity analysis on possible discounts, the second scenario estimated demand by means of a non-constrained multinomial logit model and the third scenario contemplated a segmented demand for each socio-economic group. From the first scenario, it was observed that when considering a discounts policy based on the vendor's expected profit, discounts increased market share but not necessarily generated higher profits compared to those obtained with the base model. That is, the value of the objective function decreased, compared to the base model, and in turn, demand showed an increase which can be taken as a growth strategy to increase market share in the long term.

For the second scenario, the non-constrained logit model obtained different results compared to other proposed scenarios. By omitting constraints, a significant increase in demand and in joint profits of supplier and vendor were observed. However, by not including constraints, this model is unreliable because it is not adapted to realistic consumer behavior. Consequently, the inventory level is affected by an increase of 12.6%, which can be risky due to an overestimation of demand that brings as a result a lower product turnover. In addition, it is evident that the selling price of the product obtained in this scenario differs to the maximum willingness to pay of the target population, reinforcing the low turnover of the product compared with the scenario that contemplates constraints. On the other hand, the third scenario, with segmented demand, allows to identify different target populations, letting companies to create strategies for each niche market, focused on the needs of this population and in order to develop customized products or services. Furthermore, a sales policy could be defined based on the location of a vendor and on the socio-economic environment where it is found, aimed to establish different prices that are aligned with the highest willingness to pay of the socio-economic group. Finally, future work may be focused on expanding the model to integrate multiple products, multiple vendors and/or suppliers, and to include independent prices for each vendor.

References

Ardjmand, E., Weckman, G. R., Young, W. A., Sanei Bajgiran, O., & Aminipour, B. (2016). A robust optimisation model for production planning and pricing under demand uncertainty. *International Journal of Production Research, 54*(13), 3885-3905.

Bai, T., Kan, Y. B., Chang, J. X., Huang, Q., & Chang, F. J. (2017). Fusing feasible search space into PSO for multi-objective cascade reservoir optimization. *Applied Soft Computing, 51*, 328-340.

Berger, P. D., & Bechwati, N. N. (2001). The allocation of promotion budget to maximize customer equity. *Omega, 29*(1), 49-61.

Block, H. D., & Marschak, J. (1960). Random orderings and stochastic theories of responses. *Contributions to probability and statistics, 2*, 97-132.

Bogotá, A. M. (2003). CARACTERIZACIÓN SOCIOECONÓMICA DE BOGOTÁ Y LA REGIÓN–V8.

Caicedo, F., Lopez-Ospina, H., & Pablo-Malagrida, R. (2016). Environmental repercussions of parking demand management strategies using a constrained logit model. *Transportation Research Part D: Transport and Environment, 48*, 125-140.

Castillo-López, I., & López-Ospina, H. A. (2015). School location and capacity modification considering the existence of externalities in students school choice. *Computers & Industrial Engineering, 80*, 284-294.

Castro, M., Martínez, F., & Munizaga, M. A. (2013). Estimation of a constrained multinomial logit model. *Transportation, 40*(3), 563-581.

Chen, M., & Chen, Z. L. (2015). Recent developments in dynamic pricing research: multiple products, competition, and limited demand information. *Production and Operations Management, 24*(5), 704-731.

Deng, S., & Yano, C. A. (2006). Joint production and pricing decisions with setup costs and capacity constraints. *Management Science, 52*(5), 741-756.

Ding, Y., Veeman, M. M., & Adamowicz, W. L. (2012). The influence of attribute cutoffs on consumers' choices of a functional food. *European Review of Agricultural Economics, 39*(5), 745-769.

Geunes, J., Romeijn, H. E., & Taaffe, K. (2006). Requirements planning with pricing and order selection flexibility. *Operations Research, 54*(2), 394-401.

Ghoniem, A., & Maddah, B. (2015). Integrated retail decisions with multiple selling periods and customer segments: optimization and insights. *Omega, 55*, 38-52.

Ghoreishi, M., Weber, G. W., & Mirzazadeh, A. (2015). An inventory model for non-instantaneous deteriorating items with partial backlogging, permissible delay in payments, inflation-and selling price-dependent demand and customer returns. *Annals of Operations Research, 226*(1), 221-238.

González, C., & Serna, N. (2013). The consumer's choice among television displays: A multinomial logit approach. *Lecturas de Economía, (79)*, 199-228.

Guedria, N. B. (2016). Improved accelerated PSO algorithm for mechanical engineering optimization problems. *Applied Soft Computing, 40*, 455-467.

Hashemi, S. M., Rezapour, M., & Moradi, A. (2010). An effective hybrid PSO-based algorithm for planning UMTS terrestrial access networks. *Engineering Optimization, 42*(3), 241-251.

Herrera Rojas, C. (2014). Desarrollo de un modelo de elección de ruta en metro. Master thesis. *Universidad de Chile.*

Van den Heuvel, W., & Wagelmans, A. P. (2006). A polynomial time algorithm for a deterministic joint pricing and inventory model. *European Journal of Operational Research, 170*(2), 463-480.

Jafari, H., Soltani, A., & Soltani, M. (2013). Measuring the performance of FCM versus PSO for fuzzy clustering problems. *International Journal of Industrial Engineering Computations, 4*(3), 387-392.

Karimi-Nasab, M., Modarres, M., & Seyedhoseini, S. M. (2015). A self-adaptive PSO for joint lot sizing and job shop scheduling with compressible process times. *Applied Soft Computing*, *27*, 137-147.

Kennedy, J., & Mendes, R. (2002). Population structure and particle swarm performance. In *Evolutionary Computation, 2002. CEC'02. Proceedings of the 2002 Congress on* (Vol. 2, pp. 1671-1676). IEEE.

Kim, D., & Lee, W. J. (1998). Optimal joint pricing and lot sizing with fixed and variable capacity. *European Journal of Operational Research*, *109*(1), 212-227.

Kumar, E. V., Raaja, G. S., & Jerome, J. (2016). Adaptive PSO for optimal LQR tracking control of 2 DoF laboratory helicopter. *Applied Soft Computing*, *41*, 77-90.

López-Ospina, H. A., Martínez, F. J., & Cortés, C. E. (2016). Microeconomic model of residential location incorporating life cycle and social expectations. *Computers, Environment and Urban Systems*, *55*, 33-43.

López-Ospina, H. A., Cortés, C. E., & Martínez, F. J. (2017). Residential relocation dynamics: A microeconomic model based on agents' socioeconomic change and learning. *The Journal of Mathematical Sociology*, *41*(1), 46-61.

Márquez-Díaz, L. G., Gallo-González, L. A., & Chacón-Pérez, C. A. (2011). Influence of Parking Costs on the Use of Cars in Bogota. *Ingeniería y Universidad*, *15*(1), 105-124.

Martínez, F., Aguila, F., & Hurtubia, R. (2009). The constrained multinomial logit: A semi-compensatory choice model. *Transportation Research Part B: Methodological*, *43*(3), 365-377.

Martínez, F., & Donoso, P. (2010). The MUSSA II land use auction equilibrium model. In *Residential Location Choice* (pp. 99-113). Springer Berlin Heidelberg.

Martinez, F. J., Tamblay, L., & Weintraub, A. (2011). School locations and vacancies: a constrained logit equilibrium model. *Environment and Planning A*, *43*(8), 1853-1874.

Moghadam, B., & Seyedhosseini, S. (2010). A particle swarm approach to solve vehicle routing problem with uncertain demand: A drug distribution case study. *International Journal of Industrial Engineering Computations*, *1*(1), 55-64.

McFadden, D. (1975). The revealed preferences of a government bureaucracy: Theory. *The Bell Journal of Economics*, 401-416.

Perez, J., Lopez-Ospina, H., Cataldo, A., & Ferrer, J. C. (2016). Pricing and composition of bundles with constrained multinomial logit. *International Journal of Production Research*, *54*(13), 3994-4007.

Shavandi, H., Mahlooji, H., & Nosratian, N. E. (2012). A constrained multi-product pricing and inventory control problem. *Applied Soft Computing*, *12*(8), 2454-2461.

Taleizadeh, A. A., & Noori-daryan, M. (2016). Pricing, manufacturing and inventory policies for raw material in a three-level supply chain. *International Journal of Systems Science*, *47*(4), 919-931.

Tiempo, Casa Editorial El. (2016a). "El Mercado de Televisores Mueve US$ 1.000 Millones Al Año." *Portafolio.co*. Accessed May 27.

Tiempo, Casa Editorial. (2016b). "Mundial de Brasil Cambiará El Ciclo de Ventas de Televisores." *Portafolio.co*. Accessed May 27.

Xu, J., Zeng, Z., Han, B., & Lei, X. (2013). A dynamic programming-based particle swarm optimization algorithm for an inventory management problem under uncertainty. *Engineering Optimization*, *45*(7), 851-880.

Yaghin, R., Ghomi, S. M. T., & Torabi, S. A. (2014). Enhanced joint pricing and lotsizing problem in a two-echelon supply chain with logit demand function. *International Journal of Production Research*, *52*(17), 4967-4983.

Zahara, E., & Hu, C. H. (2008). Solving constrained optimization problems with hybrid particle swarm optimization. *Engineering Optimization*, *40*(11), 1031-1049.

3

Simultaneous improvement of surface quality and productivity using grey relational analysis based Taguchi design for turning couple (AISI D3 steel/ mixed ceramic tool (Al₂O₃ + TiC))

Oussama Zerti[a]*, Mohamed Athmane Yallese[a], Abderrahmen Zerti[a], Salim Belhadi[a] and Francois Girardin[b]

[a]Mechanics and Structures Research Laboratory (LMS), Mechanical Engineering Dept., May 8th 1945 University, Guelma 24000, Algeria
[b]Laboratoire Vibrations Acoustique, INSA-Lyon, 25 bis avenue Jean Capelle, F-69621 Villeurbanne Cedex, France

CHRONICLE	ABSTRACT
Keywords: *Simultaneous improvement* *GRA* *Taguchi design* *S/N ratio* *ANOVA* *AISI D3 Steel* *Ceramic*	Current optimization strategies are based on the increase the productivity and the quality with lower cost in short time. Grey relational analysis "GRA" based on Taguchi design was proposed in this paper for simultaneous improvement of surface quality and productivity. The turning trials based on mixed Taguchi L18 factorial plan were conducted under dry cutting conditions for the machining couple: AISI D3 steel/mixed ceramic inserts (CC650). The machining parameters taken into account during this study are as follow: major cutting edge angle (χr), cutting insert nose radius (r), cutting speed (Vc), feed rate (f), and depth of cut (ap). Significant effects of machining parameters and their interactions were evaluated by the analysis of variance. Through this analysis, it have been found clearly that feed rate and cutting insert nose radius had a big significant effects on surface quality while depth of cut, feed rate followed by cutting speed had a major effect on productivity. The mathematical relationship between the machining parameters and the performance characteristics was formulated by using a linear regression model with interactions. Optimal levels of parametric combination for achieving the higher surface quality with maximum productivity were selected by grey relational analysis which is based on the high value of grey relational grade. Confirmation experiments were carried out to prove the powerful improvement of experimental results and to validate the effectiveness of the multi-optimization technique applied in this paper.

Nomenclature

ANOVA	Analysis of variance	OA	Orthogonal array
ap	Depth of cut (mm)	*Ra*	Arithmetic mean roughness (μm)
Cont %	Contribution ratio (%)	*r*	Nose radius of cutting insert (mm)
DF	Degrees of freedom	RSM	Response surface methodology
f	Feed rate (mm/rev)	SS	Sum of squares
GRA	Grey relational analysis	S/N	Signal-to-noise ratio
GRC	Grey relational coefficient	*Vc*	Cutting speed (m/min)
GRG	Grey relational grade	α	Clearance angle (degree)
MS	Mean squares	γ	Rake angle (degree)
MRR	Material removal rate (mm³/min)	λ	Inclination angle (degree)
χr	Major cutting edge angle (degree)		

* Corresponding author
E-mail: oussama_zerti@yahoo.fr (O. Zerti)

1. Introduction

Surface roughness represents an evaluation criterion of quality products and it plays an important role to estimate the manufacturing cost (Asiltürk & Akkus, 2011; Zerti et al., 2017a,b). On the other hand, the productivity is considered as a very important technological aspect that causes great effect on both product cost and series production rate (Hassan et al., 2012). For this reason, manufacturers are always committed by the good quality and high productivity in short time with low cost, because ensuring those conditions represent an index of manufacturer's qualification. Consequently, the desired surface quality with the maximum productivity is a major constraint for the choice of the optimum machining parameters in the production process. In order to achieve the desired conditions required by customers, the use of grey relational analysis as a multi-objective optimization technique based on Taguchi design is found as an efficient solution for this optimization problem. This technique has been used in different applications in because of its ease of application and reliability.

There are a number of researchers in different fields who have used grey relational analysis based on Taguchi design for a simultaneous improvement of multi-performance characteristics in order to achieve at the desired objective.

Bouzid et al. (2014) optimized cutting parameters for determining the minimum surface roughness (Ra) which corresponds to the maximum material removal rate (MRR) in turning of X20Cr13 steel with mono and multi-objective optimizations based on the L16 OA of Taguchi. Taguchi's signal-to-noise ratio was used to accomplish the objective function. Wang and Lan (2008) selected the optimum cutting conditions through the application of grey relational analysis based on Taguchi design (L9 OA) with introducing of signal to noise ratio (S/N) to get the lowest surface roughness and tool wear that correspond to the maximum of material removal rate in precision turning.

Lin (2004) reported an improvement of tool life, cutting force, and surface roughness by using Taguchi method with grey relational analysis for optimizing cutting speed, feed rate, and depth of cut during turning operations of S45C steel bars using a P20 tungsten carbide. They found that optimization of complicated multiple performance characteristics could be greatly simplified through this approach. Balasubramanian and Ganapathy (2011) solved the problem of simultaneous optimization for wire electro discharge machining (WEDM) to obtain higher material removal rate (MRR) and lower surface roughness (SR) by the use of grey relational analysis.

Hanafi et al. (2012) applied the method of grey relational analysis based on the Taguchi method for multi-objective optimization of power consumption and surface roughness when dry turning of the PEEK reinforced with 30% carbon fibers. The same technique was proposed in other several cutting process, for example: in the drilling process Noorul Haq et al. (2008) identified optimal drilling parameters namely: cutting speed, feed rate and point angle for multiple response characteristics such as surface roughness, cutting force and torque in the case of machining couple: Al/SiC metal matrix composite/TiN coated HSS twist drills under dry condition. The authors found that this technique is so reliable for improving the drilling process.

For another type of cutting process, Kuram and Ozcelik (2013) performed an experimental investigation based on the L9 OA of Taguchi method for Micro-milling of aluminium material with ball nose end mill. The authors applied Taguchi method and grey relational analysis to achieve at mono and multi-objective optimization. The works accomplished by Jailani et al. (2009) aimed to use grey relational analysis in order to optimize the sintering process parameters of Al–Si (12%) alloy/fly ash composite. The modelling of the cutting process has attracted the attention of many researchers for its great interest in the industry because it allows predicting the technological parameters without carrying out the experimental tests. An attempt was made by Gaitonde et al. (2009) to determine the link between cutting condition such as cutting speed, feed rate, and machining time and machinability aspects via response surface methodology. These considered aspects were machining force, power, specific cutting force, surface roughness, and tool wear. The study was carried out for the case of turning of high chromium AISI D2 cold work tool steel using CC650WG wiper ceramic inserts. Authors found through the response surface analysis that the surface roughness could be minimized at small values of feed rate and machining time with elevated values of cutting speed, whereas the maximum tool wear appear at Vc = 150 m/min for all values of feed rate.

Al-Ahmari (2007) formulated mathematical equations of surface roughness and cutting forces during turning of austenitic AISI 302 steel. The process parameters considered in this study were cutting speed, feed rate, depth of cut and nose radius in order to develop a machinability model. Additionally, response surface methodology (RSM) and neural networks (NN) were employed to assess the model. Zahia et al. (2015) exploited the RSM methodology that helps to formulate a reliable statistical model for monitoring the evolution of surface roughness and cutting forces according to cutting parameters such as: cutting speed, feed rate and depth of cut during the hard turning (AISI 4140) (56 HRC) with using PVD – coated ceramic insert. Zahia et al. (2013) developed a mathematical model of surface roughness that vary in function of cutting parameters, tool-nose displacements, spindle and machine tool frame. The study of Neseli et al. (2011) presented an application of response surface methodology (RSM) for modeling the average surface roughness (Ra) obtained during the turning of AISI 1040 steel, to assess the effect of tool geometry parameters on the latter. They found that the tool nose radius was the most influencing factor on the measured surface roughness.

Ceramic cutting tool is a big utilization in the machining of high alloy steel, Davim and Figueira (2007a) made a comparison between the wiper and conventional ceramics inserts to determine the influences of cutting parameters on the obtained machinability parameters (cutting forces, surface roughness, and tool wear). They found after that the use of wiper ceramics inserts allow to reach at surface roughness values less than 0.8 μm with possibility of dimensional accuracy in a work-piece, IT < 7.

Davim and Figueira (2007b) used ceramic inserts for surface finishing phase on the same material (cold work tool steel AISI D2). They revealed that obtaining surface roughness of less than 0.8 μm is feasible if the choice of cutting parameters is suitable and which also permit to eliminate cylindrical grinding operations. Aouici et al. (2014) examined the machinability behavior of cold work hard tool steel AISI D3 heat-treated (60 HRC) with a TiN doped ceramic cutting tool (SNGA120408) containing approximately 30% of TiC. The responses were estimated based on a (3^3) full factorial experimental design, where the quadratic effects were also determined. The desired optimum was set for minimum levels of surface roughness, cutting force, specific cutting force and consumed power via the statistical method (RSM) and the desirability function approach.

Singh and Dureja (2014) compared Taguchi method and RSM with a view for optimizing flank wear of tool and surface roughness during the finish operation of AISI D3 steel in hard turning. The results indicated that optimal levels of cutting parameters selected by both RSM and Taguchi method were nearly the same. Zerti et al. (2017) proposed a study with the application of Taguchi method to minimize some technological parameters (such as surface roughness, tangential force, specific cutting force, and cutting power) characterizing material machinability. They carried out 18 tests based on Taguchi design experiments during the turning of AISI D3 steel using mixed ceramic inserts (CC650) under dry cutting conditions. Bouchelaghem et al. (2010) examined the machinability behavior of AISI D3 hardened steel with CBN cutting tool for the evolution of surface roughness, cutting forces and tool wear in function of variation of cutting parameters. Bensouilah et al. (2016) conducted a comparative study to evaluate the performance of coated and uncoated mixed ceramic tools during hard turning of AISI D3 cold work tool steel. They determined the effects of cutting parameters on the machining performance through the use of ANOVA analysis of S/N ratio of the responses. The authors modeled the machining performance by linear regression for both ceramic tools CC6050 and CC650. Yallese et al. (2005) evaluated the effect of cutting parameters during the hard turning of AISI D3 steel with ceramic and CBN tool wear. They estimated the surface roughness by a power model deduced from experimental data and compared it with a theoretical model. Meddour et al. (2015) performed a statistical study to determine the significant effect of cutting speed, depth of cut, feed rate and tool nose radius on surface roughness and components of cutting force during hard turning of AISI 52100 steel by mixed ceramic cutting tool. They developed mathematical models in order to estimate those two responses. Also, they recommended that the use of big nose radius and little feed rates could improve surface quality.

The present research paper shows an experimental investigation related to the simultaneous improvement of surface quality (Ra) and productivity (MRR) using the application of grey relational analysis (GRA) based on Taguchi design (L18 OA) during the dry turning of (AISI D3 steel/ mixed ceramic inserts). Response Surface Methodology (RSM) was exploited to obtain an empiric mathematical models by regression analysis for the surface roughness and material removal rate. The ANOVA analysis of S/N ratio described the degree of influence of each of the control machining parameters and their interactions on each response. Also Pareto chart and 3D plots with their contours based on S/N ratios of responses were used to confirm the results found by ANOVA analysis. 3D surface roughness profile was made to view visualizing its topography. Confirmation tests were carried out to ensure the effectiveness of the grey relational analysis based on Taguchi design in the simultaneous improvement of the performance characteristics considered in this study.

2. Taguchi design / Grey Relational Analysis (GRA)

2.1 Taguchi design

Taguchi design is a helpful technique that has a big contribution for the improvement of the performance of systems and solving complex optimization problems (settings) during production of the product by the implementation of the design experiments that is based on the use of the orthogonal arrays which are proposed by Taguchi for minimizing the number of trials and focusing just on the essential experiments for analyzing, which lead to win the time and reducing the cost Taguchi (1986). Also this method allows controlling simultaneously controllable and uncontrollable factors by converting the responses into signal-to-noise (S/N) for identifying industrial performance of the system Zhang et al. (2007). S/N ratio is the essential criterion in the Taguchi method, it allows defining the degree of influence of the unwanted noise on the wanted signal Günay et al. (2011). Whenever the characteristic is continuous, the S/N ratios are usually divided into 3 categories given by the following equations Nalbant et al. (2007):

For (Nominal is the best): $S/N = 10\log\left(\frac{\bar{y}}{S_y^2}\right)$ (1)

For maximization (Larger-is-the better): $S/N = -10\log\left(\frac{1}{n}\sum_{i=1}^{n}\frac{1}{y_i^2}\right)$ (2)

For minimization (Smaller-is-the better): $S/N = -10\log\left(\frac{1}{n}\sum_{i=1}^{n}y_i^2\right)$ (3)

where $\bar{y}$ is the average of results obtained, S_y^2 is the variance of y, n is the number of repeat trials and y_i is the result obtained.

2.2 Grey Relational Analysis (GRA)

Grey relational analysis is a technique proposed for solving the problem of complex optimization by converting the multi-objective to a single-objective to achieve at optimal combination of parameters levels for simultaneous improvement of multiple machining characteristics Dabade (2013). The use of this method contains the steps as follow:

Step 1: Grey relational generation

According to the intended objective optimization to minimize or maximize experimental results, normalization of S/N ratio for the experimental results in the range between zero and one is necessary for grey relational generation. Depending on the objective function optimization, the normalization can be performed for two cases. If the smaller-the-better is the characteristic selected in the original sequence for minimization, then it should be normalized as given by Eq. (4).

$$x_i^*(k) = \frac{\max(x_i^0(k)) - x_i^0(k)}{\max(x_i^0(k)) - \min(x_i^0(k))}$$ (4)

If the larger-the-better is the characteristic selected in the original sequence for maximization, then it should be normalized as given by Eq. (5).

$$x_i^*(k) = \frac{x_i^0(k) - \min(x_i^0(k))}{\max(x_i^0(k)) - \min(x_i^0(k))}$$ (5)

where $x_i*(k)$ is the value after grey relational generation (normalized value), and $\max(x_i^0(k))$ and $\min(x_i^0(k))$ are the largest and smallest values of $x_i^0(k))$ for the kth response. The larger value of normalized results indicates the better performance characteristic and the best-normalized results will be equal to one.

Step 2: Grey Relational Coefficient (GRC)

Grey relational coefficient describes the correlation between the ideal and the obtained experimental results. Mathematical formula of grey relational coefficient ($\xi_i(k)$) is given as following:

$$\xi_i(k) = \frac{\Delta_{min} + \psi\Delta_{max}}{\Delta_{0i}(k) + \psi\Delta_{max}}$$ (6)

$$0 < \xi_i(k) \leq 1$$

$\Delta_{0i}(k)$ is the absolute difference between the reference sequence $x_0^k(k)$ and the S/N ratio of measured sequence $x_i^k(k)$.

$$\Delta_{0i}(k) = \left\| x_0(k) - x_i(k) \right\|$$ (7)

$$\Delta_{min} = \min_{\forall j \in i} \min_{\forall k} \left\| x_0(k) - x_i(k) \right\|$$ (8)

$$\Delta_{max} = \max_{\forall j \in i} \max_{\forall k} \left\| x_0(k) - x_i(k) \right\|$$ (9)

ψ is the distinguishing coefficient ($\psi \in [0, 1]$). In this study the value of ψ is 0.5.

Step 3: Grey Relational Grade (GRG)

Grey relational grade represents the correlation among the series, it is given by the following formula:

$$\alpha_i = \frac{1}{n}\sum_{k=1}^{n} \xi_i(k)$$ (10)

where n is the number of responses.

Step 4: Determination of optimal machining parameters

Once grey relational grade is computed, the selection of the optimal levels combination is made based on the main effects plot for (GRG). The largest value of grey relational grade that is found close to the ideal normalized value corresponds to the optimal combination. Therefore, the optimal level of the process parameters is the level with the greatest GRG value.

Step 5: Confirmation tests

Once the optimal levels are selected, the validation test occupied the final step in the optimization procedure to confirm the reliability of the optimal levels is proposed by grey relational analysis to improve system performance. This test is done by comparing the value of the S/N ratio of GRG obtained from the optimal test with that predicted $\hat{\eta}$ by the following formula with the use of optimal levels (Nalbant et al., 2007):

$$\hat{\eta} = \eta_m + \sum_{i=1}^{q}(\overline{\eta}_i - \eta_m),\tag{11}$$

where η_m is the total average of S/N ratio, $\overline{\eta}_i$ is the average of S/N ratio at the optimal level, and q is the number of the main input factors that have a significant effect on the output responses.

2.3 Grey Relational Analysis optimization based Taguchi design

Based on the above discussion, the use of the grey relational analysis coupled with Taguchi design in order to optimize the turning operations with multiple machining characteristics includes the following steps as shown in Fig. 1.

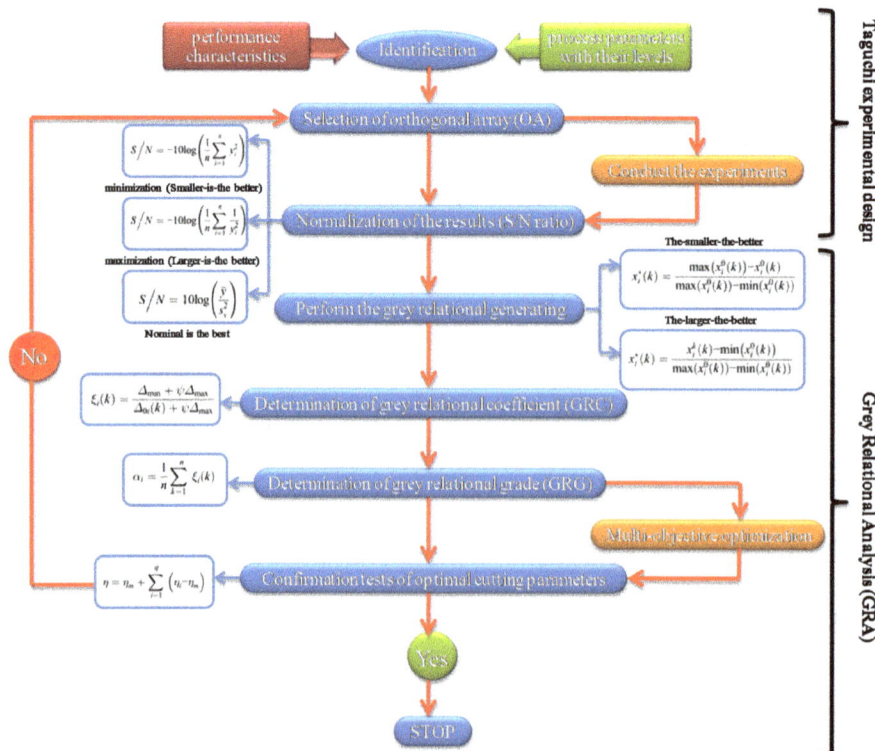

Fig. 1. Loop of multi-objective optimization of grey relational analysis (GRA) based on Taguchi experimental design.

3 Experimental set-up

During this experimental investigation, turning operations were carried out on a conventional lathe of the check company "TOS TRENCIN" SN40 model, with 6.6 kW spindle power for the turning couple (AISI D3 steel/mixed ceramic inserts CC650) under dry conditions.

3.1 Work piece material, cutting inserts and tool holders

The work piece material used was a round bar of AISI D3 steel having 70 mm in diameter and 400 mm in length. This lather is a high alloy steel that have several designation such as: DIN 1.2080, JIS SKD1, GB Cr12, AFNOR Z200Cr12. It is a tool steel with high chromium minimum risk of deformation and alteration of dimensions to thermal treatments and it has excellent wear resistance. Its chemical composition is given as follow: 2% of carbon (C), 0.30% of Manganese (Mn), 0.25% of silicon (Si), 12% of Chrome (Cr), 0.70% of Tungsten (W).

All turning operations were carried out by three mixed ceramic cutting tools CC650 were manufacturing by Sandvik Coromant and its chemical composition is as follow (Al_2O_3 (70%) + TiC (30%)). Each cutting tool is characterized by a nose radius r = 0.8mm, 0.12mm, 0.16mm and ISO geometric designation SNGA120408T01020, SNGA120412T01020, SNGA120416T01020 respectively.

Two tool holders are used in this investigation designated by ISO as PSDNN 2525 M12 and PSBNR 2525 M12, respectively. Their geometry of the active part; as shown in Fig. 2 is the same for the following angles: clearance angle (α) = 6°, rake angle (γ) = -6°, and cutting edge inclination angle (λ) = -6°, but it is different to the major cutting edge angle (χr) = 45° and 75°, respectively.

3.2 Design Experiments and cutting conditions

Mixed factorial plan reduces of Taguchi L_{18} was selected as an experimental design to study the impact of different machining parameters (Vc, f, ap) which varies at three levels (3^4) and tool geometry ($\chi r, r$) which varies at two levels (2^1) on the performance characteristics (Ra and MRR). The levels of the parameters were selected in the range of intervals advisable by manufacturer of Sandvik Coromant. The cutting parameters chosen to study with their levels are shown in Table 1.

Table 1
Process parameters and their levels.

Factor	symbol	Unit	Level 1	Level 2	Level 3
Cutting speed	Vc	m/min	220	307	440
Feed rate	f	mm/tr	0.08	0.12	0.16
Depth of cut	ap	mm	0.15	0.3	0.45
Nose radius	r	mm	0.8	1.2	1.6
major cutting edge angle	χr	Degree(°)	45	75	-

3.3 Measuring equipment

3.3.1 Surface roughness measure

The criterion measures of the surface roughness (Ra) are obtained instantly after each pass roughing by means of a Mitutoyo Surftest SJ-201 roughness meter. To prevent errors and recovery for more precision, roughness measurement was performed directly on the work-piece without dismounting it from the lathe. The measurements were repeated three times along three work-piece feed rate directions also placed at 120° (Fig. 2). The result is considered as the average of these values for each cutting condition. To properly characterize the surface roughness of the work-piece, three-dimensional topographic maps were made using an optical platform of metrology modular Altisurf 500.

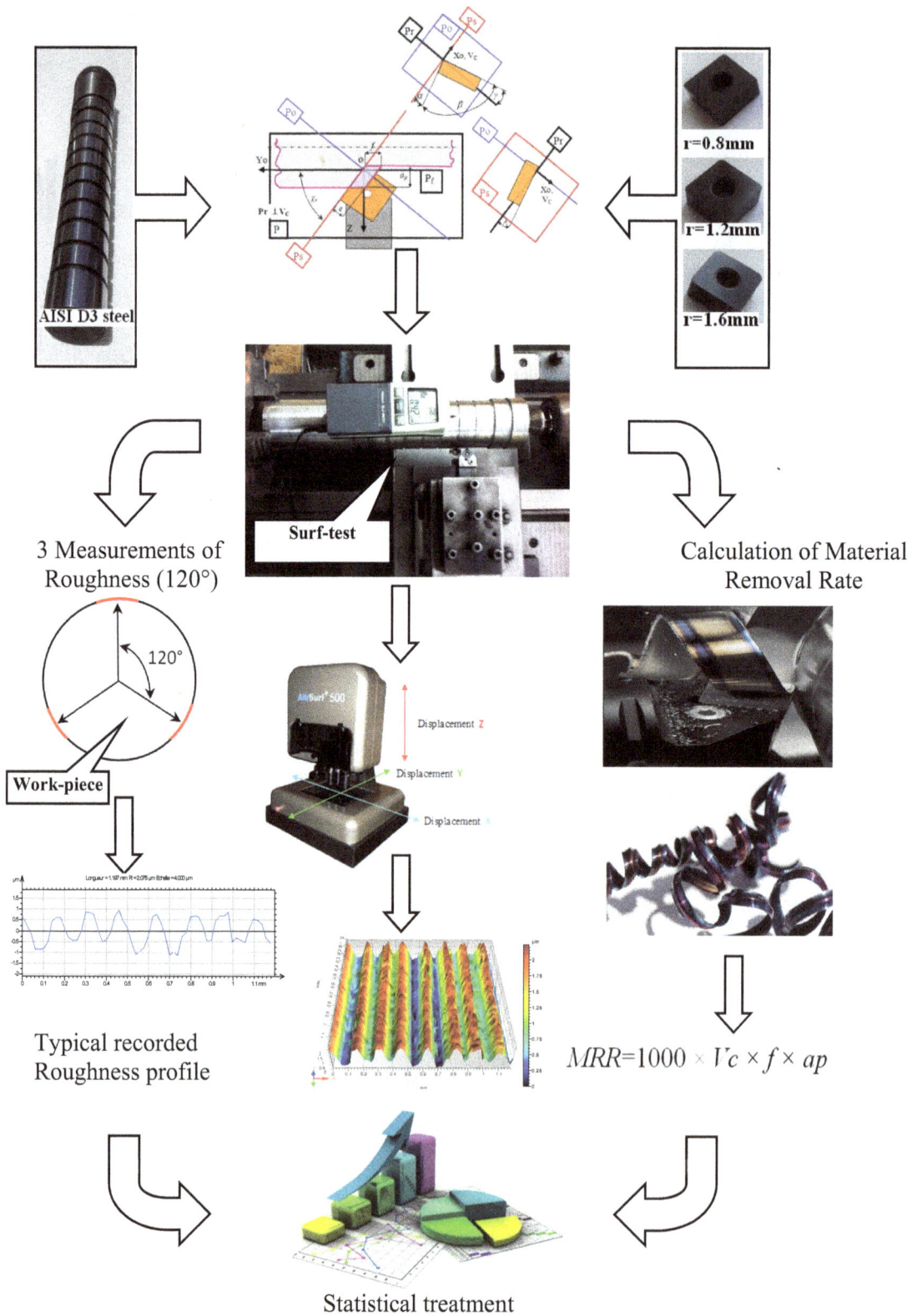

Fig. 2. Schematic diagram of the experimental set-up.

3.3.2 Formula of material removal rate

Material removal rate can be defined as the volume of material removed divided by the machining time. Another way, *MRR* is to imagine an "instantaneous" material removal rate as the rate at which the cross-section area of material being removed moves through the work-piece. This aspect of machinability is calculated using the following equation:

$$MRR = 1000 \times Vc \times f \times ap, \tag{12}$$

where *Vc* is the cutting speed (m/min), *f* is the feed rate (mm/rev), and *ap* is the depth of cut (mm) and *MRR* is the material removal rate (mm^3/min).

4. Data analysis and results

The measured values of surface roughness (*Ra*) and the calculated values of material removal rate (*MRR*) using Eq. (12) with their computed S/N ratio in this experimental study which was carried out based on various combinations of machining parameters levels proposed by Taguchi design (L$_{18}$ OA) are shown in Table 2. "The larger is the better" and "The smaller is the better" characteristics are used to calculate the S/N ratio in order to maximizing (*MRR*) and minimizing (*Ra*); i.e. Maximizing surface quality; using equations 2 and 3, respectively.

Table 2
Experimental results for surface roughness and material removal rate with their computed S/N ratios

Trail no.	Machining parameters					Response parameters			
	χr (°)	*r* (mm)	*Vc* (m/min)	*f* (mm/rev)	*ap* (mm)	*Ra* (μm)	S/N (dB)	*MRR* (mm^3/min)	S/N (dB)
1	45	0.8	220	0.08	0.15	0.47	6.56	2640	68.43
2	45	0.8	307	0.12	0.3	0.59	4.63	11052	80.87
3	45	0.8	440	0.16	0.45	0.63	3.97	31680	90.02
4	45	1.2	220	0.08	0.3	0.43	7.26	5280	74.45
5	45	1.2	307	0.12	0.45	0.55	5.14	16578	84.39
6	45	1.2	440	0.16	0.15	0.78	2.20	10560	80.47
7	45	1.6	220	0.12	0.15	0.39	8.10	3960	71.95
8	45	1.6	307	0.16	0.3	0.57	4.83	14736	83.37
9	45	1.6	440	0.08	0.45	0.40	8.03	15840	84.00
10	75	0.8	220	0.16	0.45	1.01	-0.09	15840	84.00
11	75	0.8	307	0.08	0.15	0.43	7.40	3684	71.33
12	75	0.8	440	0.12	0.3	0.65	3.70	15840	84.00
13	75	1.2	220	0.12	0.45	0.39	8.18	11880	81.50
14	75	1.2	307	0.16	0.15	0.54	5.35	7368	77.35
15	75	1.2	440	0.08	0.3	0.33	9.72	10560	80.47
16	75	1.6	220	0.16	0.3	0.51	5.79	10560	80.47
17	75	1.6	307	0.08	0.45	0.41	7.74	11052	80.87
18	75	1.6	440	0.12	0.15	0.43	7.33	7920	77.97

4.1 Analysis of variance (ANOVA)

ANOVA allows determining the significance of input parameters (χr, *r*, *Vc*, *f*, *ap*) and their interactions in order of influence on the responses (*Ra*, *MRR*). The model is based on the calculation of the sums of the squares of the (S/N) ratios of outputs. First we calculate the sum of squared deviations of the total (S/N) ratios of outputs. The sum of squared deviations SS$_T$ represents the difference between each (S/N) ratio of a measured response η_i and the total mean S/N ratio η_m, which is given as follows Lindman (1992):

$$SS_T = \sum_{i=1}^{n} (\eta_i - \eta_m)^2 \tag{13}$$

where n represents the number of trials and η_i represents the mean S/N ratio for the ith trial. The total sum of squared deviations SS_T is the sum of two terms: variance explained by the regression model (SS_d) and random residual variance (SS_e) (not explained by the model) which is written as follows:

$$SS_T = SS_d + SS_e \tag{14}$$

Another statistical tool that allows determining the significant effect of each input parameter on each output response, named (F test) by Ross (1996).

The results of variance analysis (ANOVA) for S/N (*Ra*) are shown in Table 3 and the contribution of significant terms on *Ra* are presented in Fig. 3 (a). It is seen that the feed rate occupies the first position of influencing on the quality of surface with a contribution of 50.21%, because during the feed rate of cutting tool of work-piece in turning process, the tool shape generates helicoids furrows on surface of work-piece. These furrows are deeper and broader as the feed rate increases, therefore the surface quality decreases. The second influential machining parameter is the nose radius of the tool with an impact of 20.27%. A popular established model presented by Yallese et al. (2004) to predict the surface roughness, with a cutting tool that have nose radius different of zero, is:

$$Ra = \frac{f^2}{32 \times r}, \tag{15}$$

where *Ra* is the arithmetic mean roughness (μm), *f* is the feed rate (mm/rev), *r* is the cutting tool nose radius (mm). Based on the Eq. (15) the uses of the largest tool nose radius with little feed rate improves the quality of surface. Similarly, Singh and Rao (2007); Makadia and Nanavati (2013) found that the feed rate is the most significant factor followed by tool nose radius affecting the surface roughness. The interaction $f \times ap$ comes in third position with an effect of 12.69% on quality of surface, the same interaction significance found by Aslan et al. (2007) when turning hardened AISI 4140 steel with Al_2O_3 + TiCN mixed ceramic tool. The factors (χr, Vc, ap) and the other interactions are having a slightly effect on quality of surface.

Table 3
Analysis of Variance for S/N (*Ra*)

Source	SS	DF	MS	F-value	Cont. %	Remarks
χr	1.0755	1	1.0755	116.9	1.07	Significant
r	20.4583	1	20.4583	2223.73	20.27	Significant
Vc	0.0569	1	0.0569	6.18	0.06	Insignificant
f	50.6817	1	50.6817	5508.88	50.21	Significant
ap	1.3091	1	1.3091	142.29	1.3	Significant
$\chi r \times r$	1.3651	1	1.3651	148.38	1.34	Significant
$\chi r \times Vc$	0.0179	1	0.0179	1.95	0.02	Insignificant
$\chi r \times f$	1.1715	1	1.1715	127.34	1.16	Significant
$\chi r \times ap$	3.8569	1	3.8569	419.23	3.82	Significant
$r \times Vc$	2.7503	1	2.7503	298.95	2.72	Significant
$r \times f$	0.1536	1	0.1536	16.7	0.15	Insignificant
$r \times ap$	2.9522	1	2.9522	320.89	2.92	Significant
$Vc \times f$	0.3393	1	0.3393	36.88	0.34	Significant
$Vc \times ap$	1.9258	1	1.9258	209.33	1.91	Significant
$f \times ap$	12.8056	1	12.8056	1391.91	12.69	Significant
Error	0.0183	2	0.0092		0.02	
Total	100.938	17			100	

From the analysis of Table 4 and Fig. 3 (b), it can be apparent that Vc, f, and ap have a significant effect on (*MRR*). Nevertheless, ap is the most significant factor associated with *MRR* with 54.85%. The next largest factors influencing *MRR* is f followed by Vc. Their contributions are 21.84% and 21.64%, respectively. The rest of terms do not represent any significant effects on *MRR*. The same order of significant effect of machining parameters on the *MRR* was found by Bouzid et al. (2014) when turning of X20Cr13 stainless steel.

Table 4

Analysis of Variance for S/N (*MRR*)

Source	SS	DF	MS	F-value	Cont. %	Remarks
Vc	107.757	1	7.253	10	21.64	Significant
f	108.743	1	8.151	11.24	21.84	Significant
ap	273.174	1	11.392	15.71	54.85	Significant
Vc×f	0.306	1	0.264	0.36	0.06	Insignificant
Vc×ap	0.034	1	0.027	0.04	0.01	Insignificant
f×ap	0.012	1	0.012	0.02	0	Insignificant
Error	7.978	11	0.725		1.60	
Total	498.004	17			100	

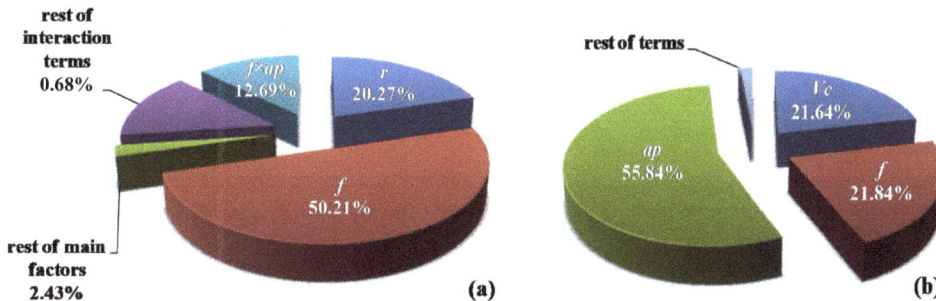

Fig. 3. Contribution of significant terms on: (a) Surface roughness (*Ra*), (b) Material removal rate (*MRR*)

4.2 Pareto analysis

Pareto analysis is a creative statistical technique which aims to identify the important causes for resolving the most problems. It is based on the Pareto principle also known as 80/20 rule which in general means that 80% of problems may be caused by as few as 20% of causes (Karuppusami & Gandhinathan, 2006). This technique was considered in this study to check and confirm the results obtained by ANOVA analysis (Fig. 4). This chart presents the ranking of the influencing machining parameters and their interactions in descending order on the *Ra* and *MRR*. The effects of factors and their interactions on the responses are standardized for a better comparison. The standardized values called (F-value) in this chart are obtained by dividing the mean squares of each factor by the error of mean squares. The more standardized the effect, the higher the factor considered influence. If the F-values which correspond to the machining parameters and their Interactions are greater than 18.51 and 4.84 for (*Ra*) and (*MRR*), respectively; the effects are significant. By against, if the values of F-values are less than 18.51 and 4.84 for (*Ra*) and (*MRR*), respectively; the effects are not significant. The confidence interval chosen is 95 % (α=0.05).

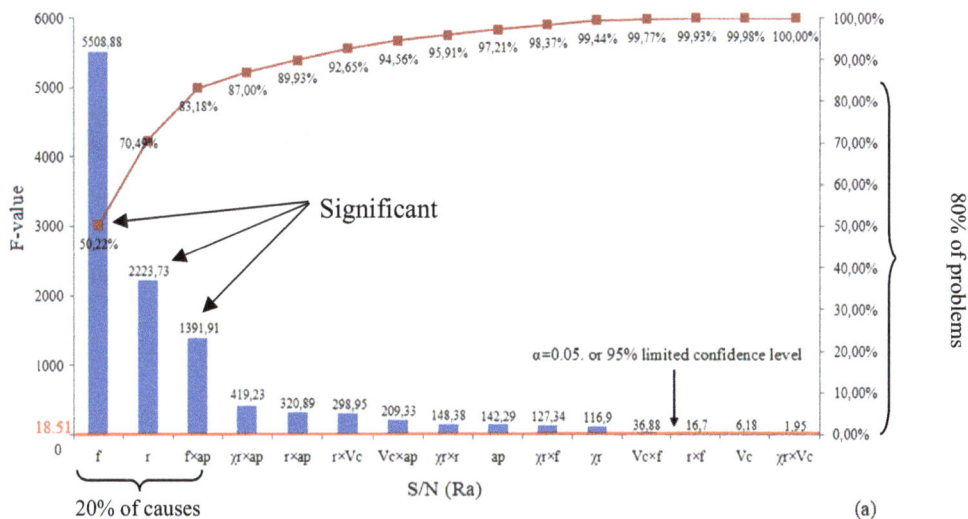

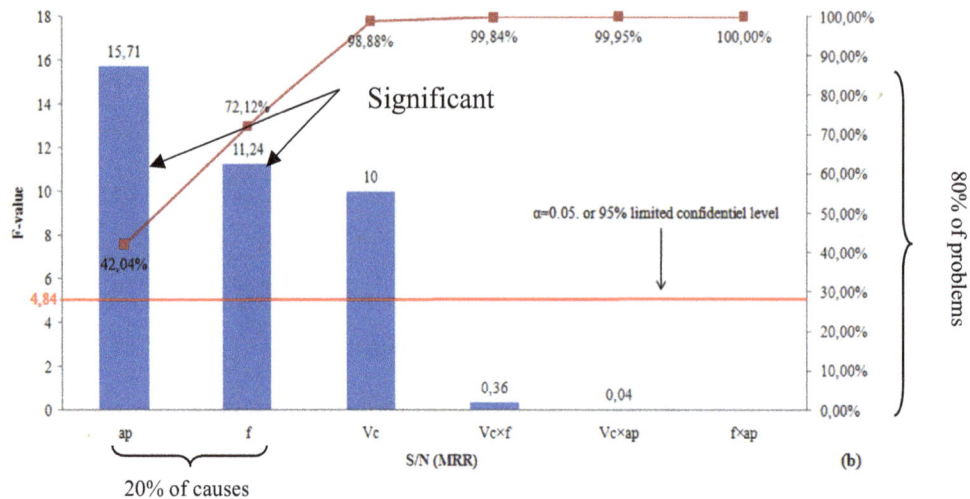

Fig. 4. Pareto analysis chart, for effect of machining parameters on: (a) surface roughness and (b) material removal rate.

4.3 Main effect factors and their interactions on responses (3D plots and contours)

The main effect plots for S/N ratios of (Ra) and (MRR) are presented in Fig. 5 (a, b), respectively. Response graphs show the evolution of S/N ratio of responses in function of variation of levels for each machining parameters. The values of the plotted points in Fig. 5 (a, b) of the S/N ratio for surface roughness and material removal rate which correspond to each level of machinability parameters are given in Table 5 (a, b) respectively. The influence of cutting parameters on the responses can be easily determined through delta value that represents the difference between max and min of S/N ratios of responses as shown in Table 5 (a, b). The higher the value of the delta, the more influential is the cutting parameter. It can be seen in Table 5 (a, b) that the significance of all main factors is ranking in descending order of influence on the responses.

It is notable from Fig. 5 (a) and Table 5 (a) that the most machining parameter affecting surface quality is the feed rate (f) followed by tool nose radius (r). The others factors represent less effects on surface quality. Also, it is clear that the surface quality deteriorates by increasing feed rate. By contrast, the increasing of cutting insert nose radius improves the surface quality (Ra).

From Fig. 5 (b) and Table 5 (b), it can be observed clearly that all main factors (Vc, f, ap) have a significant effect on material removal rate. The descending order of influence of all main factors on responses is as follow: depth of cut, feed rate followed by cutting speed. By increasing all main factors (Vc, f, ap) we can improve the productivity (MRR). It can be observed that the same ranking in descending order of influence of all main factors on responses are obtained by ANOVA analysis and is also confirmed by Pareto chart.

Table 5
S/N response table for: (a) surface roughness (Smaller is better), (b) material removal rate (Larger is better)

Table 5a

Level	γr	r	Vc	f	ap
1	5.636	4.361	5.968	7.786	6.157
2	6.125	6.308	5.85	6.181	5.989
3	-	6.973	5.823	3.675	5.496
Delta	0.489	2.611	0.145	4.11	0.661
Rank	4	2	5	1	3

Table 5b

Level	Vc	f	ap
1	76.8	76.59	74.58
2	79.69	80.11	80.61
3	82.82	82.61	84.13
Delta	6.02	6.02	9.54
Rank	3	2	1

Main Effects Plot for S/N ratios (Ra)
Data Means

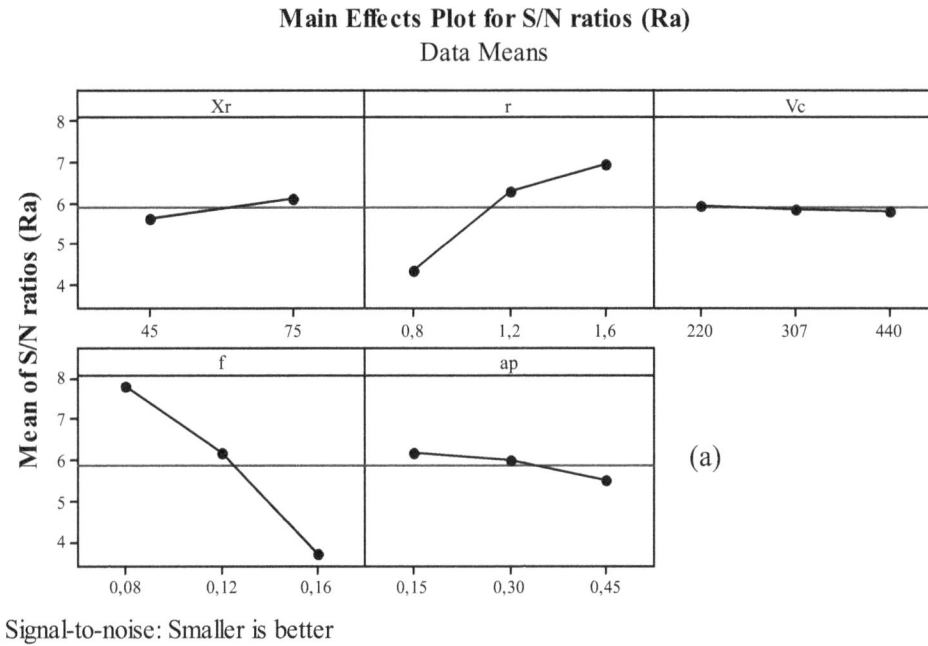

Signal-to-noise: Smaller is better

Main Effects Plot for S/N ratios (MRR)
Data Means

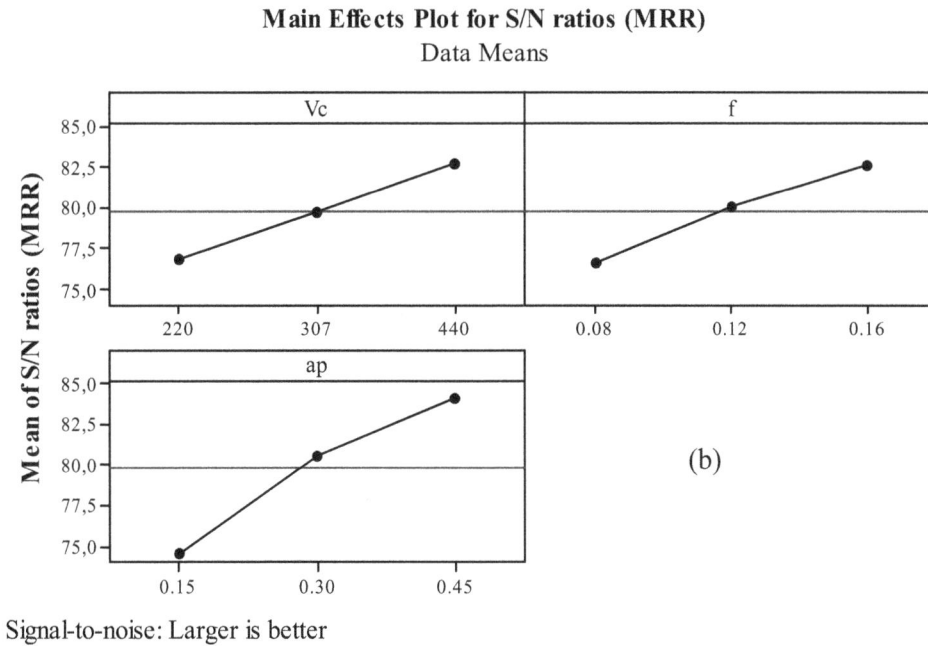

Signal-to-noise: Larger is better

Fig. 5. Main effect's plots of S/N for: (a) surface roughness, (b) material removal rate

In order to check the influence of interaction of the depth of cut and the feed rate on the surface quality (S/N ratio for Ra), 3D response surface for the effect of the interaction is drawn in Fig. 6 (a). Variables not represented in the figure are held constant at the middle level ($\chi r = 60°$, $r = 1.2$ mm, $Vc = 330$ m / min). This figure indicates that, for a given depth of cut, the surface quality is sensitive to the feed rate because the increase of this latter deteriorates quickly the surface quality. This is consistent with the conclusion of research work published by Bouzid et al. (2015) where they remarked that the surface roughness (Ra) rapidly increases by increasing feed rate. However, this decrease in surface quality becomes increasingly small with lower values of the depth of cut.

Fig. 6 (b) shows the response surface for S/N (*Ra*) in the form of a contour. It is remarkable after this figure that for any given values of depth of cut and which belongs to the interval of this study, the best surface quality is found for small values of feed rate in the study interval. Also, at higher depths of cut, better quality of surface is obtainable from 0.3 mm to 0.45mm.

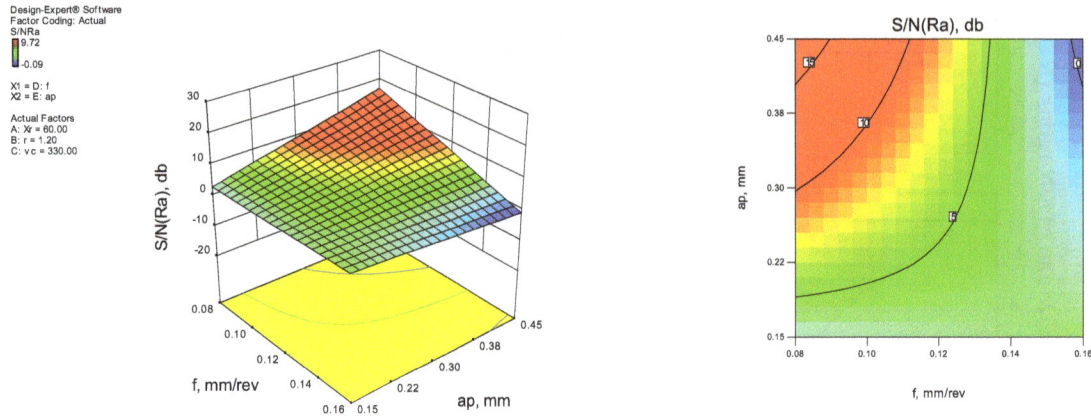

Fig. 6. 3D plot and contour for the response surface for: Effect of (*f*×*ap*) on the S/N (*Ra*)

4.4 2D profile and 3D topography of turned surface

Fig. 7 shows a representative example of 2D profile and 3D image of turned surface envisioned by means of optical platform of metrology modular Altisurf 500 with isometric view. The aim of this investigation is to examine the effect of both feed rate and nose radius on surface roughness through a comparison between 2D profile and 3D topography of turned surfaces obtained as a result of various levels combinations of machining parameters. Turning operations were conducted by the same levels of machining parameters except the feed rate and tool nose radius; for the turned surface (a): $\chi r = 75°$, $r = 0.8$ mm, $Vc = 220$ m/min, $f = 0.08$ mm/rev, $ap = 0.15$ mm, for the turned surface (b): $\chi r = 75°$, $r = 0.8$ mm, $Vc = 220$ m/min, $f = 0.16$ mm/rev, $ap = 0.15$ mm; for the turned surface (c): $\chi r = 75°$, $r = 1.6$ mm, $Vc = 220$ m/min, $f = 0.16$ mm/rev, $ap = 0.15$ mm. It clearly appears by a comparison in Fig. 7 (a, b) that the use of large feed rate yields a bad surface roughness. Because the distance between roughness asperities increases with the increase of feed rate. Whereas, by a comparison of turned surfaces (b and c) in Fig.7 it is notable that that the use of large nose radius improves the surface roughness by the crushing of the asperities. When the nose radius of cutting tool increases, the contact languor between the beak of the tool and the machined surface is increased and this leads to crushing of the asperities and traces of advance of the tool as shown in Fig. 7 (c).

5. Regression equations

Regression analysis is a computational technique that enables to found the functional relationship between the machining parameters (control factors) and the performance characteristics. The predictive equations for the performance characteristics were formulated by linear regression model with interactions given by Eq. (16).

$$Y = b_0 + \sum_{i=1}^{k} b_i X_i + \sum_{ij}^{k} b_{ij} X_i X_j + \varepsilon_i \qquad (16)$$

where b_0 is the free term of the regression equation, the coefficients b_1, b_2 … b_k and b_{12}, b_{13}, b_{k-1} are the linear and interacting terms, respectively. X_i represents the input parameters; (χr, r, Vc, f, ap) for (*Ra*) and (Vc, f, ap) for (*MRR*); and Y represents the outputs (surface roughness, material removal rate). Correlative mathematical models of *Ra* and *MRR* are given below by Eq. (17) and Eq. (18), respectively, with respective coefficients of determination R^2 of 99.72% and 99.77%. These mathematical equations are useful for the estimation of outputs parameters in the range of intervals selected in this study.

$$
\begin{aligned}
Ra = \ & -0.751523 - 0.0390656\ \chi r + 2.11283\ r + 0.00839753\ Vc + 20.5609\ f - \quad 9.98742\ ap \\
& - 0.00225299\ \chi r{\times}r + 5.98103\text{e-}005\ \chi r{\times}Vc + 0.116648\ \chi r{\times}f + 0.00051299\ \chi r{\times}ap - \\
& 0.00318008\ r{\times}Vc - 23.6922\ r{\times}f + 6.56501\ r{\times}ap - 0.0255462\ Vc{\times}f - 0.0163664\ Vc{\times}ap \quad (17) \\
& + 53.2321\ f{\times}ap
\end{aligned}
$$

$$
\begin{aligned}
MRR = \ & 12835.2 - 40.4339\ Vc - 105222\ f - 41353.3\ ap + 324.193\ Vc{\times}f + 126.452\ Vc{\times}ap + \\
& 332351\ f{\times}ap \quad\quad (18)
\end{aligned}
$$

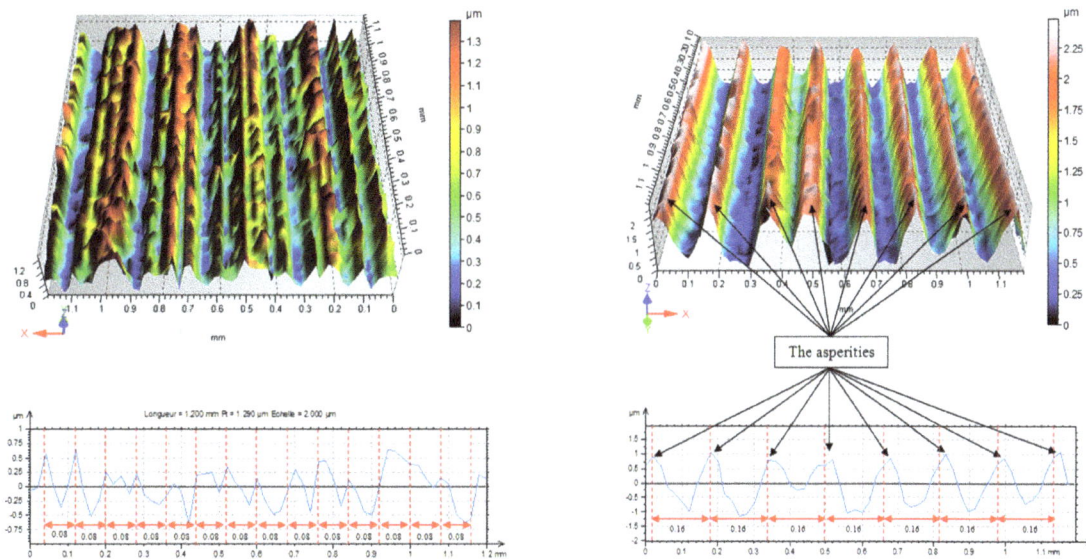

$\chi r = 75°$, $r = 0.8$ mm, $Vc = 220$ m/min, $f = 0.08$ mm/rev, $ap = 0.15$ mm

(a)

$\chi r = 75°$, $r = 0.8$ mm, $Vc = 220$ m/min, $f = 0.16$ mm/rev, $ap = 0.15$mm

(b)

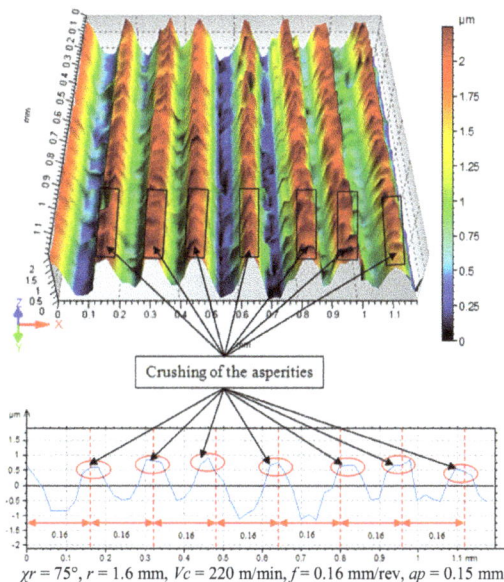

$\chi r = 75°$, $r = 1.6$ mm, $Vc = 220$ m/min, $f = 0.16$ mm/rev, $ap = 0.15$ mm

(c)

Fig. 7. Example of 2D profile and 3D topography of turned surface

To verify the reliability of correlative mathematical models of *Ra* and *MRR*, the normal probability plot vs. residuals were traced in Fig. 8 (a, b) respectively. It can be seen from Fig. 8 (a, b) that the cloud of residuals is reasonably distributed around to the straight line and there are no unusual data points in the data set. This implies that errors are distributed normally.

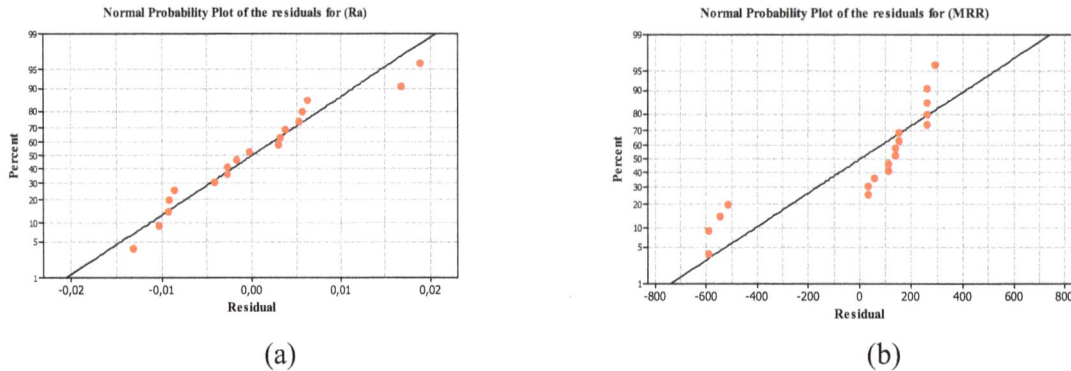

(a)   (b)

Fig. 8. Normal probability plot of residuals for: (a) surface roughness, (b) material removal rate

Fig. 9 (a, b) shows the residuals which are associated with the eighteen experimental runs of surface roughness and material removal rate, respectively. This residual is equal to the difference between the observed and predicted values of the output responses. The residuals of surface roughness belong to the interval of -0.015 to 0.020 μm versus the residuals of material removal rate, which belong at the interval of -600 to 400 mm³/min. The residuals do not represent any obvious pattern, this implies that the predictive equations are reliable for estimating the responses at any particular design points which belong at the range intervals selected in this study.

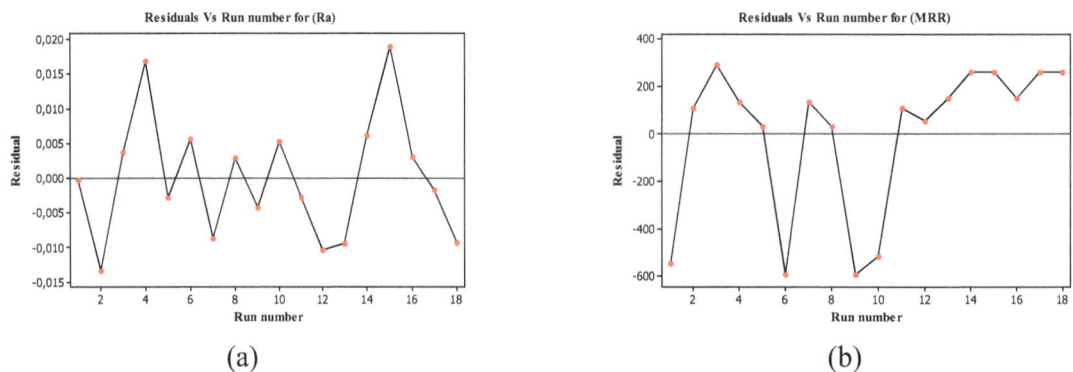

(a)   (b)

Fig. 9. Plot of residuals vs. run numbers for: (a) surface roughness, (b) removal material rate

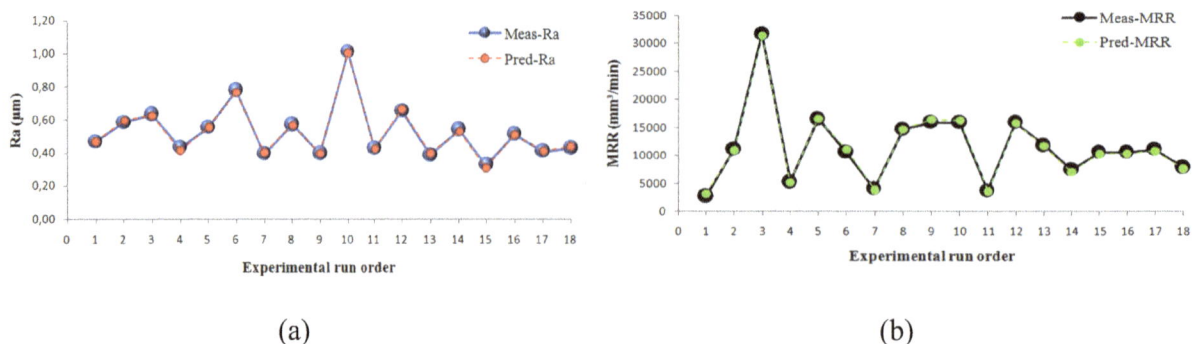

(a)   (b)

Fig. 10. Measured vs. predicted values of performance characteristics: (a) surface roughness, (b) material removal rate.

Fig. 10 (a, b) also shows a comparison between the predicted values of *Ra* and *MRR*, respectively obtained from the linear regression equations and the observed ones. This comparison proves that the estimated output responses are very close to those measured (good agreement between predicted and observed values). These figures confirm that the linear regression models are suitable for estimating the responses without carrying out experimental runs with respect to the range intervals considered in this investigation.

6. Multi-objective optimization using GRA based Taguchi design

The identification of the optimal levels combination of machining parameters using multi-objective optimization (GRA) was carried out based on the S/N ratio of measured and calculates data of surface roughness (*Ra*) and material removal rate (*MRR*), respectively. They are obtained during the dry turning operations of AISI D3 steel with ceramic cutting insert (CC650) based on Taguchi design (L_{18} OA). The simultaneous improvement in term of maximization both surface quality and productivity are proposed as an objective function. The use of this technique optimization includes the following steps which is mentioned in the second part of this paper.

Step 1: The normalization of S/N ratios for *Ra* and *MRR* in the range between zero and one using Eq. 5 (The larger-the-better) were determined for grey relational generation. The normalized data are given in Table 6.

Step 2: Calculation of $\Delta_{0i}(k)$ for normalized values of S/N ratio (*Ra, MRR*) using Eq. (7) is necessary for the computation of grey relational coefficients (GRC). Grey relational coefficients (GRC) were computed using Eq. 6 for the determination of grey relational grade (GRG). $\Delta_{0i}(k)$ and GRC values are given in Table 6.

Step 3: Grey relational grade (GRG) was computed by Eq. (10). The calculated results of GRG are shown in Table 6. This implies that the multi-objective optimization is converted to a single equivalent objective optimization.

Table 6
Results of grey relational generation, calculation of Δ_{0i} (k), grey relational coefficient and grey relational grade

Trail no.	Grey relational generation		Calculation of Δ_{0i} (k)		Grey relational coefficient		Grey relational grade
	S/N	S/N (*MRR*)	S/N	S/N (*MRR*)	S/N	S/N	GRG
	Larger-the-better						
Ideal sequence	1	1	1	1	1	1	
1	0,708	0,000	0,292	1,00	0,632	0,333	0,483
2	0,531	0,576	0,469	0,42	0,516	0,541	0,529
3	0,470	1,000	0,530	0,00	0,485	1,000	0,743
4	0,774	0,279	0,226	0,72	0,689	0,409	0,549
5	0,578	0,739	0,422	0,26	0,543	0,657	0,600
6	0,306	0,558	0,694	0,44	0,419	0,531	0,475
7	0,851	0,163	0,149	0,84	0,771	0,374	0,572
8	0,550	0,692	0,450	0,31	0,526	0,619	0,573
9	0,845	0,721	0,155	0,28	0,763	0,642	0,703
10	0,000	0,721	1,000	0,28	0,333	0,642	0,488
11	0,786	0,134	0,214	0,87	0,700	0,366	0,533
12	0,445	0,721	0,555	0,28	0,474	0,642	0,558
13	0,858	0,605	0,142	0,39	0,779	0,559	0,669
14	0,598	0,413	0,402	0,59	0,554	0,460	0,507
15	1,000	0,558	0,000	0,44	1,000	0,531	0,765
16	0,638	0,558	0,362	0,44	0,580	0,531	0,555
17	0,818	0,576	0,182	0,42	0,733	0,541	0,637
18	0,780	0,442	0,220	0,56	0,695	0,473	0,584

Step 4: Main effect plot for (GRG) was traced in order to select the optimal levels combination (Fig. 11). The mean of GRG ratio for each level of the machining parameters is presented in Table 7. In grey relational analysis the highest value of (GRG) corresponds to the best levels combination of machining parameters. Therefore, the optimal level of the machining parameters is the level with the greatest GRG value. So it can be concluded that the best level for each machining parameter was found as follow (Fig. 11): $\chi r_2 r_3 Vc_3 f_1 ap_3$, in other words, the optimal levels combination for Ra and MRR were obtained at a major cutting edge angle of 75°, insert radius of 1.6 mm, cutting speed of 440 m/min, feed of 0.08 mm/rev and depth of cut of 0.45 mm. The results of the use of this optimal levels combination are: $Ra = 0.41\,\mu m$ and $MRR = 15840$ mm³/min.

Main Effects Plot for GRG

Data Means

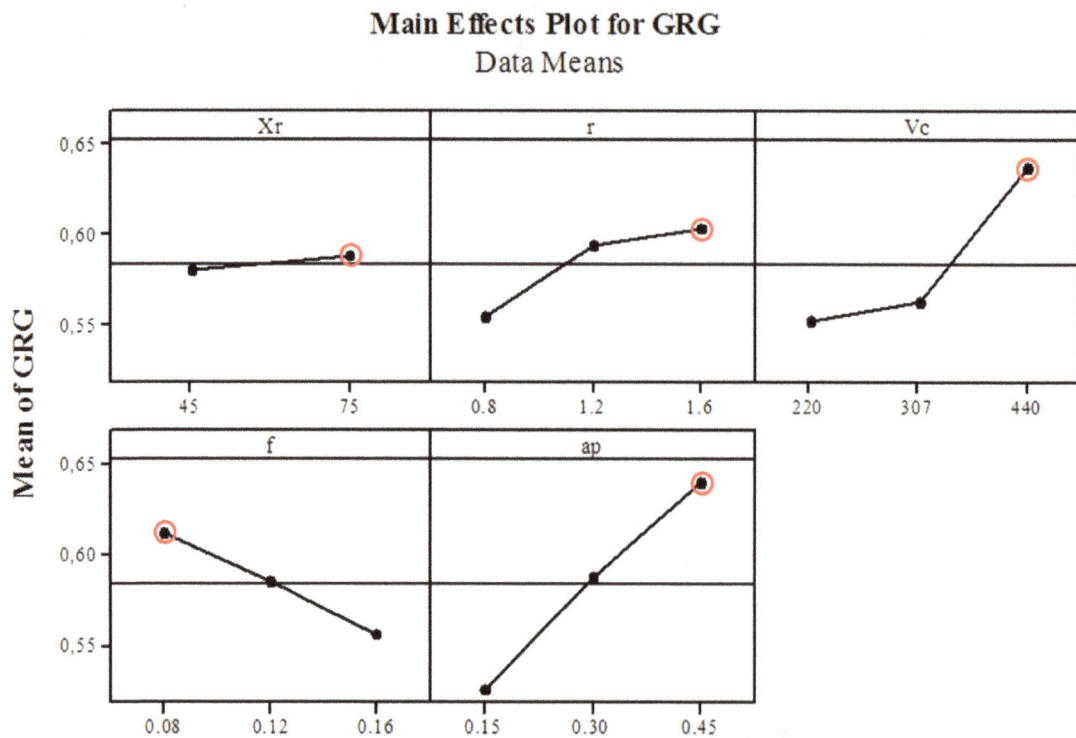

Fig. 11. Main effect's plots for GRG

Table 7

Mean of GRG ratio for each level of the machining parameters

Level	χr	r	Vc	f	ap
1	0.580	0.555	0.552	0.611	0.525
2	0.589	0.594	0.563	0.585	0.588
3	-	0.604	0.637	0.556	0.639
Delta	0.009	0.049	0.085	0.055	0.114
Rank	5	4	2	3	1

Step 5: In order to confirm the selected optimal levels combination of machining parameters through the use of GRA technique for the improvement of surface roughness and material removal rate, a comparison assessment was performed between the values of grey relational grade obtained from the optimal experiment with that predicted $\hat{\eta}$ by using the Eq. (11). Based on the results in Table 8, a notable agreement was remarked between the value found experimentally (0.69) calculated by the estimation formula (0.74). While the grey relational grade was improved from initial levels combination ($\chi r_2 r_2 Vc_2$ $f_2 ap_2$) to the optimal levels combination of machining parameters ($\chi r_2 r_3 Vc_3 f_1 ap_3$) by 0.13. According

to confirmation runs, the surface roughness and material removal rate are ameliorating approximately 1.21, 0.69 times, respectively.

Table 8
Results of the confirmation experiment

	Initial machining parameters	Optimal machining parameters	
		Prediction	Experiment
Level	$\chi r_2\, r_2\, Vc_2\, f_2\, ap_2$	$\chi r_2\, r_3\, Vc_3\, f_1\, ap_3$	$\chi r_2\, r_3\, Vc_3\, f_1\, ap_3$
Ra (μm)	0.50	-	0.41
MRR (mm³/min)	11052	-	15840
GRG	0.56	0.74	0.69
Improvement of GRG		0.13	

7. Conclusions

This investigation has detailed the procedure of applying the grey relational analysis based on Taguchi design for a simultaneous optimization in turning process. Through the results obtained in this study, the following conclusions can be drawn:

1. The mixed orthogonal array of Taguchi L_{18} is adopted in this investigation to get a small number of experiment runs for identifying the optimal levels of machining parameters using grey relational analysis.

2. Based on the ANOVA analyse of S/N ratio for (Ra), it's found that the feed rate maintains a strong effective parameter affecting the surface roughness followed by nose radius and the interaction ($f{\times}ap$) with contributions of 50.21%, 20.27%, and 12.69%, respectively.

3. The results given by ANOVA of S/N ratio for (MRR) shows that depth of cut is the most significant with the respective contribution 54.85%. The feed rate followed by cutting speed presents a statistical significance with contribution of 21.84% and 21.64%, respectively.

4. The descending order of influencing machining parameters on performance characteristics which was found by ANOVA analysis were confirmed by Pareto analysis and main effects plots for S/N ratios of (Ra) and (MRR). Effect of the interaction ($f{\times}ap$) was also confirmed by 3D plot and contour.

5. The analysis of 2D profile and 3D topographical maps of the turned surface captured by optical platform of metrology modular Altisurf 500 has presented a great importance in the examination the effect of feed rate and nose radius of cutting tool.

6. The correlative mathematical models of Ra and MRR are very efficient because of their higher coefficients of determination (R^2) values of 99.72% and 99.77%, respectively. They have an important industrial interest, since they help to make estimations of responses without carrying out experimental runs with respect to the range intervals considered in this investigation.

7. The effectiveness of these models was verified by the normal probability plot vs. residuals and the Plot of residuals vs. run numbers. It is found that the residuals are reasonably distributed around to the straight line and do not represent any obvious pattern. The plot of estimated vs. observed values of responses are also found very close to each other.

8. The grey relational analysis was used to resolve the optimization complex problem by converting the multi-objective optimization into a single equivalent objective optimization.

9. The single equivalent objective optimization in this analysis called grey relational grade (GRG). The highest value of this latter corresponds to the optimal levels combination of machining parameters. Therefore, the optimal levels combination for a simultaneous improvement of *Ra* and *MRR* was obtained at; ($\chi r_2 r_3 V c_3 f_1 a p_3$); major cutting edge angle of 75°, insert radius of 1.6 mm, cutting speed of 440 m/min, feed of 0.08 mm/rev and depth of cut of 0.45 mm. The optimized responses found by the use of optimal levels selected by grey relational analysis were (*Ra*= 0.41 μm and *MRR*= 15840 mm³/min).

10. It is notable that there was a good agreement between the value found experimentally (0.69) and the one calculated by the estimation formula (0.74). While the grey relational grade was improved from initial levels combination ($\chi r_2 r_2 V c_2 f_2 a p_2$) to the optimal levels combination ($\chi r_2 r_3 V c_3 f_1 a p_3$) by 0.13. According to confirmation runs, the surface roughness and material removal rate were improved approximately 1.21, 0.69 times, respectively.

Acknowledgments

This work was executed in the Mechanics and Structures Research Laboratory (LMS), May 8th 1945 University of Guelma, Algeria in partnership with LaMCoS (INSA-Lyon, France). Authors would like to thank the Algerian Ministry of Higher Education and Scientific Research (MESRS) and the Delegated Ministry for Scientific Research (MDRS) for granting financial support through CNEPRU Research Project, Code: J0301520140021.

References

Asiltürk, I., & Akkuş, H. (2011). Determining the effect of cutting parameters on surface roughness in hard turning using the Taguchi method. *Measurement*, *44*(9), 1697-1704.

Al-Ahmari, A. M. A. (2007). Predictive machinability models for a selected hard material in turning operations. *Journal of Materials Processing Technology*, *190*(1), 305-311.

Aouici, H., Bouchelaghem, H., Yallese, M. A., Elbah, M., & Fnides, B. (2014). Machinability investigation in hard turning of AISI D3 cold work steel with ceramic tool using response surface methodology. *The International Journal of Advanced Manufacturing Technology*, *73*(9-12), 1775-1788.

Aslan, E., Camuşcu, N., & Birgören, B. (2007). Design optimization of cutting parameters when turning hardened AISI 4140 steel (63 HRC) with Al_2O^3+ TiCN mixed ceramic tool. *Materials & design*, *28*(5), 1618-1622.

Bouzid, L., Boutabba, S., Yallese, M. A., Belhadi, S., & Girardin, F. (2014). Simultaneous optimization of surface roughness and material removal rate for turning of X20Cr13 stainless steel. *The International Journal of Advanced Manufacturing Technology*, *74*(5-8), 879-891.

Balasubramanian, S., Ganapathy, S. (2011). Grey Relational Analysis to determine optimum process parameters for Wire Electro Discharge Machining (WEDM). *International Journal of Engineering Science and Technology*, *3*(1), 95-101.

Bouchelaghem, H., Yallese, M. A., Mabrouki, T., Amirat, A., & Rigal, J. F. (2010). Experimental investigation and performance analyses of CBN insert in hard turning of cold work tool steel (D3). *Machining Science and Technology*, *14*(4), 471-501.

Bensouilah, H., Aouici, H., Meddour, I., Yallese, M.A., Mabrouki, T., Girardin, F. (2016). Performance of coated and uncoated mixed ceramic tools in hard turning process. *Measurement*, *82*, 1–18.

Bouzid, L., Yallese, M. A., Chaoui, K., Mabrouki, T., & Boulanouar, L. (2015). Mathematical modeling for turning on AISI 420 stainless steel using surface response methodology. *Proceedings of the Institution of Mechanical Engineers, Part B: Journal of Engineering Manufacture*, *229*(1), 45-61.

Davim, J. P., & Figueira, L. (2007a). Comparative evaluation of conventional and wiper ceramic tools on cutting forces, surface roughness, and tool wear in hard turning AISI D2 steel. *Proceedings of the Institution of Mechanical Engineers, Part B: Journal of Engineering Manufacture*, *221*(4), 625-633.

Davim, J. P., & Figueira, L. (2007b). Machinability evaluation in hard turning of cold work tool steel (D2) with ceramic tools using statistical techniques. *Materials & design*, *28*(4), 1186-1191.

Dabade, U.A. (2013). Multi-objective process Optimization to Improve Surface Integrity on Turned Surface of Al/SiCp Metal Matrix Composites Using Grey Relational Analysis. *Procedia CIRP, Forty Sixth CIRP Conference on Manufacturing Systems*, *7*, 299 – 304.

Gaitonde, V. N., Karnik, S. R., Figueira, L., & Davim, J. P. (2009). Analysis of machinability during hard turning of cold work tool steel (type: AISI D2). *Materials and Manufacturing Processes*, *24*(12), 1373-1382.

Günay, M., Kaçal, A., & Turgut, Y. (2011). Optimization of machining parameters in milling of Ti-6Al-4V alloy using Taguchi method. *e-Journal of New World Sciences Academy*, *6*(1), 1A0165.

Hassan, K., Kumar, A., & Garg, M. P. (2012). Experimental investigation of Material removal rate in CNC turning using Taguchi method. *International Journal of Engineering Research and Applications*, *2*(2), 1581-1590.

Hanafi, I., Khamlichi, A., Cabrera, F. M., Almansa, E., & Jabbouri, A. (2012). Optimization of cutting conditions for sustainable machining of PEEK-CF30 using TiN tools. *Journal of Cleaner Production*, *33*, 1-9.

Jailani, H. S., Rajadurai, A., Mohan, B., Kumar, A. S., & Sornakumar, T. (2009). Multi-response optimisation of sintering parameters of Al–Si alloy/fly ash composite using Taguchi method and grey relational analysis. *The International Journal of Advanced Manufacturing Technology*, *45*(3), 362-369.

Kuram, E., & Ozcelik, B. (2013). Multi-objective optimization using Taguchi based grey relational analysis for micro-milling of Al 7075 material with ball nose end mill. *Measurement*, *46*(6), 1849-1864.

Karuppusami, G., & Gandhinathan, R. (2006). Pareto analysis of critical success factors of total quality management: A literature review and analysis. *The TQM magazine*, *18*(4), 372-385.

Lin, C. L. (2004). Use of the Taguchi method and grey relational analysis to optimize turning operations with multiple performance characteristics. *Materials and manufacturing processes*, *19*(2), 209-220.

Lindman, H.R. (1992). *Analysis of variance in experimental design*. Springer-Verlag, New York, USA.

Meddour, I., Yallese, M. A., Khattabi, R., Elbah, M., & Boulanouar, L. (2015). Investigation and modeling of cutting forces and surface roughness when hard turning of AISI 52100 steel with mixed ceramic tool: cutting conditions optimization. *The International Journal of Advanced Manufacturing Technology*, *77*(5-8), 1387-1399.

Makadia, A. J., & Nanavati, J. I. (2013). Optimisation of machining parameters for turning operations based on response surface methodology. *Measurement*, *46*(4), 1521-1529.

Neseli, S., Yaldız, S., & Türkes, E. (2011). Optimization of tool geometry parameters for turning operations based on the response surface methodology. *Measurement*, *44*(3), 580–587.

Nalbant, M., Gökkaya, H., & Sur, G. (2007). Application of Taguchi method in the optimization of cutting parameters for surface roughness in turning. *Materials & design*, *28*(4), 1379-1385.

Haq, A. N., Marimuthu, P., & Jeyapaul, R. (2008). Multi response optimization of machining parameters of drilling Al/SiC metal matrix composite using grey relational analysis in the Taguchi method. *The International Journal of Advanced Manufacturing Technology*, *37*(3), 250-255.

Ross, P.J. (1996). *Taguchi techniques for quality engineering*. McGraw-Hill International Editions, Singapore.

Singh, D., & Rao, P. V. (2007). A surface roughness prediction model for hard turning process. *The International Journal of Advanced Manufacturing Technology*, *32*(11), 1115-1124.

Singh, R., & Dureja, J.S. (2014). Comparing Taguchi method and RSM for optimizing flank wear and surface roughness during hard turning of AISI D3 steel. *Proceedings of the International Conference on Research and Innovations in Mechanical Engineering*, India, 139-152.

Taguchi, G. (1986). *Introduction to Quality Engineering*. Asian Productivity Organisation, Tokyo, Japan.

Wang, M. Y., & Lan, T. S. (2008). Parametric optimization on multi-objective precision turning using grey relational analysis. *Information Technology Journal, 7*(7), 1072-1076.

Yallese, M. A., Rigal, J. F., Chaoui, K., & Boulanouar, L. (2005). The effects of cutting conditions on mixed ceramic and cubic boron nitride tool wear and on surface roughness during machining of X200Cr12 steel (60 HRC). *Proceedings of the Institution of Mechanical Engineers, Part B: Journal of Engineering Manufacture, 219*(1), 35-55.

Yallese, M. A., Boulanouar, L., & Chaoui, K. (2004). Usinage de l'acier 100Cr6 trempé par un outil en nitrure de bore cubique. *Mechanics & Industry, 5*(4), 355-368.

Zerti, O., Yallese, M.A., Belhadi, S., & Bouzid, L. (2017a). Taguchi design of experiments for optimization and modeling of surface roughness when dry turning X210Cr12 steel. *Springer International Publishing Switzerland* T. Boukharouba et al. (eds.), Applied Mechanics, Behavior of Materials, and Engineering Systems, Lecture Notes in Mechanical Engineering, DOI 10.1007/978-3-319-41468-3_22.

Zerti, O., Yallese, M. A., Khettabi, R., Chaoui, K., & Mabrouki, T. (2017b). Design optimization for minimum technological parameters when dry turning of AISI D3 steel using Taguchi method. *The International Journal of Advanced Manufacturing Technology, 89*(5-8), 1915-1934.

Zahia, H., Athmane, Y., Lakhdar, B., & Tarek, M. (2015). On the application of response surface methodology for predicting and optimizing surface roughness and cutting forces in hard turning by PVD coated insert. *International Journal of Industrial Engineering Computations, 6*(2), 267-284.

Zahia, H., Nabil, K., MA, Y., Mabrouki, T., Ouelaa, N., & Rigal, J. F. (2013). Turning roughness model based on tool-nose displacements. *Mechanics, 19*(1), 112-119.

Zhang, J. Z., Chen, J. C., & Kirby, E. D. (2007). Surface roughness optimization in an end-milling operation using the Taguchi design method. *Journal of Materials Processing Technology, 184*(1), 233-239.

4

A hybrid expert system, clustering and ant colony optimization approach for scheduling and routing problem in courier services

Eduyn López-Santana[a*], William Camilo Rodríguez-Vásquez[a] and Germán Méndez-Giraldo[a]

[a]*Sistemas Expertos y Simulación (SES), Facultad de Ingeniería, Universidad Distrital Francisco José de Caldas, Bogotá, Colombia*

CHRONICLE	ABSTRACT
Keywords: *Courier services* *Clustering* *Expert system* *Routing* *Scheduling*	This paper focuses on the problem of scheduling and routing workers in a courier service to deliver packages for a set of geographically distributed customers and, on a specific date and time window. The crew of workers has a limited capacity and a time window that represents their labor length. The problem deals with a combination of multiples variants of the vehicle routing problem as capacity, multiple periods, time windows, due dates and distance as constraints. Since in the courier services the demands could be of hundreds or thousands of packages to be delivered, the problem is computationally unmanageable. We present a three-phase solution approach. In the first phase, a scheduling model determines the visit date for each customer in the planning horizon by considering the release date, due date to visit and travel times. We use an expert system based on the know-how of the courier service, which uses an inference engine that works as a rule interpreter. In the second phase, a clustering model assigns, for each period, customers to workers according to the travel times, maximum load capacity and customer's time windows. We use a centroid based and sweep algorithms to solve the resulted problem. Finally, in the third phase, a routing model finds the order in which each worker will visit all customers taking into account their time windows and worker's available time. To solve the routing problem we use an Ant Colony Optimization metaheuristic. We present some numerical results using a case study, in which the proposed method of this paper finds better results in comparison with the current method used in the case study.

1. Introduction

The courier services are dedicated to pick up/delivery packages and/or documents that are sent from people to other people with the target of a faster and secure pick up/delivery. To manage these service systems there are different decision-making levels. In strategic level (or long-range term), the decisions state the objectives, resources, and policies of the organization in a long planning horizon, e.g., a typical problem consists in forecasting the kind of services to provide and the required capacity in the next three years. In tactical level (or medium-range term), the decisions are to determine in a gross mode the required resources to perform the service in a medium planning horizon, e.g., a typical problem consists of determining the required workforce in six months. In operational level (or short-range term), the decisions determine how to carry out the specific tasks in the service operation in short planning horizon, e.g., the scheduling problem of an operator to deliver a package to the customer's site. The operation

* Corresponding author
E-mail: erlopezs@udistrital.edu.co (E. López-Santana)

could be on all scales, from within specific towns or cities, to regional, national and global services. The complexity of the problems increases from the strategic level to operational level because the required information also increases (and more detailed) and the response time is less. The scheduling problems belong to the operational level of the decision-making process and thus is difficult to solve.

The courier services have grown in coverage and in operative complexity, although, in many of them, the traditional expertise of their workers is still used as the main input to execute operation (Rodríguez-Vásquez et al., 2016). This feature allows the chance of human error to develop inconsistencies in the process, triggered, among other reasons, by employee turnover and the constant loss of the compounded knowledge and experience about the specific tasks involved in the process. In addition, when they are faced with scheduling services manually, even the most experienced planners can only consider a limited number of possibilities and need to invest a significant amount of time to obtain a feasible schedule that it is generally far from the optimum.

The courier service generally consists in distributing thousands of packages that are managed daily to a set of customers geographically distributed by a crew of workers (Rodríguez-Vásquez et al., 2016). There is a combinatorial optimization problem for finding the best set of routes for a workforce crew known as the *vehicle routing problem* (Toth & Vigo, 2002). In a broad sense, this problem finds the best set of routes to be performed by a set of vehicles (crews) in order to serve a set of geographically-spread customers subject to some operational constraints. Thus, the courier services could be modeled as vehicle routing problem (VRP).

In the courier services, there are real-life characteristics that are related to variants of the VRP models. The VRP with time windows (VRPTW) is the most common variant to include in the courier services. This constraint arises when the appointments can be arranged before delivery and have a specific date and time to serve (Toth & Vigo, 2002). The capacitated VRP (CVRP) involves the capacity constraint as the maximum load capacity of the vehicles, (Toth & Vigo, 2002). The distance-constrained VRP (DC-VRP) involves the time or distance constraint, in addition to the capacity constraint, each vehicle has a maximum traveling distance that can be reached, usually given in terms of distance or time (Farahani et al., 2011). The multi-period VRP (MP-VRP) considers a planning horizon composed of a set of periods in which customers have to be reached at least once. Another extension of the VRP is the due dates (VRPD) and this constraint takes into account the customer's due dates (Archetti et al., 2015).

This paper presents a scheduling and routing problem in courier services that involves the variants of VRP described above: time windows, multiples periods, capacity, due date and distance-constrained. To solve this problem we propose a three-phase approach. The first phase consists of a scheduling model to find the visit date for each customer over the planning horizon by considering the release date, the due date and travel times between the customers and depot. To solve this problem, we use an expert system. The second phase is a clustering model for each period in order to assign customers to the crew according to the travel times, maximum load capacity and the customer's time windows. We use two algorithms to solve the problem: a centroid-based heuristic and sweep heuristics. Finally, the third phase consists of a routing model in order to find the sequence to visit all customers taking into account the customer's time windows and the available time of the vehicles. An Ant Colony Optimization (ACO) metaheuristic was developed to solve this problem.

The remainder of this paper is organized as follows. Section 2 reviews the background and literature related to scheduling and routing problems in courier services. Section 3 states the problem and its notation. Sections 4, 5 and 6 describe the scheduling, clustering and routing phases, correspondingly, to solve the scheduling and routing problems in courier services. Section 7 provides some numerical experiments of our proposed approach in an example inspired by a real-world case. Finally, Section 10 concludes this work and provides possible research directions.

2. Background and literature review

Scheduling and routing are very complex problems in service industries like courier companies. These problems involve decisions to allocate deliveries to a set of limited resources in a specific time under several operational constraints as time windows, capacity, due dates, among others. The complexity increases when the decision makers are faced with hundreds or thousands of packets to be scheduled in a short amount time by considering several performance measures as an operational cost, customer's service level, profitability, etc. The vehicle routing problem (VRP) is closely related with scheduling and routing problems in courier services because it involves several variants and extensions of VRP, thus these problems are NP-hard (Toth & Vigo, 2002).

Different applications of courier services tightly coupled with VRP can also be found in the literature. Malmborg (2000) states a preliminary application of scheduling in courier services. The author studies the bank messenger problems as scheduling application and it determines the starting time and the sequence of stop locations. The problem is solved using a heuristic based on a simplified criterion check processing delays. Ghiani et al. (2009) describe an approach to solve the dynamic vehicle dispatching problem with pickups and deliveries. They propose a set of anticipatory algorithms to solve the problem. Their objective function consists of minimizing the expected inconveniences of the customers and their results show better solutions compared with other approaches. Chang and Yen (2012) propose routing and scheduling strategies for city couriers in Taiwan. They seek to reduce the operational costs and improve the service level using a multi-objective multiple traveling salesman problems formulation. Their objective function minimizes the total traveled length and the unbalanced workload, simultaneously. They considered hard time windows and proposed a multi-objective scatter search framework in order to find the Pareto-optimal solutions. Yan et al. (2013) study the problem of planning and adjustment of courier routes and schedules applied to urban regions. They use VRP models that include demand and stochastic traveling times. Their approach adjusts planned routes according to actual operations in real time. Lin et al. (2014) show a prototype of a decision support system to solve the offline and online routing problems arising in courier service. They formulated the problem as a dynamic vehicle routing problem (DVRP) using fuzzy time windows in order to represent the service level. To solve the problem, a hybrid neighborhood search algorithm was developed. Their results find an improvement of the courier service level without the further expense of a longer traveling distance or a larger number of couriers. Janssens et al. (2015) present a two-phase approach to solve a courier scheduling problem. The first phase consists of partitioning a distribution region into smaller zones that are assigned to a preferred vehicle. Then, in the second phase, they modeled the problem for each zone based on VRP as a multi-objective optimization problem and develop a heuristic to solve it.

The works described above have in common the application of VRP in the courier services. Table 1 summarizes the constraints that usually are taken into consideration as capacity and time windows, also one or two simultaneous constraints are used. In this paper, we consider other constraints such as distance, and multiples periods, therefore, our solution considers four real-life characteristics at the same time, implying a high level of complexity.

Table 1

Literature of application of VRP in courier services

	Capacity	Time windows	Distance	Pickup and delivery	Multi-period	Dynamic demand	Stochastic demand
(Malmborg, 2000)	√					√	
(Ghiani et al., 2009)	√			√		√	
(Chang & Yen, 2012)	√	√					
(Yan et al., 2013)	√						√
(Lin et al., 2014)	√	√				√	
(Janssens et al., 2015)	√						
This paper	√	√	√		√		

The VRP is one of the most researched combinatorial optimization problems in the literature and the trends towards VRP includes more real-life characteristics (Braekers et al., 2016). Several variants of the VRP includes features related to scheduling and routing courier services. Braekers et al. (2016) present a survey of the trends in the VRP literature. The authors found that CVRP (capacitated VRP) and VRPTW (VRP with Time Windows) are the features that have the most publications considered and these are the most important variants used in the modeling of courier services. Another important variation of the VRP is the Distance-Constrained (DC-VRP) that involves a real-life feature as the maximum distance that a vehicle is available to drive or the labor period, however some studies are dedicated to this constraint (Almoustafa et al., 2013; Kek et al., 2008; Nagarajan & Ravi, 2012; Tlili et al., 2014). On the other hand, the pickup and delivery constraint is generally combined with VRPTW (Fikar & Hirsch, 2015; Küçükoğlu & Öztürk, 2015; Lin, 2011; Liu et al., 2013; López-Santana & Romero Carvajal, 2015). The MP-VRP (multi-period VRP) considers a planning horizon where the customers need to be visited several times with periodic or non-periodic frequency according to the service operation. Francis et al. (2006) study both cases, periodic and non-periodic frequency, and implement a service choice variable and present several heuristics for its solution. Rodriguez et al. (2015) present an iterative method to solve the PVRP through the construction of a unique visit schedule. Their approach allows the load deconsolidation, using non-regular days of visits and variable frequencies to make it a more robust and flexible tool for a real-world environment. Also, the due date is a constraint added to customers arising a variant known as VRPD (VRP with due date). Archetti et al. (2015) study this variant with multiples periods and solve it with a branch-and-cut algorithm. According to Braekers et al. (2016), it is observed that several applications have real-life characteristics individually or in some case with a limited number of other characteristics, however, the literature lacks many combinations of realistic features. In the case of courier services, multiple features must be considered simultaneously as time windows, multiples periods, capacity, due date and distance-constrained. In addition, it is possible to develop efficient solution methods to solve these problems.

On the other hand, there are a wide variety of solution methods to solve the VRP in the literature, being able to be classified in exact methods, heuristics, and metaheuristics (Eksioglu et al., 2009). Among the main exact methods used to solve the VRP are the Branch and Bound (B&B), Branch and Cut and Set-covering-based (Farahani et al., 2011). An example of a B&B can be found in Almoustafa et al. (2013), where B&B is used to solve a VRP with distance constraints. For the Branch and cut, an example of the application for the VRP with multiple periods can be found in Archetti et al. (2015). An example of the Set-covering-based is the article (Cacchiani et al., 2014).

About the heuristics, the most used are the two-phase heuristics as: the cluster-first route-second, the route-first cluster-second and the PEDAL algorithms (Toth & Vigo, 2002). In the first, the classic example is the Sweep Algorithm (Toth & Vigo, 2002), the Adaptive Large Neighborhood Search framework (ALNS) in which several types of VRP can be solved (Pillac et al., 2012). For the second family, among the route-first cluster-second methods, application of Greedy Algorithms can be found (Chu et al., 2006; Sprenger & Mönch, 2012). Other heuristics are the constructive heuristic, the Clarke and Wright's Saving Algorithm (Clarke & Wright, 1964) is the most famous heuristic. Another more recent example is presented in (Lin, 2011) using constructive heuristic as a step of the broader heuristic proposal.

The third group is the metaheuristics, which are frequently used to solve large combinatorial optimization problems. Some of the most commonly used are: Tabu search (TS), developed from heuristic from local search while additionally including a solution evaluation, local searching tactics, termination criteria and elements such as tabus in a list and the tabu length (Jia et al., 2013); Ant Colony Optimization (ACO), inspired in the feeding process of the ants and is a collective intelligence algorithm used to solve complex combinatorial optimization problems as shortest path problems (Ding et al., 2012); and Genetic Algorithm (GA), defined as adaptive search heuristic that operates over a population of solutions, based on the evolution principle that improves the solution using crossover and mutation process (Pereira & Tavares, 2009). Another metaheuristic used to solve VRPs using neighborhoods is the work of Rincon-Garcia et al. (2017) that uses Large Neighborhood Search approaches and Variable Neighborhood Search techniques to guide the search. Likewise, Galindres-Guancha et al. (2018) present a methodology of three stages to solve the multi-depot VRP. In the first one, the starting solutions are built up using constructive heuristics. The second stage consists of improving each starting solution with an iterated local search multi-objective metaheuristic (ILSMO). In the last stage, a single front is found using concepts of dominance and taking as a base for the previous results.

In the field of Expert Systems (ES), which is a knowledge-based system and uses the reasoning methods that emulate the behavior and performance of human expert to solve problems. The ES is a specific field of artificial intelligence (Díez et al., 2001; Méndez-Giraldo et al., 2013). An ES is a computer program that reasons and uses a knowledge base to solve complex problems in a particular domain (Krishnamoorthy & Rajeev, 1996; Kusiak, 1990). Sahin et al. (2012) show a statistical analysis of hybrid ES approaches and their applications. The publications have an increasing trending as an indicator of the popularity of hybrid ES. Dios and Framinan (2016) present a review of case studies in manufacturing scheduling tools, while Wagner (2017) shows a case study using ES in different areas and state future trends. Both studies affirm that the ESs play a strategic role in the scheduling methods and techniques since they allow to involve the human expertise in the algorithms and to hybrid with other techniques. However, in service systems the application of ES is scarce, López-Santana and Méndez-Giraldo (2016) present a proposal of an ES for scheduling in service systems. They state that the application of knowledge-based systems and ES for scheduling in services systems are scarce and propose a structure to determine the service system and identify the tools to solve a specific scheduling problem.

When examining the literature, one should notice the lack of modeling approaches for scheduling and routing problems in courier services because of its complexity in the multiple real-life characteristics and the large-scale of the problems. Indeed, most of the papers deal with applications of VRP and its variants individually. The novelty of this work consists of exploring and combining multiple real-life characteristics as time windows, due dates, multiples periods, distance-constrained and capacity constraints in the VRP model applied to courier services. According to the literature trends, we propose a hybrid solution approach that integrates an expert system to scheduling deliveries, two heuristics to clustering customers and a metaheuristic based in ACO to solve a routing problem.

3. Problem statement

We consider a set of customers $V_c = \{1, 2, ..., N\}$ geographically distributed, in which each customer expects to receive a package. The set of workers (vehicles) who visit the customers are identicaly denoted by $K = \{1, 2, ..., m\}$ and belong to a central hub $\{0\}$ that is considered to be the starting point and the end point of all workers. We assume that a worker is assigned to a vehicle, thus we use worker and vehicle, interchangeably. All customers must be visited once during the planning horizon. Then, we define the problem in a directed complete graph $G = \{V, A\}$ where $V = V_c \cup \{0\}$ is the set of vertices and $A = \{(i, j): i, j \in V, i \neq j\}$ is the set of arcs. Each arc (i, j) has a non-negative associated value t_{ij} that represents the traveling time from i to j. The objective is to establish the set of routes to visit all customers seeking to minimize the total routing distance for the vehicles. The assumptions and conditions of the problem are summarized as follow:

- The service time is the same for all customers and all vehicles.
- Each customer imposes a hard time-window in which the delivery must start. This means that the vehicle must arrive to start the service before the end of that time window, and it can arrive before the beginning of the time window, but the customer will not be serviced before this earliest time.
- The vehicles are homogeneous, i.e., they have the same capacity but they have different time windows for their labor period. It means that each vehicle has an availability limitation for the traveling time and service time.
- The vehicles start and finish in a central depot.
- The travel times are deterministic and fulfill the triangle inequality.

Table 2introduces the notation used in the mathematical models of our approach. Fig. 1 shows our approach that for solving three models:

Table 2

Notation used for scheduling, clustering and routing models

Sets:

V_c:	set of customers
V:	set of vertices
A:	set of arcs
P:	set of periods in the planning horizon
K:	set of vehicles
V_p:	set of vertices to visit in period p
A_p:	set of arcs that connect the set of vertices to visit in a period p
V_{pk}:	set of vertices to visit in period p for vehicle k
A_{pk}:	set of arcs that connect the set of vertices to visit in a period p for vehicle k

Parameters:

N:	number of customers
m:	number of vehicles
Tv:	service time in a customer
Q_k:	capacity of vehicle k
d_j:	demand of customer j
e_j:	lower bound of time window for customer j
l_j:	upper bound of time window for customer j
n_k:	lower bound for the available time of vehicle k
p_k:	upper limit for the available time of vehicle k
R_k:	maximum work time for vehicle k
t_{ij}:	travel time for arc $(i,j) \in A$
h_j:	release date of the visit with customer j
G_j:	due date to visit customer j
D:	average visit capacity of the vehicles per hour
H:	Total available hours of all vehicles

Binary Variables:

z_{jp}:	is 1 if the customer j is visited on the period p and 0 in otherwise
y_{jk}:	is 1 if the customer j is visited by the vehicle k and 0 in otherwise
x_{ij}:	is 1 if the arc (i,j) is used for the optimal solution and 0 in otherwise
y_k:	is 1 if the vehicle k is used and 0 in otherwise

Integer Variables:

N_p:	amount of customers to visit on period p

Continuous Variables:

S_j:	time in which the visit of customer j starts
u_j:	accumulated load delivered up to reaching customer j

1. Scheduling model: The objective function, input, output and solution method for this phase is the set of parameters shown in Fig. 1. This model defines the visit date of each vertex from the set V_C in the planning horizon of p periods, by considering the release date, due date and the travel times t_{ij} for all $(i,j) \in A$. To solve this model, we use an expert system based on the know-how of the courier service as the knowledge base. The inference engine works as a rule interpreter. The output consists of two sets: a set of customers allocated for each period and its respective sub set of arcs for each period.

2. Clustering model: This model takes as input the outputs of the scheduling model, the demand and times windows of the customers, and the capacity and labor schedule for the workers (see Fig. 1). The clustering model consists of grouping all vertices of the set V_p according to the traveling time t_{ij}, that the arc $(i,j) \in A_p$, and allocates it to a vehicle k by considering its labor schedule $[n_k, p_k]$, its maximum load capacity Q_k and the time windows of the customers $[e_j, l_j]$. The outputs are a set of customers allocated for each vehicle in each period and its associated sub set of arcs.

3. Routing model: The inputs consists of the outputs of clustering model, the travel times, customer's time windows and vehicle's time windows. The routing model is an individual optimization process to determine the order in which each vehicle will visit all customers assigned during the clustering process, taking into account the time windows of the customers and the available time of the vehicles. The output is a set of routes for each vehicle, with the scheduled detailed in the start time of service at each customer.

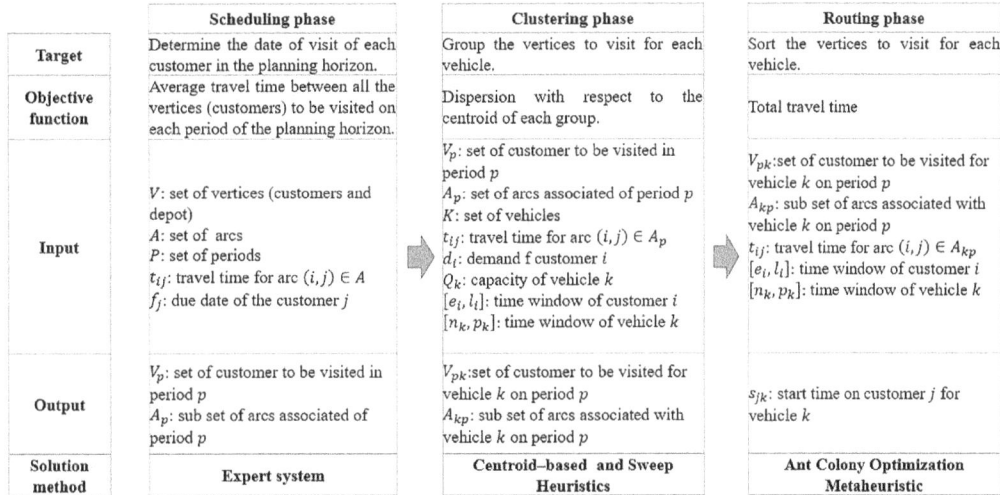

	Scheduling phase	Clustering phase	Routing phase
Target	Determine the date of visit of each customer in the planning horizon.	Group the vertices to visit for each vehicle.	Sort the vertices to visit for each vehicle.
Objective function	Average travel time between all the vertices (customers) to be visited on each period of the planning horizon.	Dispersion with respect to the centroid of each group.	Total travel time
Input	V: set of vertices (customers and depot) A: set of arcs P: set of periods t_{ij}: travel time for arc $(i,j) \in A$ f_j: due date of the customer j	V_p: set of customer to be visited in period p A_p: set of arcs associated of period p K: set of vehicles t_{ij}: travel time for arc $(i,j) \in A_p$ d_i: demand f customer i Q_k: capacity of vehicle k $[e_i, l_i]$: time window of customer i $[n_k, p_k]$: time window of vehicle k	V_{pk}:set of customer to be visited for vehicle k on period p A_{kp}: sub set of arcs associated with vehicle k on period p t_{ij}: travel time for arc $(i,j) \in A_{kp}$ $[e_i, l_i]$: time window of customer i $[n_k, p_k]$: time window of vehicle k
Output	V_p: set of customer to be visited in period p A_p: sub set of arcs associated of period p	V_{pk}:set of customer to be visited for vehicle k on period p A_{kp}: sub set of arcs associated with vehicle k on period p	s_{jk}: start time on customer j for vehicle k
Solution method	Expert system	Centroid–based and Sweep Heuristics	Ant Colony Optimization Metaheuristic

Fig. 1. Proposed solution approach to scheduling and routing in courier services

In the next three sections, we will present the components of our solution approach. First, we present the scheduling model to allocate the customers for each period; second, we develop the clustering model to determine the set of customers to each worker and finally, we present the routing model to find the sequence to visit the customers for each worker.

4. Scheduling model

We have a set of customers to be scheduled in a period within the planning horizon. Since the number of customer is too large, then we need to solve a scheduling model by taking into account the appointments arranged with them, the due dates of the customers, the maximum load capacity of the vehicles, the number of vehicles available, the average visiting capacity of the vehicles by hour and the total available hours of the vehicle. The first phase consists of allocating customers to periods. The next sections explain the mathematical formulation of the problem and its solution procedure that consists of an expert system.

4.1 Problem formulation

Given a set V_C as the set of customers that requires to be visited during the planning horizon P, we need to allocate each customer for a period to be visited. This problem uses a binary variable z_{jp} that take the value of 1 only if the customer $j \in V_c$ is visited on period $p \in P$; and 0, otherwise.

The scheduling model can be formulated mathematically as follows:

$$\min \frac{\sum_{p \in H} \sum_{(i,j) \in A} t_{ij} z_{jp}}{N} \tag{1}$$

subject to:

$$z_{jp} = f\big(h_j, G_j, t_{ij}, Q, m, H, D\big) \qquad\qquad j \in V, p \in P \tag{2}$$

$$\sum_{p \in P} z_{jp} = 1 \qquad\qquad p \in P \tag{3}$$

$$\sum_{j\in V} z_{jp} \leq HD \cdot D \qquad\qquad p \in P \qquad\qquad (4)$$

$$z_{jp} \in \{0,1\} \qquad\qquad j \in V, p \in P \qquad\qquad (5)$$

The objective function (1) seeks to minimize the average distance between all customers to visit each period in the planning horizon, allowing to maximize the concentration in the visiting groups. Constraints (2) represents the relation between assigning a specific customer to a period depending on the appointments arranged with them, the due dates of the customers, the maximum load capacity of the vehicles, the amount of total vehicles available, the average visiting capacity of the vehicles by hour and the total available hours of the vehicles. In the scheduling model, this relation is represented by a set of rules that belongs to an expert system which also includes basic rules to comply with the constraints, e.g., the date of the appointments arranged with customers, and priority rules as ranking customers that are closer to the due date, customers who do not have an appointment and are closer than one who does so they can be visited in the same period. Constraint (3) assures that each customer will have a determined visiting date in the planning horizon. Constraint (4) controls the maximum amount of deliveries that can be programmed to be visited according to the total load capacity. And finally, Constraint (5) defines the binary variable used.

4.2 Solution procedure

Regarding the courier service, and referring to the definition in this paper, scheduling is performed by the workers using mainly the know-how and expertise of the business to define the period in the planning horizon in which the visit of each customer will be executed. Our plan consists of providing a model which allows complying with the efficacy of the scheduling process and to prevent the chance of human error. An expert system (ES) is a well-designed system which replicates the cognitive process that experts use to solve particular problems (Turban, 1989). In addition, from the designed rules that came from workers experience, the ES will use as a performance function, the objective function established in Eq. (1), that consists in the minimization of the average distance between all customers to visit each period of the planning horizon.

The ES is specialized in a specific field and aims to solve problems through reasoning methods that emulates the performance of a human expert (Díez et al., 2001; Méndez-Giraldo et al., 2013). Fig. 2 shows the architecture of a traditional ES.

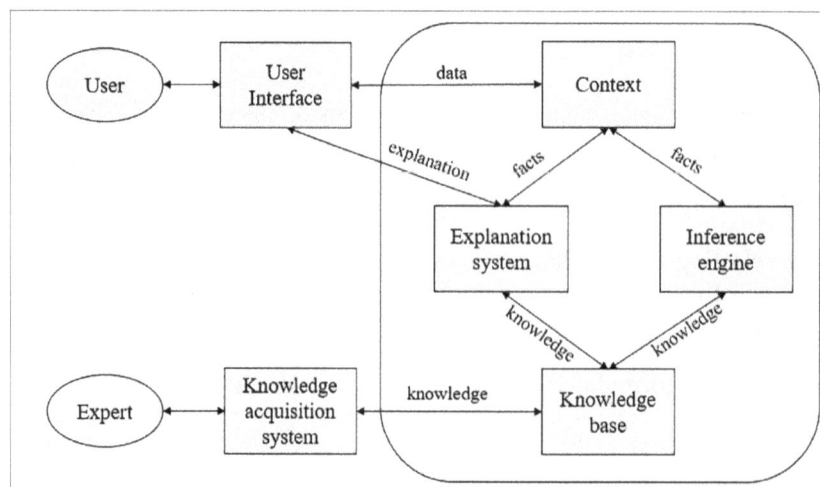

Fig. 2. Architecture of a traditional ES (source: modified of Krishnamoorthy and Rajeev (1996))

The ES operates according to López-Santana and Méndez-Giraldo (2016). When the ES starts the process of inference, it needs a context or working memory, which represents the set of established facts. The explanation system simulates the process of the answer of an expert to the questions: *How* is a decision arrived at? and *Why* do we need a data? Since the knowledge base has to be continuously updated and/or appended depending on the growth of knowledge in the domain, an interface between expert and ES is necessary. The knowledge acquisition system executes this interface, and is not an on-line component, it can be implemented in many ways. In addition, another interface is necessary between the user and the ES. The user interface allows the user interacts with the ES giving data, defining facts and monitoring the status of the problem-solving.

The knowledge base is the component where all knowledge provided by experts is stored in an orderly and structured way under a set of relationships, such as rules or probability distributions, facts and heuristics that represent the thinking of the expert. The major task of ES rules the basis of its operation, but we can also use representation schemes such as semantic networks, frames, among others. The knowledge base is independent of the mechanisms of inference and search methods.

The inference engine performs two main tasks (López-Santana & Méndez-Giraldo, 2016): in the first phase, it examines the facts and rules, and if possible, add new facts; in the second one, it chooses the order in which inferences are made. Notably, the findings obtained by the inference engine, are made based on deterministic or stochastic information data and can be simple or compound.

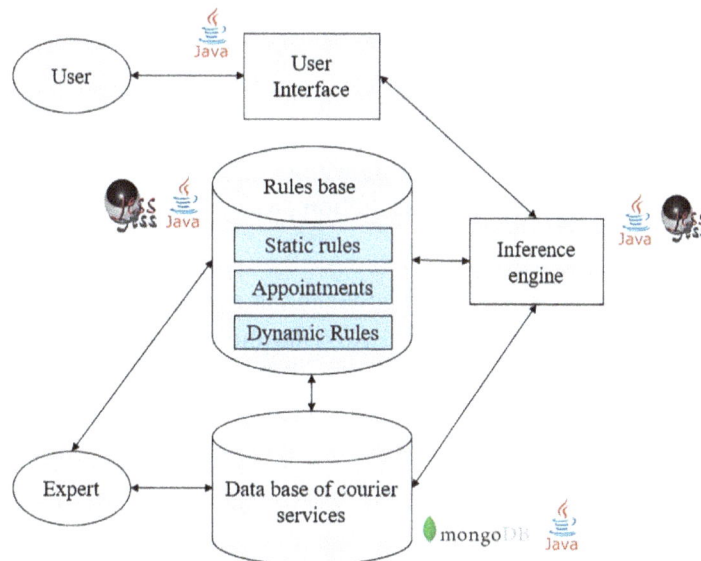

Fig. 3. Architecture of knowledge base and inference engine

The main components that make up our proposed ES are the base of knowledge and the inference engine. Fig. 3 presents the structure of our ES. The *knowledge base* is represented by the data base of courier service through a process of documentary review of internal procedures, service agreement and also through interviews with employees who carry out scheduling activities, which may be considered as pseudo-expert. They will be formulated the rules of the type if (condition) then (action), which together create rules base. Both, data base and rules base, are the knowledge base of our ES and were built in mongoDB® and Java®.

The *inference engine*: this is a rule interpreter who decides when to apply the rules. This was built in Jess (Java expert system shell) and Java®. From this definition, the rules of the knowledge base will be clustered into three modules, to be applied in an orderly and sequential way as follows:

Module of Static rules: The first module contains static rules since they are applied to every customer's package to visit of the available stock. These rules are based on the internal procedures and service agreements with the customers. The result is a feature modification named route status, this feature can assume three values: *pending*, *dismissed* and *scheduled*. If some rule is fulfilled, the feature is changed to "*dismissed*". If none of the module rules is fulfilled, the feature is remained with the value "pending".

Module of Appointments: The second module contains only one rule, however, it is the major and application basis of the next module. The rule consists of evaluating if the customer has a booked appointment within the planning horizon if so, it changes the route to be "*scheduled*" and the same route date as the appointment date.

Module of Dynamic rules: The third module contains a set of rules that are activated only if they are required. These rules were designed to take advantage of ES, which consists of intensive searches, in other words, instead of re-executing computational efforts by repeatedly applying different rules to the same ones, efforts are directed towards searching on a large basis of facts, those that activate the rules. These rules are built based on the experience and know-how of the pseudo-experts who program the routes.

The set of rules that make up the base of knowledge and the inference engine that was developed for the case study are listed in the Appendix A. The user interface and all components are integrated into Java®. The outputs of scheduling model are the sets of customers that are scheduled to visit for each period. The next section describes the clustering model in order to build a determined number of clusters for each period associate each one for a worker.

5. Clustering model

With the results of scheduling model, we have for each period in the planning horizon, a set of customers is scheduled to be visited, however, the scale of these sets are too large, thus we use a strategy for grouping in sets for each vehicle using a clustering model. To solve this model we propose two algorithms: a centroid-based and sweep algorithms. The next sections describe the mathematical formulation of the clustering model and its solution procedures.

5.1 Problem formulation

Given the set of customers V_p scheduled to be visited in each period p, we use a binary variable y_{jk} that takes the value of 1 only if the vehicle $k \in K$ visits the customer $j \in V_p$; and 0, otherwise. For each period p, the clustering model can then be stated as follows,

$$\min \frac{\sum_{k \in K} \sum_{i \in V_p} t_{ij} y_{jk}}{N_p} \tag{6}$$

subject to:

$$\sum_{j \in V_p} d_j y_{jk} \leq Q_k \qquad\qquad k \in K \tag{7}$$

$$\sum_{k \in K} y_{jk} = 1 \qquad\qquad j \in V_p \tag{8}$$

$$e_j - n_k y_{jk} \geq 0 \qquad\qquad j \in V_p, k \in K \tag{9}$$

$$y_{jk} \in \{0, 1\} \qquad\qquad j \in V_p, k \in K \tag{10}$$

The objective function (6) seeks to minimize the average distance among all the customers to be visited by the vehicle k. Constraints (7) limit the maximum load capacity of each vehicle. Constraints (8) state that a customer can be allocated only for one vehicle. Constraints (9) ensure that the customers with time windows are allocated to vehicles that start their shift before. Constraints (10) define the binary variable used.

5.2 Solution procedure

There are several methods that allow clustering customers in groups according to distances, costs, etc. (Patiño Chirva et al., 2016). We selected two procedures: Centroid-Based and Sweep algorithms. Given $d(v_j)$ as the demand of customer $j \in V_p$, v_j denotes the vertex of the customer j, $[e, l]_{v_j}$ is the time window of customer j and $[e, l]_{l_i}$ is the schedule labor period of cluster l_i, i.e., the time window when a vehicle is available to deliver the packages. We assume that a cluster corresponds to a vehicle.

The Centroid-Based algorithm starts from the geometry of the geometric centers, around which the cluster environment is generated. Algorithm 1 shows the pseudocode of the centroid-based algorithm for clustering the set of customers to be visited for each period. This method is divided into two steps according to Shin and Han (2012) and we modify to adapt the time windows for customers and vehicles.

Algorithm 1. Centroid-Based algorithm
1: **First step** // Clustering construction
2: $i = 0$ and Q =vehicle capacity
3: **while** $(V_p \neq \emptyset)$ // all customers are allocated
4:    v_j: the farthest-to-origin customer among all un-clustered customers
5:    **if** $([e,l]_{v_i} \in [e,l]_{l_i})$
6:       Generate cluster l_i with customer v_j
7:       $Q_{l_i} = Q$ // State cluster capacity
8:       **while** $(d(v_j) \leq Q_{l_i})$
9:          Add v_j to l_i
10:          $Q_{l_i} = Q_{l_i} - d(v_j)$
11:          Compute $CG(l_i)$
12:          $V_p = V_p - j$ and $V_p' = V_p$
13:          **for all** $k \in V_p'$
14:             v_k: node closest to $CG(l_i)$
15:             **if** $([e,l]_{v_k} \in [e,l]_{l_i})$
16:                $v_j = v_k$
17:                **Break**
18:             **Else**
19:                $V_p' = V_p' - k$
20:             **end if**
21:          **end for**
22:       **end while**
23:       $i = i + 1$
24:    **end if**
25: **end while**
26: **End first step**
27: **Second step** // Cluster adjustment
28: State $L = \{l_0, l_1, ..., l_m\}$, which is the set of cluster generated in the first step
29: **for** $i = 0$ to m
30:    **for all** $v_k \in l_i$
31:       **for all** $l_j \in L$.
32:          **if** $([e,l]_{v_k} \in [e,l]_{l_i}$ and $i \neq j$ and v_k is closer to $CG(l_j)$ than to $CG(l_i)$ and $d(v_k) \leq Q_{l_j})$
33:             Move customer v_k from l_i to l_j
34:             Compute $CG(l_j)$ and $CG(l_i)$
35:             $Q_{l_i} = Q_{l_i} + d(v_k)$
36:             $Q_{l_i} = Q_{l_i} - d(v_k)$
37:          **end if**
38:       **end for**
39:    **end for**
40:    $i = i + 1$
41: **end for**
42: **End Second step**

In the first step is the clustering construction, the algorithm selects the node farthest from the source, within nodes that have not been previously allocated and a cluster is generated; then the geometric center calculated with Eq. (11) must be computed between these nodes, where w_i^x and w_i^y are coordinates in x and y, respectively, and the nodes belonging to the cluster.

$$CG(l_i) = \left(\sum_{i=0}^{m} w_i^x / m \ , \ \sum_{i=0}^{m} w_i^y / m \right) \tag{11}$$

To add nodes to the first cluster (l_0), the cluster construction algorithm finds v_j among un-clustered nodes, which is located closest from $CG(l_0)$, and includes v_j to l_0 only if the demand of v_j does not exceed the available capacity of l_0 and its time window belongs to the labor schedule of l_0. If v_j is added to cluster l_0, the capacity of the cluster is reduced by the demand of v_j and $CG(l_0)$ is recalculated. The same processes above are conducted until the available capacity of l_0 becomes smaller than the demand of the closest node from $GC(l_0)$. When the demand of v_j exceeds the available capacity of l_0, the algorithm stops to expand l_0, and finds the farthest node among un-clustered nodes again in order to generate another cluster, l_1. These processes are repeated until no unvisited node exists, i.e., when $V_p \neq \emptyset$. In summary, the nodes closest to the geometric center are added taking into account the defined capability of each cluster, which corresponds to the capacity of a vehicle, the time windows associated with the customers and the available schedules of the vehicles. The time complexity of this step is $O(n^2)$.

In the second step, the clusters generated in step one are adjusted. Cluster adjustment means that if customer v_k, which belongs to cluster l_i, is closer to $CG(l_j)$ than $C_G(l_i)$, the demand of v_k does not exceed the available capacity of l_j, and the time window of v_k belongs of the available labor time of l_j then move v_k from l_i to cluster l_j. If a customer moves from l_i to l_j , $CG(l_i)$ and $CG(l_j)$ are also recalculated. The time complexity of this step is also $O(n^2)$.

Algorithm 2. Sweep algorithm

1:      **First step** // Initialization
2:      Set $i = 0$ and $Q = $ vehicle capacity
3:      Compute polar coordinates of customers (ρ_i, θ_i)
4:      Sort cluster L_k by schedule
5:      Order customers according value θ_i
6:      **if** $(\theta_i = \theta_{i+1})$
7:              Sort customers in ascending order according to the value ρ_i
8:      **end if**
9:      **End First Step**
10:    **Second step** //Cluster generation
11:    **for all** $k \in l_k$ //from first to last cluster in l_k
12:            Select an initial customer v_j
13:            $k = 0$ and $l_k = \{v_j\}$
14:            $Q_{l_k} = Q_{l_k} - d(v_j)$
15:            $N_P = N_p - j$
16:            **while** $(N_p \neq \emptyset)$
17:                    Select the next customer v_{j+1}
18:                    **if** $([e,l]_{v_{j+1}} \in [e,l]_{l_k}$ and $d(v_{j+1}) \leq Q_{l_k})$
19:                            $l_k = l_k \cup \{v_{j+1}\}$
20:                            $Q_{l_k} = Q_{l_k} - d(v_j)$
21:                            $N_p = N_p - j$
22:                    **end if**
23:            **end while**
24:            $k = k + 1$
25:    **end for**
26:    **End Second Step**

The Sweep algorithm is based on polar coordinates (Gillett & Miller, 1974; Toth & Vigo, 2002). Algorithm 2 shows pseudocode for clustering the customers based in sweep algorithm. The first step is the initialization process, which consists of computing the polar coordinates and sort them in ascending order. Moreover, the clusters $L = \{l_0, l_1, \ldots, l_K\}$ are sorted by schedule time. The second step is the cluster construction. It starts from an origin point a straight line which is rotating and creating a zone in which the customers must be assigned. The area addressed in the sweeping process constitutes the cluster, as long as it complies with the capacity constraint indicated for each one of them. The algorithm selects v_j among un-clustered nodes includes v_j to l_0 only if the demand of v_j does not exceed the available capacity of l_0 and its time window belongs to the labor schedule of l_0. If v_j is added to cluster l_0, the capacity of it is reduced by the demand of v_j. The same processes above are conducted until the available capacity of l_0 becomes smaller than the demand of the closest node from $GC(l_0)$. Then we select the next cluster, l_1 and the process is repeated until no unvisited node exists, i.e. $V_p \neq \emptyset$. Like centroid-based algorithm, the sweeping heuristics were modified to take into account to the time windows associated with the customers and the available schedules of the vehicles. The time complexity of this algorithm is $O(n^2)$.

We have two algorithms to compute the clusters, then we will compare the average distances to select the clusters with the less average of distance between all customers of set V_p. The outputs of this model are the sets of clusters that represent the workers or vehicles for each period that need to be visited. The next section describes the routing model to determine the sequence in which the customers are visited.

6. Routing model

With the results of the clustering phase, we have a set of customers to be allocated in paths in order to be severed for the crew of vehicles. This situation arises with a special routing model that consists in a traveling salesman problem with time windows (TSPTW) because each vehicle is solved separately. The next sections explain the mathematical formulation of the routing model and the solution procedure that is based on the Ant Colony Optimization (ACO) metaheuristic.

6.1 Problem formulation

Given a set of customer V_{kp} allocated to a period p and a vehicle k for the clustering model, the vehicle must serve the customers, thus we are faced to solve a routing model. The problem is defined on the auxiliary directed graph $G' = (V'_{kp}, A'_{kp})$, with $V'_{kp} = V_{kp} \cup \{0, n+1\}$ and $A'_{kp} = \{(0,i): i \in V_{kp}\} \cup \{(i, n+1): i \in V_{kp}\} \cup \{(i,j): i,j \in V_{kp}, i \neq j\}$, where vertices 0 and $n+1$ denote the depot. The node $n+1$ is a copy of the depot 0 and is introduced for the sake of clarity in the mathematical formulation presented hereafter. We assume a zero service time at the depot, i.e. $T_{v_0} = T_{v_{n+1}} = 0$.

This problem consists of designing a set of routes for each vehicle such that:

- Each customer is visited exactly once,
- The time window $[e_j, l_j]$ is met, i.e., the vehicle must arrive to start service before time l_j and no later than e_j, but the customer will not be serviced before the beginning of the time window, and
- The time window $[n_k, p_k]$ is met, i.e., the vehicle has a time window that represents its labor period.

The problem is formulated as a TSPTW. This problem uses binary variables x_{ij} that takes the value of 1 only if the vehicle traverses the arc $(i,j) \in A_{kp}$, where $i \neq j$; and 0, otherwise. The variable s_j defines the start time of a service $j \in V_{kp}$. Finally, the variables u_j represents the accumulated load in customer $j \in V_{kp}$. For each vehicle k and each period p in the planning horizon, the mathematical model is formulated as follows:

$$\min \sum_{(i,j)\in A_{kp}} t_{ij} x_{ij} \tag{12}$$

subject to:

$$\sum_{j \in V_{kp} \setminus j \neq i} x_{ij} = 1 \qquad\qquad i \in V_{kp} \qquad (13)$$

$$\sum_{i \in V_{kp} \setminus i \neq j} x_{ij} - \sum_{i \in V_{kp} \setminus i \neq j} x_{ji} = 0 \qquad\qquad j \in V_{kp} \qquad (14)$$

$$\left(u_i + d_j - u_j \right) \leq M \left(1 - x_{ij} \right) \qquad\qquad (i,j) \in A_{kp} \qquad (15)$$

$$\left(u_i + d_j - u_j \right) \geq -M \left(1 - x_{ij} \right) \qquad\qquad (i,j) \in A_{kp} \qquad (16)$$

$$u_i \leq Q \qquad\qquad i \in V_{kp} \qquad (17)$$

$$s_i + t_{ij} + Tv - M \left(1 - x_{ij} \right) \leq s_j \qquad\qquad (i,j) \in A_{kp} \qquad (18)$$

$$e_j \sum_{i \in V_{kp}} x_{ji} \leq s_j \leq l_j \sum_{i \in V_{kp}} x_{ji} \qquad\qquad j \in V_{kp} \qquad (19)$$

$$n_k \leq s_j \leq p_k \qquad\qquad j \in \{0, n+1\} \qquad (20)$$

$$x_{ij} \in \{0,1\} \qquad\qquad (i,j) \in A_{kp} \qquad (21)$$

$$u_j \geq 0 \qquad\qquad j \in V_{kp} \qquad (22)$$

The objective function (11) aims to minimize the total travel time related to the visits performed for each vehicle. Constraints (12) ensure a visit each customer only once. Constraints (13) allow keeping fluency between trips in a route, i.e., after visiting one customer, a vehicle must immediately start another trip. Constraints (14) and (15) allocate values to the accumulated load variables and prevent sub-tours. Constraints (16) state the limited capacity of the vehicle. Constraints (17) establish the relationship between the vehicle's departure time from a customer and its immediate successor. Constraints (18) and (19) enforce the time windows at the customers and the vehicles. Constraints (20) impose binary conditions on the flow variables. Finally, constraints (21) defines the accumulated load variables as non-negative.

6.2 Solution procedure

With the previous procedure we allocate a set of customers for each period in the planning horizon, as a result we can simplify the problem as a TSPTW, capacity constraints, and schedule capacity. To solve this problem an ACO metaheuristic is proposed since TSPTW is considered an NP-complete, i.e., belong to the combinatorial optimization problems that are considered difficult to solve. However, according to Glover and Kochenberger (2003), ACO has had several successful implementations in a wide variety of combinatorial optimization problems. In these problems, generally, ACO algorithms are linked with additional features, such as local optimizers against specific problems, that take ant solutions for local optimum (Glover & Kochenberger, 2003). Some recent examples of successful applications of the ACO metaheuristic for the TSPTW can be found in the literature (e.g. Kara & Derya, 2015; López-Ibáñez & Blum, 2010). In addition, if we consider that ACO metaheuristic has also been used for the VRPTW, more applications can be found in the literature (e.g. Cheng & Mao, 2007; Ding et al., 2012; Pureza et al., 2012; Yu et al., 2011).

Cheng and Mao (2007) propose an ACO to solve the TSPTW. The authors developed a modified ant algorithm called ACS-TSPTW (Ant Colony System-TSPTW) based on the ACO technique to solve the TSPTW. Two local heuristics are integrated in ACS-TSPTW algorithm to manage the time windows limitations. Table 3 summarizes the main notation used for the proposed ACO.

Table 3
Notation used for proposed ACO to solve the TSPTW

$\tau_{ij}(t)$: pheromone from node i to j during the iteration t
$\Delta\tau_{ij}(t)$: pheromone update from node i to j during iteration t
A: set of arcs (i, j)
$\mathcal{N}_i$: paths that are missing from the current ant
g_{ij}: heuristic that increases the importance of nodes near their closing time
h_{ij}: heuristic that increases the importance of nodes that do not require timeout
$\mathcal{T}^*$: best current path
$\mathcal{L}^*$: travel time of the best current road

According to Cheng and Mao (2007), the ACO algorithm defines the amount of pheromone deposited for each path $(i, j), \forall (i, j) \in A$ with global and local update. The local update is computed as follows:

$$\tau_{ij}(t) = (1 - \varpi)\tau_{ij}(t - 1) + \varpi\Delta\tau_{ij}(t), \tag{23}$$

where $\tau_{ij}(t)$ is the pheromone quantity on the way i, j for the time t, $0 < \varpi < 1$ is the local evaporation rate and $\Delta\tau_{ij}(t)$ is the pheromone increase on the way i, j given by:

$$\Delta\tau_{ij}(t) = \begin{cases} \mathcal{L}^*, & \text{if } (i, j) \in \mathcal{T}^* \\ 0, & \text{otherwise} \end{cases}, \tag{24}$$

where $\mathcal{L}^*$ is the distance traveled by the best path $\mathcal{T}^*$ and, the global update is given by:

$$\tau_{ij}(t) = (1 - \theta)\tau_{ij}(t - 1) + \theta \Delta\tau_{ij}(t), \tag{25}$$

where $0 < \theta < 1$ is the global evaporation rate. The transition node rule consists in choice to move to the next node j is being taken with the Eq. (26).

$$j = \begin{cases} \arg_{z \in N_i}^{max} \tau_{ij}(t)(g_{ij})^\beta (h_{ij})^\gamma, & \text{if } q \leq q_0 \\ S & , \text{ otherwise} \end{cases}, \tag{26}$$

Two local heuristics are defined by Eq. (28) and Eq. (29). The parameters (β, γ) are user defined that determine the importance of g_{ij} and h_{ij} and S is s the transition probability rule defined in Eq. (27).

$$P_{ij} = \begin{cases} \dfrac{\tau_{ij}(t)(g_{ij})^\beta (h_{ij})^\gamma}{\sum_{z \in N_i} \tau_{iz}(t)(g_{iz})^\beta (h_{iz})^\gamma}, & \text{if } j \in \mathcal{N}_i, \\ 0 & , \text{ otherwise} \end{cases} \tag{27}$$

where $\mathcal{N}_i$ is the set of nodes that the ant has not visited in this tour.

The local heuristic g_{ij} increases the importance of the nodes near their closing time and is defined by:

$$g_{ij} = \begin{cases} \dfrac{1}{1 + \exp\left(\delta(G_{ij} - \mu)\right)}, & \text{if } G_{ij} \geq 0 \\ 0 & , \text{ otherwise} \end{cases} \tag{28}$$

where G_{ij} is the remaining time to the node j to close and is defined as $G_{ij} = b_j - t_j$ where b_j is the time to the node j close and t_j is the current time 1, δ controls the probability curve and μ is defined as the average of $G_{ij} \geq 0$. The intuition behind this heuristic is that the ant should visit those nodes whose arrival times are closer to their upper time-window constraints before those nodes with later upper time-window constraints in order to avoid the risk of lateness (Cheng & Mao, 2007). The local heuristic h_{ij} increases the importance of nodes near to its input time and is defined by:

$$h_{ij} = \begin{cases} \dfrac{1}{1 + \exp\left(\lambda(H_{ij} - v)\right)}, & \text{if } H_{ij} \geq 0 \\ 1 & , \text{ otherwise} \end{cases} \tag{29}$$

where H_{ij} is the remaining time to the node j to open and is defined as $H_{ij} = a_j - t_j$ where a_j is the time to the node j to open and t_j is the current time, λ controls the probability curve and v is defined as the average of $H_{ij} \geq 0$. Note that $h_{ij} = 1$, when H_{ij} is negative, i.e., the open nodes are given the highest priority.

Algorithm 3 introduces the basic processing steps for the proposed ACO. The first step consists of starting the variables with initial values. The second step builds a path according to local heuristics g_{ij} and h_{ij}. The third step saves the best path found so far and his travel time, and updates the pheromone. The fourth one evaluates the stopping criterion, if it is met the algorithm stops and reports the solution, else the second step is run again and the iterative process continues until the stopping criterion is reached.

Algorithm 3. Pseudocode ACO-TSPTW

1:	**First step** // Initialization
2:	Set $t = 0$ or $t =$ route start time.
3:	**for all** $(i, j) \in A$ $\rightarrow$ Set initial values $\tau_{ij}(t)$
4:	Put m ants in the tank or at a random node.
5:	**for all** $(i, j) \in A$ $\rightarrow$ Set $\Delta\tau_{ij}(t) = 0$
6:	**End first step**
7:	**Second Step** //Path construction
8:	**for** $i = 0$ to n: // n is the number of nodes
9:	$\quad$ **for** $k = 1$ to m:
10:	$\quad\quad$ **if** $\mathcal{N}_i \neq \emptyset$
11:	$\quad\quad\quad$ //Local heuristics
12:	$\quad\quad\quad$ **for all** $j \in \mathcal{N}_i$ $\rightarrow$ Compute g_{ij} and h_{ij} // equations (28) and (29)
13:	$\quad\quad\quad$ **if** $g_{ij} = 0 \,\forall\, j \in \mathcal{N}_i$
14:	$\quad\quad\quad\quad$ $k = k + 1$ // no more feasible paths
15:	$\quad\quad\quad$ **end if**
16:	$\quad\quad\quad$ //Move:
17:	$\quad\quad\quad$ $j \rightarrow$ equations (26) and (27)
18:	$\quad\quad\quad$ Delete j from $\mathcal{N}_i$
19:	$\quad\quad$ **end if**
20:	$\quad$ **end for**
21:	**end-for**
22:	**End second step**
23:	**Third step**
24:	Save $\mathcal{T}^*$ and $\mathcal{L}^*$ // the best path found so far and his travel time, respectively
25:	**for all** $i \in \{1..m\}$ $\mathcal{N}_i = n$ //for each ant
26:	**for all** $(i, j) \in A$://Local update
27:	$\quad$ $\tau_{ij}(t) \rightarrow$ equation (23) //update the pheromone
28:	**end for**
29:	**for all** $(i, j) \in \mathcal{T}^*$:
30:	$\quad$ $\tau_{ij}(t) \rightarrow$ equation (25) //update the pheromone
31:	**end for**
32:	**End third step**
33:	**Fourth step** // Stopping criterion
34:	**if** the stopping criterion is met
35:	$\quad$ **Exit**
36:	**Else**
37:	$\quad$ **Go to** Second step
38:	**end if**
39:	**End fourth step**

For the initial amount of pheromone on each arc, Cheng and Mao (2007) suggest to determine it based on an approximate estimate of the tour length and, the stopping criterion is set to a maximum acceptable number of iterations to run the algorithm.

7. Results

In this section, we illustrate the results of our proposed method for scheduling and routing in courier services obtained in a real world case. All tests in this work were run using Java 8 on a Windows 8 64-bit machine, with an Intel i5 3337 processor (2×1.8 GHz) and 6 GB of RAM.

7.1 Case study

We illustrate the performance of the proposed method with an example based on a real-world scenario, where a courier company in Bogotá has a crew of more than two hundred motorcycle operators with delivery package in the city. The company has three operating centers, from which approximately 2.100 packages per day are delivered. That is 700 packages on average for each center, with 20 to 32 motorcycle operators, these can carry from 10 to 60 packages per day according to a schedule. We assume a planning horizon of three days.

The packages that must be delivered to the customer have a set of attributes, which determine the selection according to priorities in order to be scheduled. In the real case study, the packages have more than 60 attributes, however, many of them are informative and do not affect the rules of our ES. Table 4 lists the set of attributes considered in the case study. For the sample, a database with the packages that were in its system was requested by the courier company on September 20, 2016. In the sample received there were 17.492 packages. Fig. 4 shows the geographical dispersion of the packets on the map of Bogota. We assume a capacity of 2.700 visits per day and 75 vehicles.

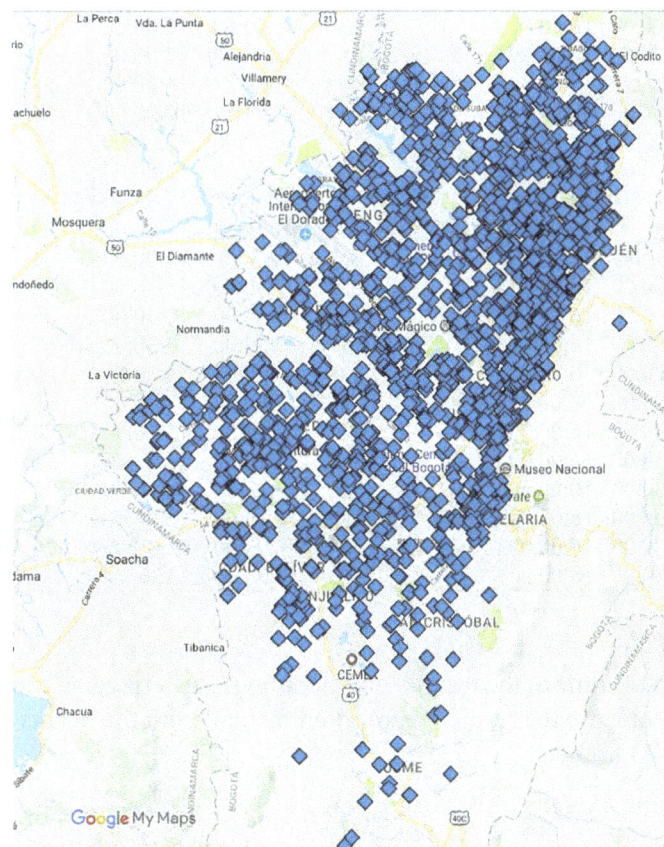

Fig. 4. Geographical dispersion of sample of case study

Table 4

List of rules attributes of the case study

- Identification of the document
- Type of delivery
- Appointment date
- Appointment start time
- End date quotes
- Real state
- Manageable actions
- Real reason
- Telephone management qualification
- Number of calls
- Number of visits
- Due date
- CX (Longitude)
- CY (Latitude)

Table 5 shows the capacity Q_k as the number of visits per day and the time windows $[n_k, p_k]$ for each vehicle k.

Table 5

Capacity and time window of vehicles in case study

k	Q_k	n_k	p_k	k	Q_k	n_k	p_k	k	Q_k	n_k	p_k
1	60	8:00:00	18:00:00	26	30	10:30:00	16:00:00	51	38	8:00:00	13:30:00
2	40	8:00:00	15:00:00	27	25	8:00:00	12:00:00	52	38	8:00:00	14:00:00
3	25	9:00:00	14:00:00	28	35	10:00:00	17:00:00	53	38	8:00:00	14:00:00
4	20	9:30:00	14:00:00	29	35	10:30:00	17:00:00	54	38	8:00:00	14:00:00
5	30	10:00:00	16:00:00	30	25	8:00:00	12:00:00	55	41	8:00:00	14:30:00
6	30	10:00:00	16:00:00	31	25	8:00:00	12:00:00	56	41	8:00:00	14:00:00
7	30	8:00:00	14:00:00	32	35	10:00:00	17:00:00	57	41	8:00:00	14:30:00
8	30	8:00:00	14:00:00	33	30	10:00:00	16:00:00	58	41	8:00:00	14:30:00
9	30	10:00:00	16:00:00	34	35	10:00:00	17:00:00	59	44	8:00:00	14:30:00
10	30	10:00:00	16:00:00	35	25	8:00:00	12:00:00	60	44	8:00:00	15:30:00
11	30	11:30:00	16:00:00	36	25	8:00:00	12:00:00	61	50	8:00:00	16:00:00
12	40	8:00:00	14:30:00	37	35	10:30:00	17:00:00	62	60	8:00:00	18:00:00
13	30	10:30:00	16:00:00	38	35	10:00:00	17:00:00	63	60	8:00:00	18:00:00
14	40	8:00:00	15:00:00	39	35	10:30:00	17:00:00	64	60	8:00:00	18:00:00
15	30	10:00:00	16:00:00	40	25	9:30:00	13:30:00	65	60	8:00:00	17:00:00
16	30	10:00:00	16:00:00	41	25	8:30:00	12:30:00	66	60	8:00:00	17:00:00
17	30	10:00:00	16:00:00	42	25	9:30:00	13:30:00	67	60	8:00:00	17:00:00
18	30	10:00:00	16:00:00	43	29	10:00:00	14:30:00	68	60	8:00:00	18:00:00
19	30	10:30:00	16:00:00	44	32	8:00:00	12:00:00	69	60	8:00:00	18:00:00
20	30	10:30:00	16:00:00	45	32	8:00:00	13:00:00	70	60	8:00:00	18:00:00
21	30	10:30:00	16:00:00	46	32	8:00:00	13:00:00	71	60	8:00:00	18:00:00
22	30	10:30:00	16:00:00	47	32	9:30:00	14:30:00	72	60	8:00:00	18:00:00
23	30	10:00:00	16:00:00	48	32	8:00:00	13:00:00	73	60	8:00:00	17:30:00
24	30	10:30:00	16:00:00	49	35	8:00:00	13:30:00	74	60	8:00:00	17:30:00
25	30	10:30:00	16:00:00	50	35	9:30:00	15:00:00	75	60	8:00:00	17:30:00

7.2 Results of case study

In this section, we report some of the results for scheduling, clustering and routing models of the proposed method. Finally, we compare the results of proposed method with the results of the current method in the case study.

i. Results of scheduling model

Section 4 described the structure of our ES for scheduling phase. For the ES we use the Jess rule engine library (available in http://www.jessrules.com/jess/index.shtml). We apply the scheduling model over a planning horizon of three days. The performance measure is the average distance between all customers for each day. Table 6 presents the results of scheduling phase for three days for the case study. Fig. 5 shows the geographical distribution of the customers for each day.

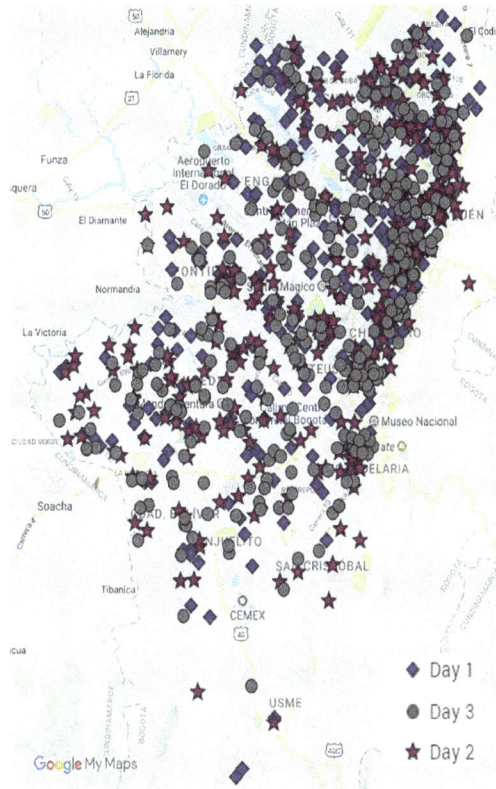

Fig. 5. Geographical distribution of customers allocated in scheduling phase

Table 6
Results of scheduling phase for three days in the case study

p	Number of customers	Average distance (km)
Day 1	2700	10.020
Day 2	2700	9.710
Day 3	2700	9.740

ii. Results of clustering model

For the clustering phase, we apply both algorithms, centroid-based and sweep, for each day in the planning horizon. Table 7 shows the results of the average of distance between all customers. The average distance obtained for the centroid-based algorithm is in average 62% less than sweep algorithm, which indicates the centroid-based algorithm found better results in cooperation with the sweep algorithm for all days. To illustrate our results, Fig. 6 shows ten clusters generated with both algorithms, in *a)* centroid-based and *b)* sweep. We can observe that the centroid-based algorithm's cluster are more concentrated while the sweep algorithm's clusters are scattered. In addition, Table 8 shows the average distance of each cluster generated with the centroid-based algorithm for Day 1.

Table 7
Results of clustering phase for three days in the case study

p	Number of customers	Average distance (km)		Difference Centroid-based vs Sweep	% difference Centroid-based vs Sweep
		Sweep	Centroid-based		
Day 1	2700	3.160	1.200	-1.960	-62%
Day 2	2700	3.020	1.140	-1.870	-62%
Day 3	2700	2.960	1.150	-1.810	-61%
Average	**2700**	**3.040**	**1.160**	**-1.880**	**-62%**

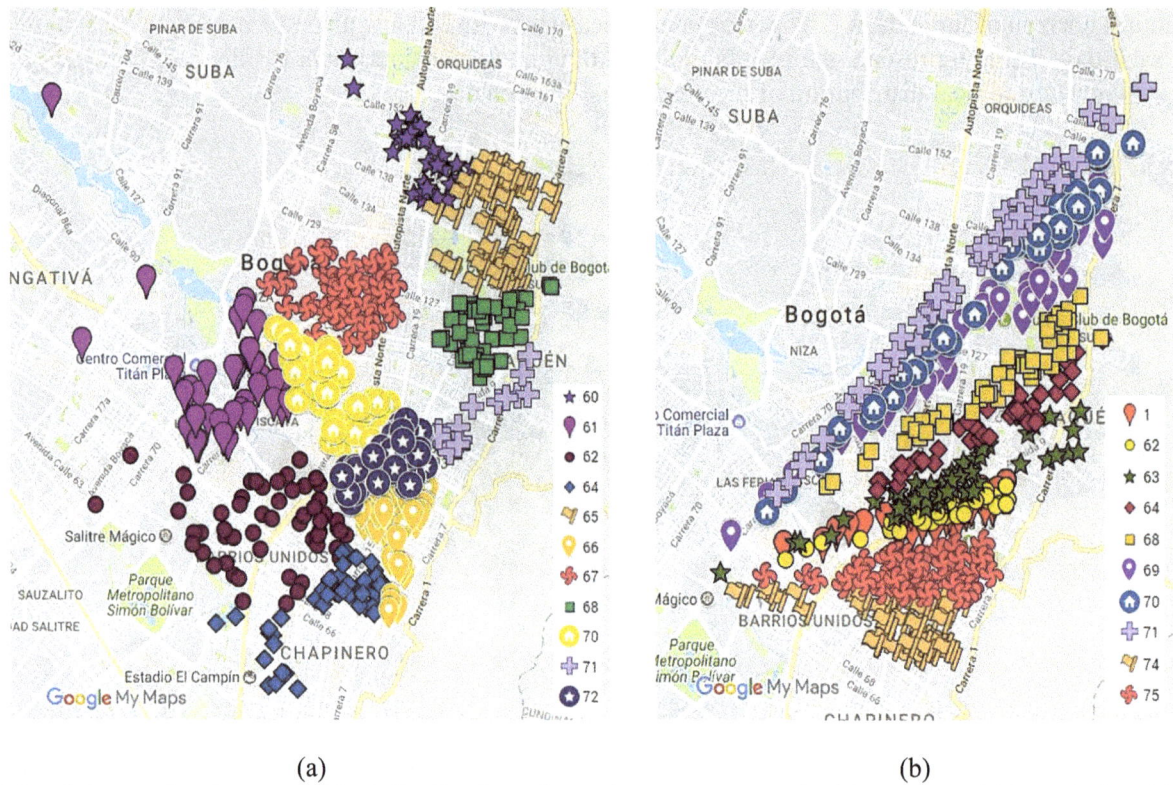

(a)   (b)

Fig. 6. Example of geographical distribution of ten clusters generated with *a)* centroid-based algorithm, and *b)* sweep algorithm

Table 8
Results of centroid-based algorithm for Day 1

Cluster k	Average distance (km)	Cluster k	Average distance (km)	Cluster k	Average distance (km)
1	3.95	26	0.87	51	1.39
2	1.60	27	1.40	52	1.24
3	1.21	28	1.22	53	1.05
4	1.34	29	1.12	54	1.39
5	1.36	30	1.47	55	0.74
6	1.63	31	0.68	56	0.88
7	1.34	32	1.29	57	1.36
8	1.50	33	1.18	58	1.33
9	1.12	34	1.20	59	1.24
10	1.62	35	0.76	60	0.91
11	1.41	36	1.23	61	1.51
12	1.80	37	1.71	62	1.57
13	1.50	38	1.06	63	0.78
14	1.42	39	1.42	64	1.16
15	1.48	40	0.63	65	1.05
16	1.43	41	0.71	66	0.95
17	1.23	42	1.00	67	1.01
18	1.32	43	1.51	68	0.80
19	1.25	44	1.15	69	1.44
20	1.05	45	0.88	70	1.10
21	1.11	46	1.12	71	0.46
22	0.94	47	0.46	72	0.73
23	1.13	48	0.85	73	0.73
24	1.21	49	1.04		
25	0.69	50	1.30		

iii.   Results of routing model

For the routing phase we apply the ACO algorithm described in Section 6 for each cluster. For setting the parameters of ACO algorithm, we use the numbers suggest by Cheng and Mao (2007) as follows,

$$m = 3, q_0 = 0.99, \theta = \omega = 0.1, \beta = 0.5, \gamma = 3 \text{ and } \delta = \lambda = 0.05.$$

Table 9 shows the total routing distance for each day, the number of customers scheduled and the average distance between customers in the route. A total of 2700 customers did not schedule because the 5% of the paths were infeasible since the time windows.

Table 9

Results of ACO algorithm in routing phase for three days in the case study

p	Total routing distance (km)	Scheduled customers	Average distance between customers (km)
Day 1	2512	2613	0.96
Day 2	2237	2273	0.98
Day 3	2567	2553	1.01
Average	**2438**	**2480**	**0.98**

Table 10 shows the detailed results for the 73 vehicle's paths in Day 1. To illustrate our results, Fig. 7 shows the path 68 in Day 1. The visiting order is represented by colors and numbers.

Table 10

Results of ACO algorithm in routing phase for three days in the case study

k	Routing distance (km)	Scheduled customers	Average distance between customers (km)	k	Routing distance (km)	Scheduled customers	Average distance between customers (km)
1	0.00	0	Infeasible	38	20.19	35	0.58
2	75.18	40	1.88	39	23.59	35	0.67
3	10.67	25	0.43	40	5.34	25	0.21
4	13.30	20	0.67	41	7.97	25	0.32
5	45.21	30	1.51	42	8.75	25	0.35
6	18.50	30	0.62	43	50.92	29	1.76
7	18.29	30	0.61	44	19.62	32	0.61
8	16.54	30	0.55	45	10.89	32	0.34
9	35.05	30	1.17	46	13.60	32	0.43
10	25.14	30	0.84	47	16.92	32	0.53
11	35.67	30	1.19	48	11.76	32	0.37
12	36.06	40	0.90	49	17.66	35	0.50
13	48.75	30	1.63	50	52.07	35	1.49
14	27.35	40	0.68	51	18.90	38	0.50
15	52.78	30	1.76	52	24.23	38	0.64
16	22.01	30	0.73	53	16.25	38	0.43
17	45.82	30	1.53	54	25.17	38	0.66
18	46.45	30	1.55	55	33.33	41	0.81
19	40.38	30	1.35	56	19.61	41	0.48
20	18.32	30	0.61	57	67.15	41	1.64
21	15.93	30	0.53	58	18.44	41	0.45
22	34.39	30	1.15	59	61.89	44	1.41
23	27.25	30	0.91	60	43.23	44	0.98
24	42.52	30	1.42	61	82.77	50	1.66
25	7.83	30	0.26	62	111.54	60	1.86
26	13.76	30	0.46	63	57.84	60	0.96
27	30.77	25	1.23	64	82.65	60	1.38
28	53.58	35	1.53	65	82.97	60	1.38
29	18.63	35	0.53	66	68.23	60	1.14
30	13.15	25	0.53	67	72.80	60	1.21
31	6.68	25	0.27	68	60.51	60	1.01
32	49.14	35	1.40	69	41.89	60	0.70
33	43.17	30	1.44	70	52.51	60	0.88
34	52.70	35	1.51	71	27.25	60	0.45
35	17.26	25	0.69	72	53.70	60	0.90
36	12.83	25	0.51	73	23.92	27	0.89
37	58.32	35	1.67				

iv. Summary and comparison

We compare the average distance between the customers for each phase. Table 11 shows the results of each phase. The average distance between the customers decreases in each phase for all days. This result indicates that the customers are concentrated, thus the routing distance is reduced.

Table 11
Summary of average distance between customers for each phase in the case study

p	Scheduling phase	Clustering phase	Routing phase
Day 1	10.02	1.20	0.96
Day 2	9.71	1.14	0.98
Day 3	9.74	1.15	1.01
Average	**9.82**	**1.16**	**0.98**

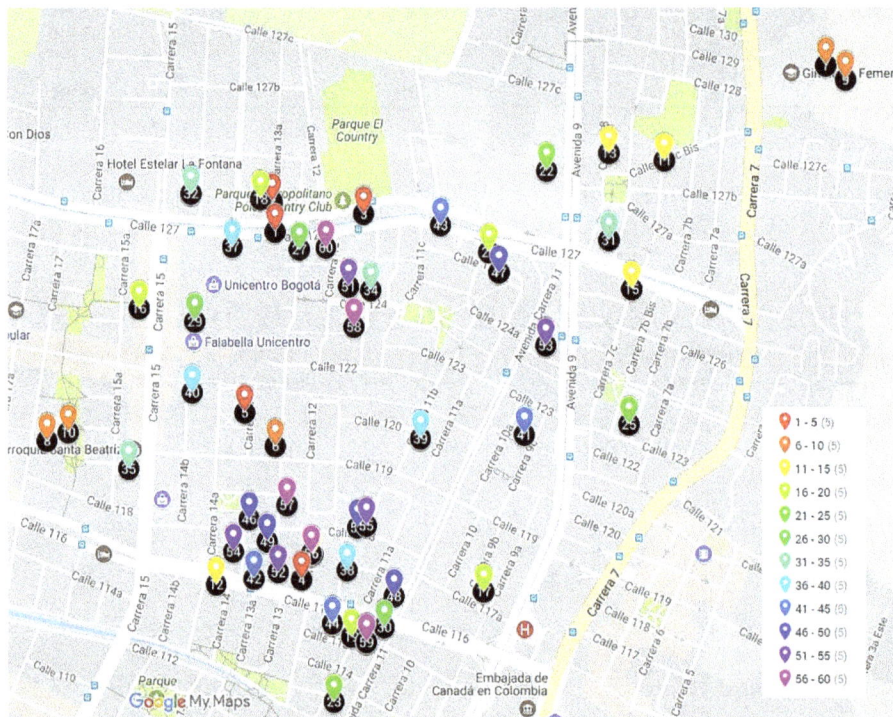

Fig. 7 Example of geographical distribution of path 68 in Day 1

Likewise, we compare the service level as the fulfillment of the customers with the delivery date. In the company, the actual service level is between 80% and 85%. Our approach obtains an average of 93%. The improvement is 11%. Table 12 summarizes these results for each day.

Table 12
Summary of service level of solution approach and actual results in the case study

p	Customers with delivery date	Customers with delivery date scheduled	% service level of our approach	% service level of company's method	% improvement our approach vs company's method
Day 1	546	524	96%	81%	15%
Day 2	433	367	85%	82%	3%
Day 3	333	324	97%	82%	15%
Average	**437**	**405**	**93%**	**82%**	**11%**

8. Conclusions and future work

This work has presented a hybrid expert system, clustering, and ant colony optimization approach to solve a scheduling and routing problem in courier services where a set of customers geographically distributed are scheduled to be severed in a planning horizon and a crew of operators is routing in order to visit the customers' sites and deliver a package or documents. The contribution of this paper is twofold: for the scheduling and routing point view, the problem is closer to VRPs but the problem exhibits some distinctive aspects like the presence of time windows in the customers and the vehicles, the due dates for the customers, multiples periods and distance-constrained; for the solution method point view, the courier services are faced with a large-scale problems thus an exact method is not available to find a good solution in a short time, then we propose an integration of knowledge based model and heuristics/metaheuristics techniques in order to solve the problem using the knowledge in the courier service to identify and classify the customers and clustering and routing method to allocate the available crew to attend the customers.

We proposed a solution approach which consists of the integration of knowledge based model and heuristics/metaheuristics techniques in a three phase procedure. In the first phase, a scheduling model determined the visit date for each customer over the planning horizon considering the release date, the due date and travel times between the customers and depot. We have used an expert system to solve this problem seeking to minimize the average distance between the customers. In the second phase, a clustering model used two algorithms: a centroid-based and a sweep algorithms. For each period, the model built a set of groups the customers according to the travel times, maximum load capacity and the customer's time windows. Finally, in the third phase, a routing model assigned and scheduled the customers for each cluster to one operator and found the sequence to visit them taking into account the customer's time windows and the available time of the operators. We have solved a traveling salesman problem with time windows using an ACO algorithm.

Results over a real-world case inspired in a courier company have illustrated the application of our proposed procedure in order to allocate. Our proposed method looked for reducing the average of distance between the customers as a performance measure of the concentration of the customers. These results suggest that as the average distance is reduced, the construction is increased; thus the routing distances is less and the customers are severed with a better service level. The results of the case study suggest that our proposed method could outperform a procedure-based in the intuition of the planners, which does not include the expert system, and clustering algorithms.

Future work should focus on the extension of the problem to a dynamic setting, in which unexpected customer's orders or operator's features occur and it is necessary to reschedule to attend the customers. Also, it is possible to incorporate additional constraints such as the technical skills for operators, uncertainty in the travel times and visits, etc. Another opportunity to improve the solution is to incorporate other features in the expert system likes a learning module in order to add more information in the knowledge base, other classification methods based on fuzzy logic in order to manage imprecision and uncertainty, for instance the priority could be managed as linguistic variable and be represented as a fuzzy number and use a fuzzy inference system to determine the scheduling package, etc. In addition, to improve the execution time, it is possible to develop a parallelizable algorithm, which could be distributed among several computers (or multiple cores) thus reducing the computation time for all phases in our approach.

Acknowledgements

We would like to acknowledge to Centro de Investigaciones y Desarrollo Científico at Universidad Distrital Francisco José de Caldas (Colombia) by supporting partially under Grant No. 2-602-468-14.

Appendix A. Rules of the Proposed Expert System

Section 4 introduced the knowledge base and the inferences system of our expert system. These were been built with the internal procedures and the service agreement in a real world company of courier services. For the Module of Statics Rules, the attributes that allow discarding the packages that do not have to get on the route are:

Appointment: Date of the Next appointment
Real status: Process state in which the package is located
Manageable Stock: internal rating
Real reason: qualification of the last visit that was made
Telephone management rating: last call rating

Table A1 lists the set of rules that are used in Module of Static Rules for the expert system in scheduling phase.

Table A1
List of rules of Module of Static Rules

If has an appointment date after the planning horizon then change from pending to dismissed
If has real status Route Assigned then change from pending to dismissed
If has real status Charged and verified then change from pending to dismissed
If has real status Packed then change from pending to dismissed
If has real status Delivered then change from pending to dismissed
If has real status City Forwarded then change from pending to dismissed
If has Not manageable stock then change from pending to dismissed
If has Real reason Non covered city then change from pending to dismissed
If has Real reason Customer has not ready documents then change from pending to dismissed
If has Real reason Incomplete address 2 then change from pending to dismissed
If has Real reason Incomplete delivery address then change from pending to dismissed
If has Real reason Address 2 does not exist then change from pending to dismissed
If has Real reason Delivery address does not exist then change from pending to dismissed
If has Real reason Complete management then change from pending to dismissed
If has Real reason Deficient identification of the final user then change from pending to dismissed
If has Real reason Non localized delivery then change from pending to dismissed
If has Real reason Moved then change from pending to dismissed
If has Real reason Entity request then change from pending to dismissed
If has Real reason Final user request then change from pending to dismissed
If has Real reason Rejected then change from pending to dismissed
If has Real reason No request rejection then change from pending to dismissed
If has Real reason Stock rejection then change from pending to dismissed
If has Real reason Destroyed rejection then change from pending to dismissed
If has Real reason Phone rejection then change from pending to dismissed
If has Real reason Declined then change from pending to dismissed
If has Real reason Unspecified time leaving then change from pending to dismissed
If has Real reason Holder without details then change from pending to dismissed
If has Real reason Uncovered zones then change from pending to dismissed
If has telephonic management rating Customer calls entity then change from pending to dismissed
If has telephonic management rating Customer has not ready documents then change from pending to dismissed
If has telephonic management rating Wrong customer names then change from pending to dismissed
If has telephonic management rating Returned to entity then change from pending to dismissed
If has telephonic management rating No ID then change from pending to dismissed
If has telephonic management rating Rejected then change from pending to dismissed
If has telephonic management rating Rejection because no needs then change from pending to dismissed
If has telephonic management rating Unspecified time leaving then change from pending to dismissed
If has telephonic management rating Return request then change from pending to dismissed
If has telephonic management rating Minor child owner then change from pending to dismissed

For Module of Appointments, only the following rule applies: "If the appointment date is contained in the planning horizon then change route status to scheduled and make route date equal to appointment date".

The rules of Module 3 that were built from the experience of the pseudo-experts use the concept of onward chaining (Forward Chaining), which is why they were ordered from the rule with conditions to comply from the most strict to the more flexible, this in order to prioritize the most important and urgent packages to manage. Table A2 describes the rules of Module of Dynamic Rules.

Table A2

List of rules of Module of Dynamic Rules

If there is available capability and the customer is closer to 5 customers with appointment or more and the package is expired and has 0 visits or less and has more than 5 calls then change from pending to scheduled.
If there is available capability and the customer is closer to 4 customers with appointment or more and the package is expired and has 0 visits or less and has more than 4 calls then change from pending to scheduled.
If there is available capability and the customer is closer to 3 customers with appointment or more and the package is expired and has 1 visit or less and has more than 3 calls then change from pending to scheduled.
If there is available capability and the customer is closer to 2 customers with appointment or more and the package is expired and has 1 visit or less and has more than 2 calls then change from pending to scheduled.
If there is available capability and the customer is closer to 1 customer with appointment or more and the package is expired and has 2 visits or less and has more than 1 call then change from pending to scheduled.
If there is available capability and the customer is closer to 0 customers with appointment or more and the package is expired and has 2 visits or less and has more than 0 calls then change from pending to scheduled.

References

Almoustafa, S., Hanafi, S., & Mladenović, N. (2013). New exact method for large asymmetric distance-constrained vehicle routing problem. *European Journal of Operational Research*, *226*(3), 386–394. Journal Article. https://doi.org/10.1016/j.ejor.2012.11.040

Archetti, C., Jabali, O., & Speranza, M. G. (2015). Multi-period Vehicle Routing Problem with Due dates. *Computers & Operations Research*, *61*, 122–134. https://doi.org/10.1016/j.cor.2015.03.014

Braekers, K., Ramaekers, K., & Van Nieuwenhuyse, I. (2016). The vehicle routing problem: State of the art classification and review. *Computers & Industrial Engineering*, *99*, 300–313. https://doi.org/10.1016/j.cie.2015.12.007

Cacchiani, V., Hemmelmayr, V. C., & Tricoire, F. (2014). A set-covering based heuristic algorithm for the periodic vehicle routing problem. *Discrete Applied Mathematics (Amsterdam, Netherlands : 1988)*, *163*(Pt 1), 53–64. https://doi.org/10.1016/j.dam.2012.08.032

Chang, T.-S., & Yen, H.-M. (2012). City-courier routing and scheduling problems. *European Journal of Operational Research*, *223*(2), 489–498. https://doi.org/10.1016/j.ejor.2012.06.007

Cheng, C.-B., & Mao, C.-P. (2007). A modified ant colony system for solving the travelling salesman problem with time windows. *Mathematical and Computer Modelling*, *46*(9–10), 1225–1235. Journal Article. https://doi.org/10.1016/j.mcm.2006.11.035

Chu, F., Labadi, N., & Prins, C. (2006). A Scatter Search for the periodic capacitated arc routing problem. *European Journal of Operational Research*, *169*(2), 586–605. https://doi.org/10.1016/j.ejor.2004.08.017

Clarke, G. u, & Wright, J. W. (1964). Scheduling of vehicles from a central depot to a number of delivery points. *Operations Research*, *12*(4), 568–581. Journal Article. https://doi.org/10.1287/opre.12.4.568

Díez, R. P., Gómez, A. G., & Martínez, N. de A. (2001). *Introduction to Artificial Intelligence: Expert Systems, Artificial Neural Networks, and Evolutionary Computation*. Universidad de Oviedo.

Ding, Q., Hu, X., Sun, L., & Wang, Y. (2012). An improved ant colony optimization and its application to vehicle routing problem with time windows. *Neurocomputing*, *98*, 101–107. Journal Article. https://doi.org/10.1016/j.neucom.2011.09.040

Dios, M., & Framinan, J. M. (2016). A review and classification of computer-based manufacturing scheduling tools. *Computers and Industrial Engineering*, *99*, 229–249. https://doi.org/10.1016/j.cie.2016.07.020

Eksioglu, B., Vural, A. V., & Reisman, A. (2009). The vehicle routing problem: A taxonomic review. *Computers & Industrial Engineering*, *57*(4), 1472–1483. Journal Article. https://doi.org/10.1016/j.cie.2009.05.009

Farahani, R. Z., Rezapour, S., & Kardar, L. (2011). *Logistics Operations and Management*. Book, Elsevier. https://doi.org/10.1016/C2010-0-67008-8

Fikar, C., & Hirsch, P. (2015). A matheuristic for routing real-world home service transport systems facilitating walking. *Journal of Cleaner Production*, *105*, 300–310. https://doi.org/10.1016/j.jclepro.2014.07.013

Francis, P., Smilowitz, K., & Tzur, M. (2006). The Period Vehicle Routing Problem with Service Choice. *Transportation Science*, *40*(4), 439–454. https://doi.org/10.1287/trsc.1050.0140

Galindres-Guancha, L. F., Toro-Ocampo, E. M., & Gallego- Rendón, R. A. (2018). Multi-objective MDVRP solution considering route balance and cost using the ILS metaheuristic. *International Journal of Industrial Engineering Computations*, *9*(1), 33–46. https://doi.org/10.5267/j.ijiec.2017.5.002

Ghiani, G., Manni, E., Quaranta, A., & Triki, C. (2009). Anticipatory algorithms for same-day courier dispatching. *Transportation Research Part E: Logistics and Transportation Review*, *45*(1), 96–106. https://doi.org/10.1016/j.tre.2008.08.003

Gillett, B. E., & Miller, L. R. (1974). A Heuristic Algorithm for the Vehicle-Dispatch Problem. *Operations Research*, *22*(2), 340–349. https://doi.org/10.1287/opre.22.2.340

Glover, F. W., & Kochenberger, G. A. (2003). *Handbook of metaheuristics*. (G. Fred & Gary A. Kochenberger, Eds.) (Vol. 57). Book, Springer US. https://doi.org/10.1007/b101874

Janssens, J., Van den Bergh, J., Sörensen, K., & Cattrysse, D. (2015). Multi-objective microzone-based vehicle routing for courier companies: From tactical to operational planning. *European Journal of Operational Research*, *242*(1), 222–231. https://doi.org/10.1016/j.ejor.2014.09.026

Jia, H., Li, Y., Dong, B., & Ya, H. (2013). An Improved Tabu Search Approach to Vehicle Routing Problem. *Procedia - Social and Behavioral Sciences*, *96*, 1208–1217. https://doi.org/10.1016/j.sbspro.2013.08.138

Kara, I., & Derya, T. (2015). Formulations for Minimizing Tour Duration of the Traveling Salesman Problem with Time Windows. *Procedia Economics and Finance*, *26*, 1026–1034. https://doi.org/10.1016/S2212-5671(15)00926-0

Kek, A. G. H., Cheu, R. L., & Meng, Q. (2008). Distance-constrained capacitated vehicle routing problems with flexible assignment of start and end depots. *Mathematical and Computer Modelling*, *47*(1–2), 140–152. https://doi.org/10.1016/j.mcm.2007.02.007

Krishnamoorthy, C. S., & Rajeev, S. (1996). *Artificial intelligence and expert systems for engineers*. Boca Raton: CRC Press.

Küçükoğlu, İ., & Öztürk, N. (2015). An advanced hybrid meta-heuristic algorithm for the vehicle routing problem with backhauls and time windows. *Computers & Industrial Engineering*, *86*, 60–68. https://doi.org/10.1016/j.cie.2014.10.014

Kusiak, A. (1990). *Intelligent manufacturing systems*. London: Prentice Hall International.

Lin, C., Choy, K. L., Ho, G. T. S., Lam, H. Y., Pang, G. K. H., & Chin, K. S. (2014). A decision support system for optimizing dynamic courier routing operations. *Expert Systems with Applications*, *41*(15), 6917–6933. https://doi.org/10.1016/j.eswa.2014.04.036

Lin, C. K. Y. (2011). A vehicle routing problem with pickup and delivery time windows, and coordination of transportable resources. *Computers & Operations Research*, *38*(11), 1596–1609. https://doi.org/10.1016/j.cor.2011.01.021

Liu, R., Xie, X., Augusto, V., & Rodriguez, C. (2013). Heuristic algorithms for a vehicle routing problem with simultaneous delivery and pickup and time windows in home health care. *European Journal of Operational Research*, *230*(3), 475–486. https://doi.org/10.1016/j.ejor.2013.04.044

López-Ibáñez, M., & Blum, C. (2010). Beam-ACO for the travelling salesman problem with time windows. *Computers & Operations Research*, *37*(9), 1570–1583. Journal Article. https://doi.org/10.1016/j.cor.2009.11.015

López-Santana, E. R., & Méndez-Giraldo, G. A. (2016). A Knowledge-Based Expert System for Scheduling in Services Systems. In J. C. Figueroa-García, E. R. López-Santana, & R. Ferro-Escobar (Eds.), *Applied Computer Sciences in Engineering WEA 2016* (pp. 212–224). Springer International Publishing AG. https://doi.org/10.1007/978-3-319-50880-1_19

López-Santana, E. R., & Romero Carvajal, J. de J. (2015). A hybrid column generation and clustering approach to the school bus routing problem with time windows. *Ingeniería*, *20*(1), 111–127. https://doi.org/http://dx.doi.org/10.14483/udistrital.jour.reving.2015.1.a07

Malmborg, C. J. (2000). Current modeling practices in bank courier scheduling. *Applied Mathematical Modelling*, *24*(4), 315–325. https://doi.org/10.1016/S0307-904X(99)00044-X

Méndez-Giraldo, G., Álvarez, L., Caicedo, C., & Malaver, M. (2013). *Expert system for scheduling production-research and development of a prototype* (1st ed.). Colombia: Universidad Distrital Francisco José de Caldas.

Nagarajan, V., & Ravi, R. (2012). Approximation algorithms for distance constrained vehicle routing problems. *Networks, 59*(2), 209–214. https://doi.org/10.1002/net.20435

Patiño Chirva, J. A., Daza Cruz, Y. X., & López-Santana, E. R. (2016). A Hybrid Mixed-Integer Optimization and Clustering Approach to Selective Collection Services Problem of Domestic Solid Waste. *Ingeniería, 21*(2), 235–247. https://doi.org/http://dx.doi.org/10.14483/udistrital.jour.reving.2016.2.a09

Pereira, F. B., & Tavares, J. (2009). *Bio-inspired Algorithms for the Vehicle Routing Problem.* (F. B. Pereira & J. Tavares, Eds.). Springer-Verlag Berlin Heidelberg. https://doi.org/10.1007/978-3-540-85152-3

Pillac, V., Guéret, C., & Medaglia, A. L. (2012). A parallel matheuristic for the technician routing and scheduling problem. *Optimization Letters*, 1–11. https://doi.org/10.1007/s11590-012-0567-4

Pureza, V., Morabito, R., & Reimann, M. (2012). Vehicle routing with multiple deliverymen: Modeling and heuristic approaches for the VRPTW. *European Journal of Operational Research, 218*(3), 636–647. https://doi.org/10.1016/j.ejor.2011.12.005

Rincon-Garcia, N., Waterson, B. J., & Cherrett, T. J. (2017). A hybrid metaheuristic for the time-dependent vehicle routing problem with hard time windows. *International Journal of Industrial Engineering Computations, 8*(1), 141–160. https://doi.org/10.5267/j.ijiec.2016.6.002

Rodríguez-Vásquez, W. C., López-Santana, E. R., & Méndez-Giraldo, G. A. (2016). Proposal for a Hybrid Expert System and an Optimization Model for the Routing Problem in the Courier Services. In J. C. Figueroa-García, E. R. López-Santana, & R. Ferro-Escobar (Eds.), *Applied Computer Sciences in Engineering WEA 2016* (pp. 141–152). Springer International Publishing AG. https://doi.org/10.1007/978-3-319-50880-1_13

Rodriguez, S., Correa, D., & López-Santana, E. (2015). An Alternative Iterative Method to Periodic Vehicle Routing Problem. In S. Cetinkaya and J. K. Ryan (Ed.), *IIE Annual Conference and Expo 2015* (pp. 2001–2010).

Sahin, S., Tolun, M. R., & Hassanpour, R. (2012). Hybrid expert systems: A survey of current approaches and applications. *Expert Systems with Applications, 39*(4), 4609–4617. https://doi.org/10.1016/j.eswa.2011.08.130

Shin, K., & Han, S. (2012). A Centroid-based Heuristic Algorithm for the Capacitated Vehicle Routing Problem. *Computing and Informatics, 30*(4), 721–732. Retrieved from http://www.cai.sk/ojs/index.php/cai/article/view/192

Sprenger, R., & Mönch, L. (2012). A methodology to solve large-scale cooperative transportation planning problems. *European Journal of Operational Research, 223*(3), 626–636. https://doi.org/10.1016/j.ejor.2012.07.021

Tlili, T., Faiz, S., & Krichen, S. (2014). A Hybrid Metaheuristic for the Distance-constrained Capacitated Vehicle Routing Problem. *Procedia - Social and Behavioral Sciences, 109*, 779–783. https://doi.org/10.1016/j.sbspro.2013.12.543

Toth, P., & Vigo, D. (2002). *The vehicle routing problem. Optimization* (Vol. 9). Philadelphia: SIAM. https://doi.org/10.1137/1.9780898718515

Turban, E. (1989). *Decision support and expert systems: management support systems* (2nd ed.). Book, Prentice Hall PTR.

Wagner, W. P. (2017). Trends in expert system development: A longitudinal content analysis of over thirty years of expert system case studies. *Expert Systems with Applications, 76*, 85–96. https://doi.org/10.1016/j.eswa.2017.01.028

Yan, S., Lin, J.-R., & Lai, C.-W. (2013). The planning and real-time adjustment of courier routing and scheduling under stochastic travel times and demands. *Transportation Research Part E: Logistics and Transportation Review, 53*, 34–48. https://doi.org/10.1016/j.tre.2013.01.011

Yu, B., Yang, Z. Z., & Yao, B. Z. (2011). A hybrid algorithm for vehicle routing problem with time windows. *Expert Systems with Applications, 38*(1), 435–441.

5

Estimation and optimization of flank wear and tool lifespan in finish turning of AISI 304 stainless steel using desirability function approach

Lakhdar Bouzid[ab*], Sofiane Berkani[a], Mohamed Athmane Yallese[a], François Girardin[c] and Tarek Mabrouki[d]

[a]Mechanical Engineering Department, Mechanics and Structures Research Laboratory (LMS), May 8th 1945 University, P.O. Box 401, Guelma 24000, Algeria
[b]University of Larbi Ben M'Hidi, Oum el Bouaghi, Algeria
[c]Laboratoire Vibrations Acoustique, INSA-Lyon, 25 bis avenue Jean Capelle, F-69621 Villeurbanne Cedex, France
[d]University of Tunis El Manar, ENIT, Tunis, Tunisia

CHRONICLE	ABSTRACT
Keywords: *Flank wear* *Surface roughness* *Lifespan* *Modeling* *DFA* *Optimization*	The wear of cutting tools remains a major obstacle. The effects of wear are not only antagonistic at the lifespan and productivity, but also harmful with the surface quality. The present work deals with some machinability studies on flank wear, surface roughness, and lifespan in finish turning of AISI 304 stainless steel using multilayer Ti(C,N)/Al$_2$O$_3$/TiN coated carbide inserts. The machining experiments are conducted based on the response surface methodology (RSM). Combined effects of three cutting parameters, namely cutting speed, feed rate and cutting time on the two performance outputs (i.e. VB and Ra), and combined effects of two cutting parameters, namely cutting speed and feed rate on lifespan (T), are explored employing the analysis of variance (ANOVA). The relationship between the variables and the technological parameters is determined using a quadratic regression model and optimal cutting conditions for each performance level are established through desirability function approach (DFA) optimization. The results show that the flank wear is influenced principally by the cutting time and in the second level by the cutting speed. In addition, it is indicated that the cutting time is the dominant factor affecting workpiece surface roughness followed by feed rate, while lifespan is influenced by cutting speed. The optimum level of input parameters for composite desirability was found Vc_1-f_1-t_1 for VB, Ra and Vc_1-f_1 for T, with a maximum percentage of error 6.38%.

1. Introduction

The tool wear, especially the flank wear, is one of the most important aspects that affect lifespan and product quality in machining. It is a major form of tool wear in metal cutting, which adversely affects the dimensional accuracy and product quality, is the main hurdle in the wide implementation of coated carbide tools to machining of stainless steel in the industry. Practically the lifespan is evaluated by the measure of the flank wear. If it increases quickly, the lifespan becomes very short and vice versa. In finish turning, tool life is measured by the machining time taken by the same insert until the flank wear reaches its allowable limit of 0.3 mm. Wear is an important technological parameter of control in the machining process. It is the background for the evaluation of the tool life and surface quality (Yallese et

* Corresponding author
E-mail: issam.bouzid@yahoo.com (L. Bouzid)

al., 2008; Uvaraja & Natarajan, 2012; Çaydas, 2010). Therefore, development of a reliable flank wear progression model will be extremely valuable. Significant efforts have been devoted by several researchers in understanding and modeling the tool wear progression, wear mechanisms, tool lifespan and surface quality in metal cutting. In recent years, a significant emphasis has been placed in the development of predictive models in metal cutting. Analytical models are easy to implement and can give much more insight about the physical behavior in metal cutting. Kramer (1986) developed a model for prediction of the wear rates of coated tools in high-speed machining of steel. The abrasive wear and the chemical dissolution were considered as dominant wear mechanisms. Singh and Rao (2010) developed flank wear prediction model of ceramic inserts in hard turning. Flank wear rate was modeled considering abrasion, adhesion, and diffusion as dominant wear mechanisms. Normal load/force incurred on the flank face was modeled using experimental results. However, increase in the normal load with the progress in flank wear was not considered in the model. It is widely reported that cutting forces influence more with the progress in flank wear, which appeared as one of the most promising techniques for monitoring tool wear. Singh and Vajpayee (1980) developed a flank wear model considering abrasion as the dominant wear mechanism. Yallese et al. (2009) have shown that for the 100Cr6 steel, the machined surface roughness is a function of the local damage form and the wear profile of a CBN tool. When augmenting cutting speed tool wear increases and leads directly to the degradation of the surface quality. In spite of the evolution of flank wear up to the allowable limit VB =0.3 mm, arithmetic roughness Ra did not exceed 0.55 µm. A relation between VB and Ra in the form $Ra = k.e^{\beta(VB)}$ is proposed. Coefficients k and β vary within the ranges of 0.204–0.258 and 1.67–2.90, respectively. It permits the follow-up of the tool wear.

A common problem in product or process design is the selection of design variable setting which meets a required specification of quality characteristics. For this purpose, among global approximation approaches, the response surface methodology (RSM) has recently attracted the most attention since it has performed well in comparison to other approaches (Garcia et al., 1981; Smith, 1973). RSM consists of the following three steps: (1) data gathering, (2) modeling, and (3) optimization (Aouici et al., 2014). Neşeli et al. (2011) applied response surface methodology (RSM) to optimize the effect of tool geometry parameters on surface roughness in hard turning of AISI 1040 with P25 tool. Yallese et al. (2009) found that a cutting speed of 120 m/min is an optimal value for machining X200Cr12 using CBN7020. In an original work carried out by Çaydaş (2009), the effects of the cutting speed, feed rate, depth of cut, workpiece hardness, and cutting tool type on surface roughness, tool flank wear, and maximum tool–chip interface temperature during an orthogonal hard turning of hardened/tempered AISI 4340 steels were investigated. Dureja et al. (2009) applied the response surface methodology (RSM) to investigate the effect of cutting parameters on flank wear and surface roughness in hard turning of AISI H11 steel with a coated-mixed ceramic tool. The study indicated that the flank wear is influenced principally by feed rate, depth of cut and workpiece hardness. When turning hardened 100Cr6, Banga and Abrão (2003) found that cutting speed is the most factor influencing tool lifespan. These authors have shown that PCBN cutting tools provide longer tool lifespan than both mixed and composite ceramics. A model built to evaluate the machinability of Hadfield steel using RMS and ANOVA techniques was presented by (Horng et al., 2008). The study revealed that the flank wear is influenced by the cutting speed while the interaction effect of the feed rate with the nose radius and the corner radius of the tool have statistical significance on obtained surface roughness.

The current study investigates the influence of cutting parameters (cutting speed, feed rate and cutting time, with a constant cutting depth ap = 0.15 mm) in relation to flank wear (VB), lifespan (T) and surface roughness (Ra) on machinability. The processing conditions are turning of stainless steel (AISI 304) with CVD coated carbide tools using both response surface methodology (RSM) and ANOVA. This latter is a computational technique that enables the estimation of the relative contributions of each of the control factors to the overall measured response. In this work, only the significant parameters will be used to develop mathematical models using response surface methodology. The latter is a collection of mathematical and statistical techniques that are useful for the modeling and analysis of problems in which response of interest is influenced by several variables and the objective is to optimize the response.

2. Experimental procedures

2.1. Material

Straight turning operations were carried out on 100 mm diameter and 400 mm length bars made of AISI 304 stainless steel with the chemical specification given in Table 1. Machining operations were achieved on a 6.6 KW power TOS TRENCIN model SN40 lathe.

Table 1
Chemical composition of AISI 304 steel (*wt %*)

C	Cr	Ni	Si	Mn	Mo	Fe	Other components
0.02	16.91	7.69	0.33	1.44	0.41	72.10	1.1

2.2. Cutting tool and tool holder

Cutting inserts used are "Ti(C, N)/Al$_2$O$_3$/TiN" CVD coated carbide referenced as GC2015 (SNMG 12-04-08-MF). The cutting inserts were clamped on a right-hand tool holder with designation PSBNR25x25M12. The geometry of the right-hand tool holder is characterized by the following angles: $\chi_r = +75°$, $\lambda = -6°$, $\gamma = -6°$ and $\alpha = +6°$.

2.3. Measurement setup

A roughness meter (2D) *Surftest 201 Mitutoyo* was employed to measure surface roughness *Ra*. The length examined is 4 mm with a cut-off of 0.8 mm and the measured values of *Ra* are within the range 0.55 – 3.2 μm. Flank wear *VB* is usually observed in the flank face of a cutting insert. Among the different forms of tool wear, flank wear is the important measure of the lifespan as it affects the surface quality of the workpiece. Long-term wear tests have been carried out through straight turning to evaluate CVD coated carbide tool flank wear for various cutting conditions. Flank wear is measured using a binocular microscope (*Visuel Gage 250*) equipped with (*Visual Gage 2.2.0*) software Fig. 1.

2.4. Planning of experiments

In order to develop the mathematical model based on RSM, two full factorials design (3^3 and 3^2) are adopted as the experimental design method. In the current study, cutting speed, feed rate and cutting time are identified as the factors which affect the responses such as surface roughness, flank wear and lifespan. Three levels are defined for each factor to investigate surface roughness and flank wear behavior. On the other hand, to investigate the lifespan behavior three levels are defined for two factors (cutting speed and feed rate) (Table 2).

Table 2
Attribution levels of cutting parameters

Control parameters	Unit	Symbol	Levels		
			Level 1	Level 2	Level 3
Flank wear (VB) and surface roughness (Ra)					
Cutting speed	m/min	Vc	280	330	400
Feed rate	mm/rev	f	0.08	0.11	0.14
Cutting time	min	t	4	10	16
Lifespan (T)					
Cutting speed	m/min	Vc	280	330	400
Feed rate	mm/rev	f	0.08	0.11	0.14

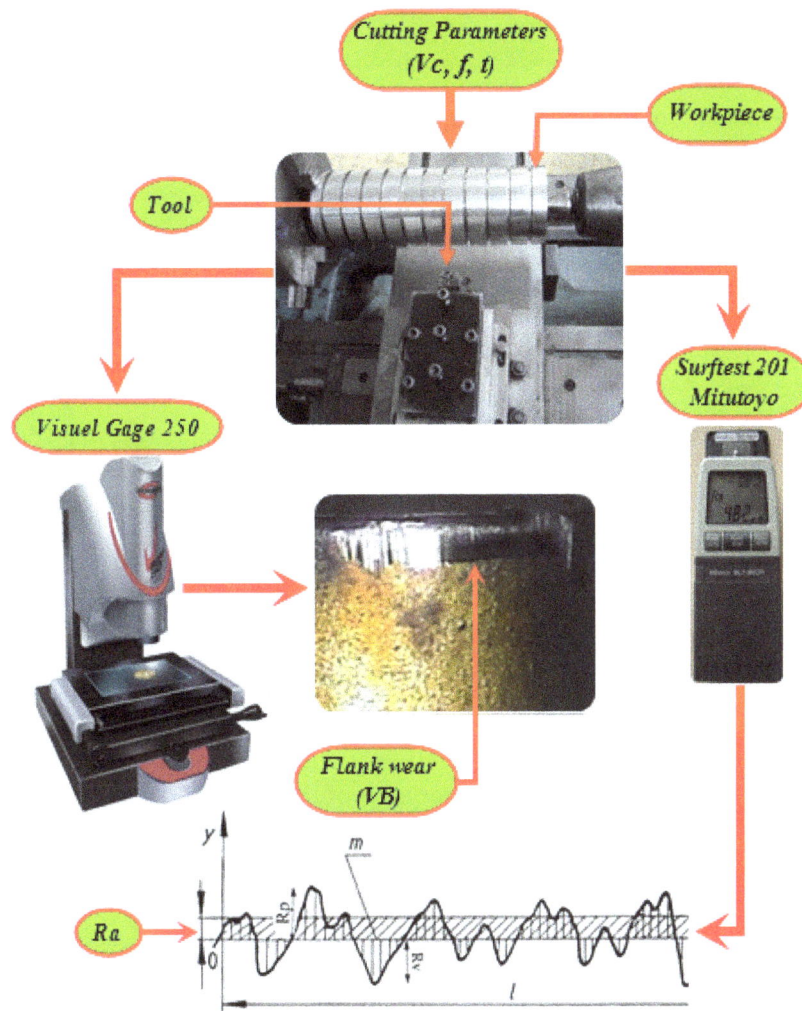

Fig. 1. Illustration of measured surface roughness and flank wear

2.5. RSM-technique

In the present investigation, the second-order RSM-based mathematical models for flank wear (*VB*), surface roughness (*Ra*) and lifespan (*T*) were developed with cutting speed (*Vc*), feed rate (*f*), and cutting time (*t*) as the process parameters. RSM technic is recognized as a statistical technique based on simple multiple regressions. Using this technique, the effect of two or more factors on quality criteria can be investigated and optimum values could be obtained. The results are expressed in 3D series or counter map. In the procedure of analysis, the approximation of response (*Y*) was proposed using the fitted second-order polynomial regression model which is commonly called the quadratic model. The quadratic model of *Y* can be written as follow Eq. (1):

$$Y = a_0 + \sum_{i=1}^{3} a_i X_i + \sum_{i=1}^{3} a_{ii} X_i^2 + \sum_{i<j}\sum a_{ij} X_i X_j + \varepsilon \qquad (1)$$

where a_0 is constant, a_i, a_{ii} and a_{ij} represent respectively the coefficients of linear, quadratic and cross product terms. X_i reveals the coded variables that correspond to the studied machining parameters such as cutting speed (*Vc*), feed rate (*f*) and cutting time (*t*), and ε is a random experimental error. The analysis of variance (ANOVA) has been applied to check the adequacy of the developed machinability models (Bouzid et al., 2015; Berkani et al. 2015; Zahia et al. 2015; Keblouti et al. 2017). The ANOVA table consists of sum of squares and degrees of freedom. The sum of squares is performed into contributions from the polynomial model and the experimental value and was calculated by the following Eq. (2):

$$SS_f = \frac{N}{N_{nf}} \sum_{i=1}^{N_{nf}} (\bar{y}_i - \bar{y})^2 \qquad (2)$$

The mean square is the ratio of sum of squares to degrees of freedom was calculated by the following Eq. (3):

$$Ms_i = \frac{SS_f}{df_i} \qquad (3)$$

The F-value is the ratio of mean square of regression model to the mean square of the experimental error was calculated by the following Eq. (4):

$$F_i = \frac{Ms_i}{Ms_e} \qquad (4)$$

This analysis was out for a 5 % significance level, i.e., for a 95 % confidence level. The last column of the tables shows the percentage of each factor contribution (Cont. %) on the total variation, then indicating the degree of influence on the result, was calculated by the following Eq. (5):

$$Cont.\% = \frac{SS_f}{SS_T} \times 100 \qquad (5)$$

3. Results and discussion

Table 3 and Table 4 show all the values of the response factors: flank wear (*VB*), surface roughness (*Ra*) and lifespan (*T*), and were made with the objective of analysing the influence of the cutting speed (*Vc*), feed rate (*f*), and cutting time (*t*) on the total variance of the results. The surface roughness was obtained in the range of 0.55–3.2 μm, flank wear and lifespan were obtained in the range of 0.025-0.51 mm, and 10-44 min, respectively.

Table 3
Experimental results for surface roughness and flank wear

Run	Factors			Responses	
	Vc, m/min	f, mm/rev	t, min	Ra, μm	VB, mm
1	280	0.08	4	0.56	0.025
2	280	0.08	10	0.61	0.050
3	280	0.08	16	0.74	0.100
4	280	0.11	4	0.81	0.030
5	280	0.11	10	1.17	0.074
6	280	0.11	16	1.25	0.110
7	280	0.14	4	1.32	0.045
8	280	0.14	10	1.34	0.069
9	280	0.14	16	1.35	0.110
10	330	0.08	4	0.55	0.040
11	330	0.08	10	0.62	0.115
12	330	0.08	16	0.80	0.190
13	330	0.11	4	0.79	0.060
14	330	0.11	10	1.21	0.135
15	330	0.11	16	1.60	0.170
16	330	0.14	4	1.31	0.06
17	330	0.14	10	1.47	0.185
18	330	0.14	16	1.92	0.350
19	400	0.08	4	0.80	0.050
20	400	0.08	10	1.24	0.200
21	400	0.08	16	1.99	0.410
22	400	0.11	4	0.87	0.065
23	400	0.11	10	1.55	0.290
24	400	0.11	16	2.95	0.460
25	400	0.14	4	1.16	0.070
26	400	0.14	10	1.70	0.300
27	400	0.14	16	3.20	0.510

Table 4
Experimental results for lifespan

Run	Factors		Response
	Vc, m/min	f, mm/rev	T, min
1	280	0.08	44
2	280	0.11	42
3	280	0.14	39
4	330	0.08	27
5	330	0.11	20
6	330	0.14	15
7	400	0.08	15
8	400	0.11	11
9	400	0.14	10

3.1. ANOVA results

Table 5 shows the results of ANOVA analysis for flank wear surface roughness and lifespan. In addition, the same Table 5 shows the degrees of freedom, sum of square, mean of square, F-value and P-value.

Table 5
ANOVA for response surface quadratic models

Source	df	SS	Ms	F	P	Cont,%	Remarques
a) Flank wear (VB)							
Model	9	0.49	0.05	65.46	< 0.0001	97.2	Significant
Vc	1	0.17	0.17	203.66	< 0.0001	33.6	Significant
f	1	0.02	0.02	18.91	0.0004	3.12	Significant
t	1	0.23	0.23	279.92	< 0.0001	46.18	Significant
Vc×f	1	0.00	0.00	2.50	0.1326	0.41	Not Significant
Vc×t	1	0.08	0.08	98.26	< 0.0001	16.21	Significant
f×t	1	0.00	0.00	4.44	0.0503	0.73	Significant
Vc²	1	0.00	0.00	0.10	0.7539	0.02	Not Significant
f²	1	0.00	0.00	0.19	0.6723	0.03	Not Significant
t²	1	0.00	0.00	0.01	0.9294	0	Not Significant
Residual	17	0.01	0.00				
Total	26	0.50					
b) Surface roughness (Ra)							
Model	9	10.51	1.17	31.75	< 0.0001	94.34	Significant
Vc	1	2.21	2.21	60.14	< 0.0001	19.84	Significant
f	1	2.59	2.59	70.52	< 0.0001	23.25	Significant
t	1	3.56	3.56	96.88	< 0.0001	31.96	Significant
Vc×f	1	$1.7.10^{-3}$	$1.7.10^{-3}$	0.046	0.8319	0.02	Not Significant
Vc×t	1	1.91	1.91	51.85	< 0.0001	17.15	Significant
f×t	1	0.094	0.094	2.55	0.129	0.84	Not Significant
Vc²	1	0.17	0.17	4.55	0.0479	1.53	Significant
f²	1	0.055	0.055	1.49	0.2389	0.49	Not Significant
t²	1	0.086	0.086	2.33	0.1455	0.77	Not Significant
Residual	17	0.63	0.037				
Total	26	11.14					
c) Lifespan (T)							
Model	5	1477.93	295.59	50.30	0.0043	98.82	Significant
Vc	1	1320.17	1320.17	224.67	0.0006	88.27	Significant
f	1	79.83	79.83	13.59	0.0346	5.34	Significant
Vc×f	1	0.15	0.15	0.03	0.8833	0.01	Not Significant
Vc²	1	147.89	147.89	25.17	0.0153	9.89	Significant
f²	1	0.89	0.89	0.15	0.7233	0.06	Not Significant
Residual	3	17.63	5.88				
Total	8	1495.56					

The ration of contribution of different factors and their interactions were also presented. The main purpose was to analyse the influence of cutting speed (Vc), feed rate (f), and cutting time (t) on the total variance of the results. From the analysis of Table 5, it can be apparent seen that the cutting time (t), cutting speed (Vc), interactions ($Vc \times t, f \times t$) and feed rate (f) all have significant effect on the flank wear

(*VB*). But, the effect of cutting time is the most significant factor associated for flank wear with 46.18 %. The next largest factor influencing *VB* is the cutting speed. Its contribution is 33.60 % to the model. The interaction (*Vc*×*t*) were less significant, while (*Vc*×*f*) interaction and productions *(Vc², f², t²)* were found to be negligible. However, the value of "P" in Table 5 for surface roughness (*Ra*) model is less than 0.05 which indicates that the model is significant, which is desirable as it indicates that the terms in the model have a significant effect on the response. In the same manner, the main effect of cutting time (*t*), feed rate factor (*f*), cutting speed (*Vc*), the interaction of cutting speed and cutting time (*Vc*×*t*), and the product (*Vc²*) are significant model terms. It can be seen that the cutting time (*t*) is the most important factor affecting *Ra*. Its contribution is 31.96 %. The second important factor affecting *Ra* is the feed rate, because its increase generates helicoid furrows, the result of tool shape and helicoid movement tool-workpiece. These furrows are deeper and broader as the feed rate increases. Its contribution is 23.25%. The next factors influencing *Ra* are the cutting speed, the interaction (*Vc*×*t*) and the product (*Vc²*). Other model terms can be said to be not significant.

Finality, it can be apparently seen in Table 5 that the cutting speed factor (Cont. ≈ 88 %), the feed rate factor (Cont. ≈ 5 %) and the product *Vc²* (Cont. ≈ 9.9 %) have statistical significance on the lifespan (*T*), especially the cutting speed.

3.2. Influential factors

To better view the results of the analysis of variance, a Pareto graph is built (Fig. 2 a, b, and c). This figure ranks the cutting parameters and their interactions of their growing influence on the flank wear (*VB*), surface roughness (*Ra*) and lifespan (*T*). Effects are standardized (F-value) for a better comparison. If the (F-value) values are greater than (F-table = 4.45) for *VB* and *Ra*; and greater than (F-table = 10.13) for lifespan, the effects are significant. By cons, if the values of F-value are less than (4.45; 10.13) the effects are not significant. The confidence interval chosen is 95 %.

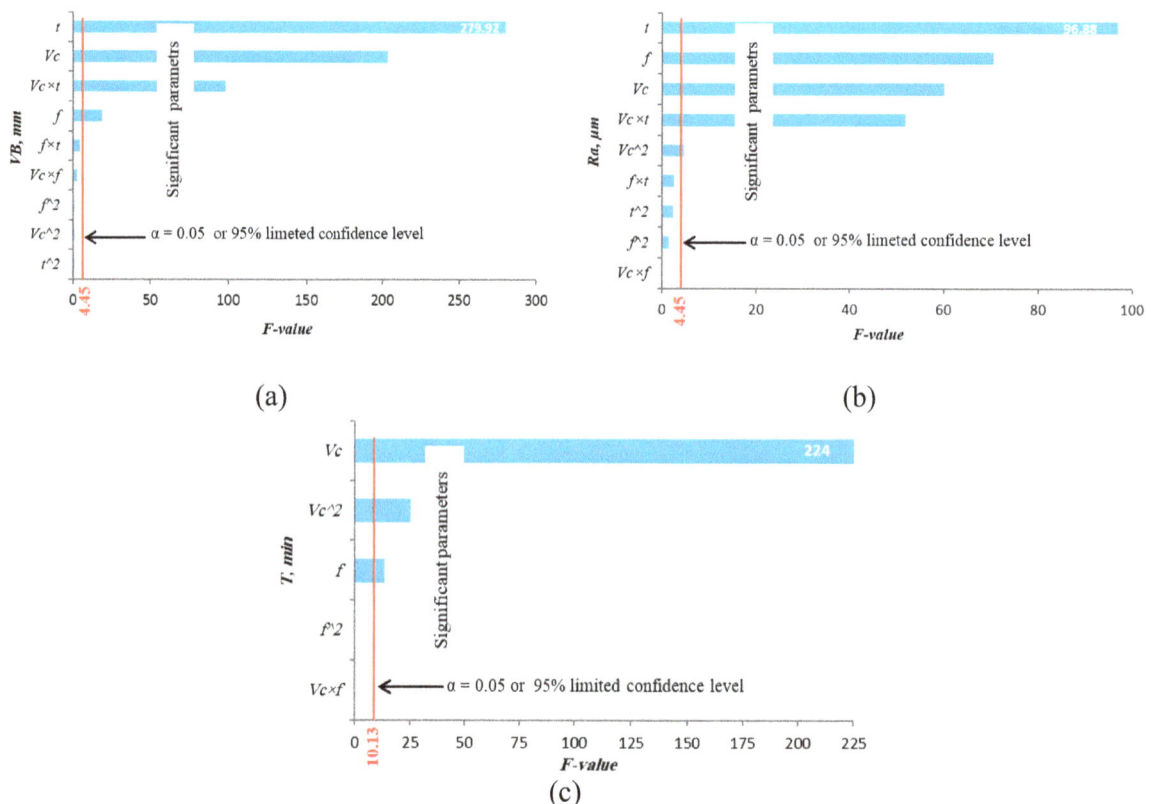

(a)

(b)

(c)

Fig. 2. Pareto graphs of: **a)** flank wear, **b)** surface roughness and **c)** lifespan

3.3. Regression equations

The relationship between the inputs (cutting speed, feed rate and cutting time) and outputs (*VB, Ra* and *T*) was modelled by quadratic regression equations. The different quadratic models obtained from statistical analysis can be used to predict the flank wear, surface roughness and lifespan according to the studied factors. The models and its determination coefficients obtained for different cutting phenomena are presented in Eqs. (6-8) respectively to (flank wear, surface roughness and lifespan).

$$VB = 0.65 - 0.0021Vc - 3.68f - 0.069t + 1.074.10^{-6}Vc^2 + 0.0072\ Vc \times f + 0.00022Vc \times t$$
$$+ 5.61f \times f + 0.097f \times t + 2.93.10^{-5}\ t^2 \tag{6}$$
$R^2 = 97.20\%$

$$Ra = 5.68 - 0.037\ Vc + 33.37f - 0.421\ t + 4.79.10^{-5}Vc^2 - 6.6.10^{-3}Vc \times f + 0.0011Vc \times t$$
$$- 106.173\ f^2 + 0.49f \times t + 0.0033\ t^2 \tag{7}$$
$R^2 = 94.39\%$

$$T = 413,26 - 1,94Vc - 321,22\ f + 0,0024Vc^2 + 0,11Vc \times f + 740,741f^2 \tag{8}$$
$R^2 = 98.82\%$

In order to reduce the models, only the significant parameters will be conserved.

$$VB = 0.19 - 6.59.10^{-4}Vc - 0.011f - 0.069t + 2.27.10^{-4}Vc \times t + 0.097222f \times t \tag{9}$$
$R^2 = 96.73\%$

$$Ra = 6.35 - 0.037\ Vc + 12.703\ f - 0.3\ t + 1.102.10^{-3}\ Vc \times t + 4.79.10{-5}\ Vc^2 \tag{10}$$
$R^2 = 92.27\%$

$$T = 400.77 - 1.92\ Vc - 122.22\ f - 2.47.10^{-3}\ Vc^2 \tag{11}$$
$R^2 = 98.75\%$

3.4. Models validation

3.4.1. Graphical Validation

The above models can be used to predict flank wear, surface roughness and lifespan at the particular design points. The differences between measured and predicted responses are illustrated in Figs. (3-5). These figures indicate that the quadratic models are capable to representing the system under the given experimental domain.

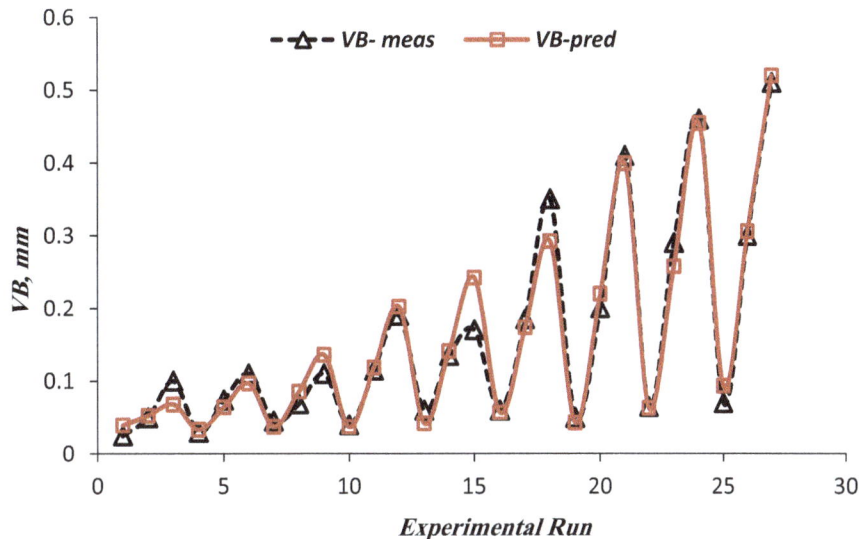

Fig. 3. Comparison between measured and predicted values for flank wear

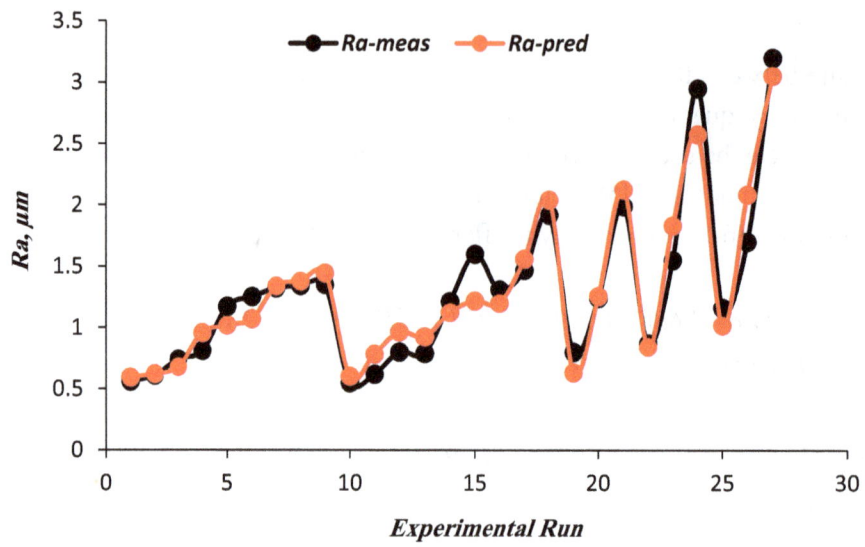

Fig. 4. Comparison between measured and predicted values for surface roughness

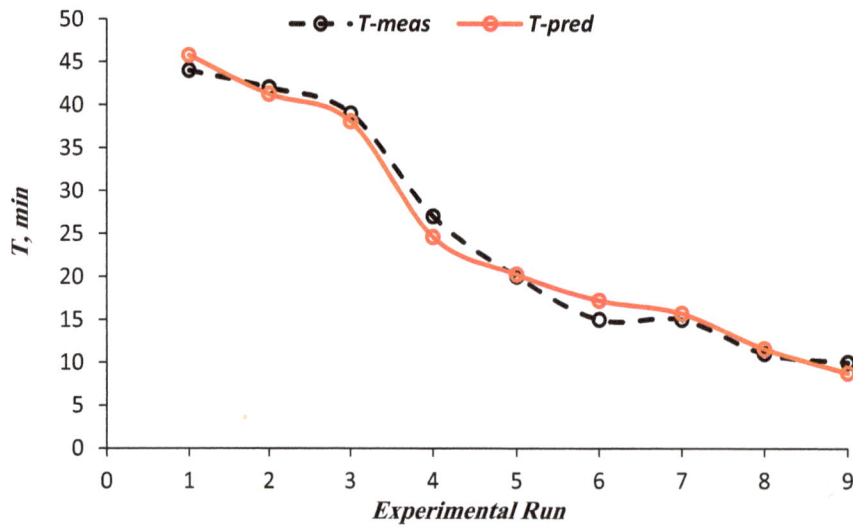

Fig. 5. Comparison between measured and predicted values for lifespan

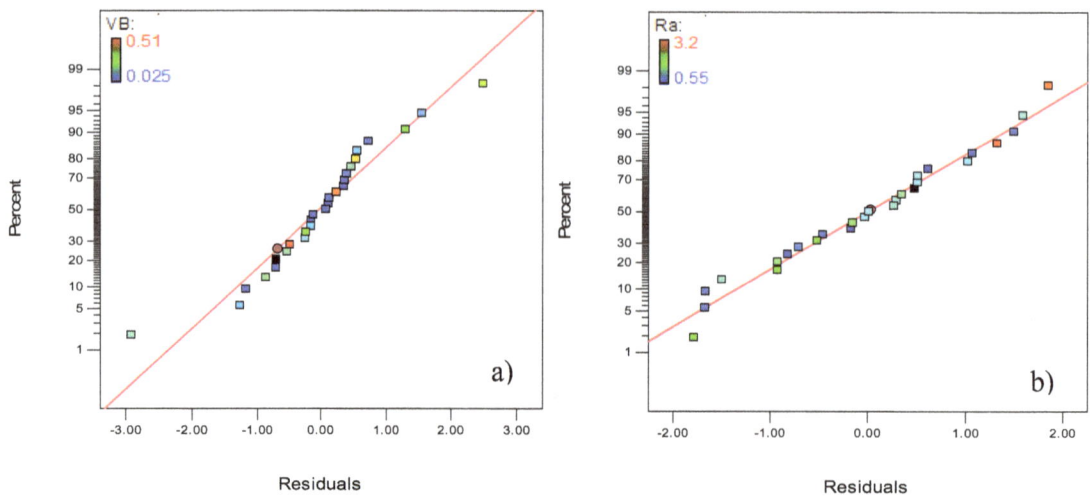

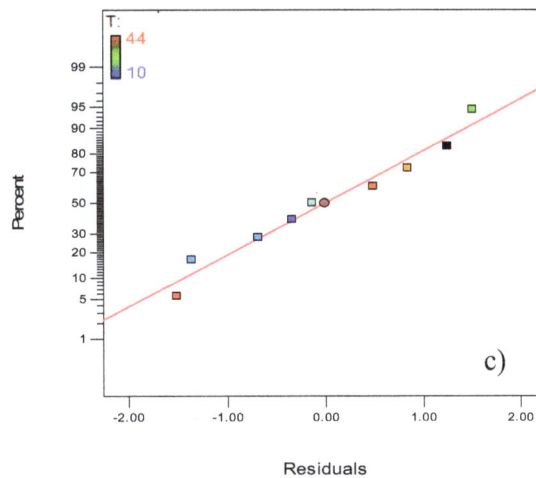

Fig. 6. Normal probability plots of predicted response for: **a)** flank wear, **b)** surface roughness and **c)** lifespan

The Anderson–Darling test and normal probability plots of predicted response for: surface roughness, flank wear and tool lifespan respectively, are presented in Figs.6 *(a, b, c)*. The data closely follows the straight line. The null hypothesis is that the data distribution law is normal and the alternative hypothesis is that it is non-normal. Using the P-value which is greater than alpha of 0.05 (level of significance), the null hypothesis cannot be rejected (i.e., the data don't follow a normal distribution). It implies that the models proposed are adequate.

3.4.2. Mathematical validation

The analysis of variance (ANOVA) was used to check the adequacy of developed models for a given confidence interval. The ANOVA table consists of sum of squares and degrees of freedom. In order to perform an ANOVA, the sum of squares is usually completed into contributions from regression model and residual error. As for this technique, if the calculated value of F-ratio of model is more than the standard tabulated value of table (F-table) for a given confidence interval, then the model is adequate within the confidence limit (Meddour et al., 2015; Muthukrishnan & Davim, 2009; Elbah et al., 2013). The adequacy of developed mathematical models is presented in Tables 6. The model accuracy (Δ) is commonly given by the following Eq.12 (Kaddeche et al., 2012):

$$\Delta = \frac{100}{n} \sum_{i=1}^{n} \left| \frac{y_{i,expt} - y_{i,pred}}{y_{i,pred}} \right|, \tag{12}$$

where $y_{i,expt}$ is the measured value of response corresponding to i^{th} trial, $y_{i,pred}$ is the predicted value of response corresponding to i^{th} trial and n is the number of trials. Eqs. (9-11) are used to test the accuracy of the models using the experimental data. The prediction errors of these models are illustrated in Table 6 together with determination coefficients. It is concluded that the correlations are valid and can be used for predictions when turning AISI304 stainless steel.

Table. 6
ANOVA analysis, percent prediction error of the experimental data and R^2 values for *VB*, *Ra* and *T*

Responses	SS M	SS R	D.f M	D.f R	Ms M	Ms R	F-test	F-table	P-value
VB	0.48	0.014	9	17	0.054	0.0008	65.45	2.49	<0.0001
Ra	10.51	0.63	9	17	1.17	0.037	31.75	2.49	<0.0001
T	1477.9	17.62	5	3	259.58	5.87	50.3	9.01	0.0043

M: model; R: residual

Responses	% Prediction error of the experimental data	R^2 (%) Values of models
VB	14.31	96.73
Ra	11.51	92.27
T	6.14	98.75

3.5. Responses surface analysis

3.5.1. Flank wear

Fig. 7 illustrates the evolution of the flank wear according to the cutting speed, cutting time and feed rate. It is found that tool wear increases with increasing effects of both cutting time and speed. It can be concluded that the cutting time exhibits maximum influence on flank wear. The maximum value of flank wear is found with height level of cutting time and cutting speed.

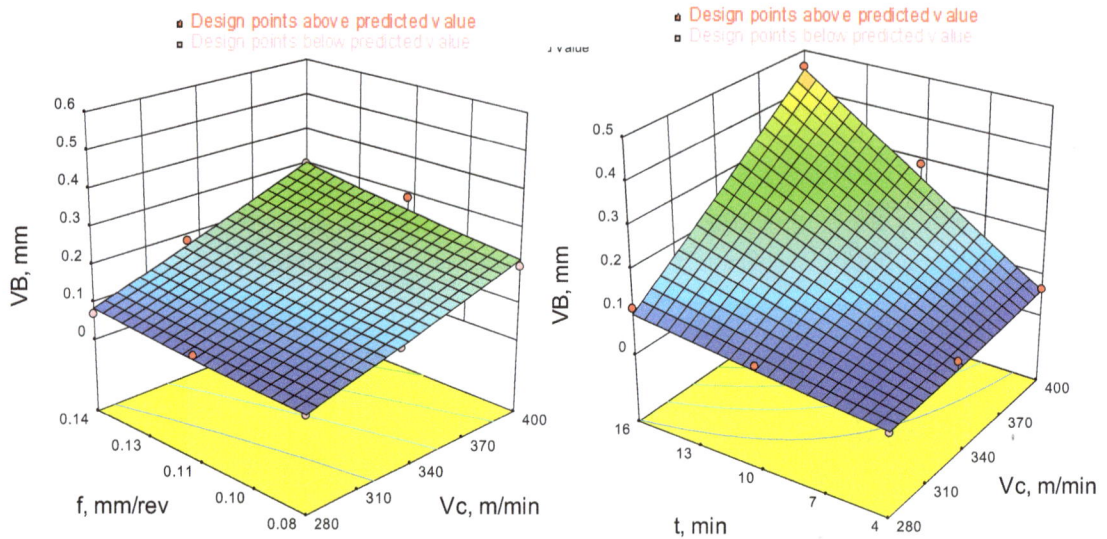

Fig. 7. Effect of cutting speed, feed rate and cutting time on flank wear

3.5.2. Workpiece surface roughness

The estimated response surface for the surface roughness in relation to the cutting parameters (Vc, f and t) presented in (Fig. 8.) shows that the cutting speed had a significant influence on machined surface roughness.

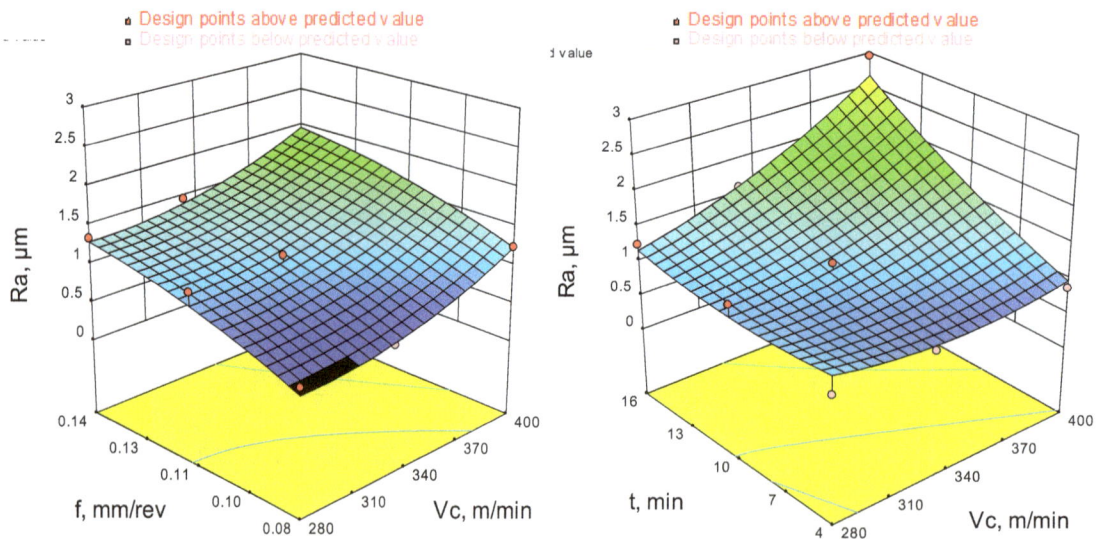

Fig. 8. Effect of cutting speed, feed rate and cutting time on surface roughness

A high values of surface roughness noted in small value of cutting speed that can be explained by the presence of built up edge (Fig. 9.) on the surface due to the high ductility of austenitic stainless steel. With the increasing of cutting speed the surface roughness values decrease until a minimum value reached beyond which they increase. The decrease in surface roughness when increasing of cutting speed to 340 m/min can be explained by the presence of micro-welds on machined surface due to high heat at cutting zone and the height of built-up-edge which lead to the breaking of BUE and carried away on the machined surface as seen in (Fig. 9). Further, increasing the cutting speed causes an increase in surface roughness because the cutting tool nose wear increases causing the poor surface finish (Ciftci, 2005). In the other hand, the roughness (Ra) tends to increase, considerably with increase in feed rate (f) and cutting time (t).

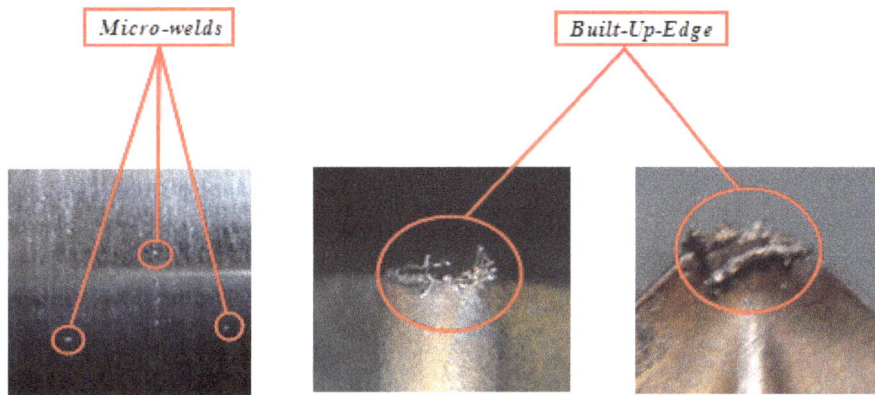

Fig. 9. Micro-Weld on machined surface and Built-Up Edge on cutting insert

3.5.3. Lifespan

The effect of feed rate (f) and cutting speed (Vc) on the tool life (T) is shown in Fig. 10. This figure displays that the value of tool life (T) decrease with the increase of cutting speed and feed rate. The decrease is approximately 77.27% of T.

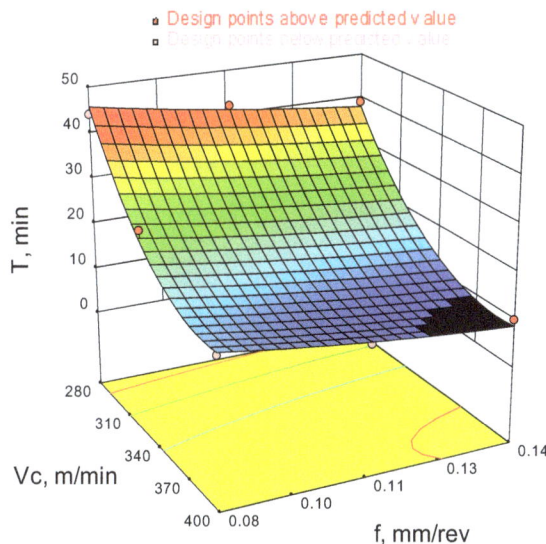

Fig. 10. Effect of cutting speed and feed rate on tool lifespan

3.6. Micrographs for flank wear VB of the GC2015 tool

For the considered regime (*Vc* = 280 m/min, *ap* = 0.15 mm and *f* = 0.08 mm/rev), flank wear *VB* of the coated carbide tool GC2015 spreads regularly. Figure 11 shows the micrographs for *VB* of GC2015 insert, its lifetime is 44 min.

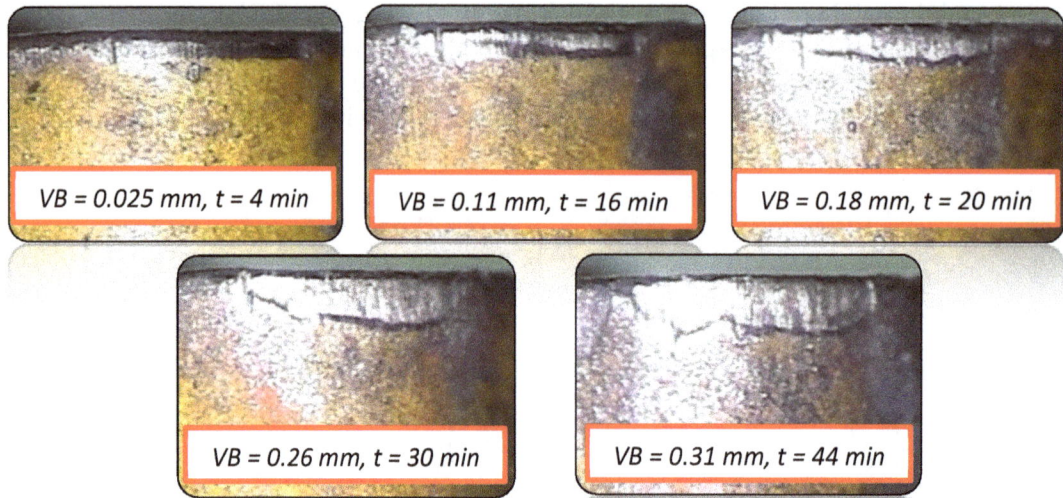

VB = 0.025 mm, t = 4 min

VB = 0.11 mm, t = 16 min

VB = 0.18 mm, t = 20 min

VB = 0.26 mm, t = 30 min

VB = 0.31 mm, t = 44 min

Fig. 11. Micrographs for *VB* of GC2015 at *ap* = 0.15 mm; *f* = 0.08 mm/rev and *Vc* = 280 m/min

4. Optimisation of machining performance using desirability function approach (DFA)

The desirability function approach to simultaneously optimizing multiple responses was originally proposed by Derringer and Suich. Essentially, the approach is to translate the functions to a common scale [0, 1], combine them using the geometric mean and optimize the overall metric. DFA continue to be a commonly preferred method because it can solve multi response optimization problem by converting it into single response optimization problem which enable to sort the computational work (Bouzid et al., 2014). The concept of desirability function involves the translation of the responses from its individual desirability to scale of composite desirability (overall desirability function). The DFA optimization process is used to perform through the following steps:

Step 1: Define the independent input parameters of the experimental design and the desired responses (y_i) to be optimized.

Step 2: Adopt an experimental design plan and conduct experiments for the designed parameter combinations.

Step 3: Calculate individual desirability (d_i) for each response (y_i) using desirability functions.

- *Larger–the–Better:* For a goal to find a maximum, the individual desirability is shown as follows:

$$di = \begin{cases} 0 & \text{if} & y_i \leq y_{i\min} \\ \dfrac{y_i - y_{i\min}}{y_{i\max} - y_{i\min}} & \text{if} & y_{i\min} \leq y_i \leq y_{i\max} \\ 1 & \text{if} & y_i \geq y_{i\min} \end{cases} \quad (13)$$

- *Smaller–the–Better*: For a goal to find a minimum, the individual desirability is shown as follows:

$$di = \begin{cases} 0 & \text{if} \quad y_i \leq y_{i\,min} \\ \dfrac{y_{i\,max} - y_i}{y_{i\,max} - y_{i\,min}} & \text{if} \quad y_{i\,min} \leq y_i \leq y_{i\,max} \\ 1 & \text{if} \quad y_i \geq y_{i\,min} \end{cases} \tag{14}$$

Step 4: Select the parameter combination that will maximize composite desirability (*Dc*), and determine the optimal parameter and its level combination based on the higher value of composite desirability (*Dc*).

$$Dc = \left(\prod_{i=1}^{p} d_i \right)^{\frac{1}{P}}, \tag{15}$$

where (*di*) is the individual desirability of the response (*i*) and (*P*) is the number of response in the measure. The desirability ranges from zero to one.

Step 5: The last step is performing ANOVA to indicate the most significant factor that affects the multiple performance characteristics.

4.1. DFA optimization for VB, Ra and T

During the optimisation process, the aim was to find the optimal values of machining parameters in order to produce the lowest flank wear, surface roughness (*VB* and *Ra*) and the highest lifespan (*T*).

Table 7

Evaluated Individual desirability and composite desirability with their rank for *VB* and *Ra*

Runs	Factors			Individual desirability (di)		Composite desirability (Dc) (%)	Rank
	Vc (m/min)	f (mm/rev)	t (min)	VB (%)	Ra (%)		
1	280	0.08	4	100	99.62	99.81	1
2	280	0.08	10	94.85	97.74	96.28	3
3	280	0.08	16	84.54	92.83	88.59	9
4	280	0.11	4	98.97	90.19	94.48	4
5	280	0.11	10	89.90	76.60	82.98	11
6	280	0.11	16	82.47	73.58	77.90	15
7	280	0.14	4	95.88	70.94	82.47	12
8	280	0.14	10	90.93	70.19	79.89	14
9	280	0.14	16	82.47	69.81	75.88	18
10	330	0.08	4	96.91	100	98.44	2
11	330	0.08	10	81.44	97.36	89.05	8
12	330	0.08	16	65.98	90.57	77.30	16
13	330	0.11	4	92.78	90.94	91.86	6
14	330	0.11	10	77.32	75.09	76.20	17
15	330	0.11	16	70.10	60.38	65.06	21
16	330	0.14	4	92.78	71.32	81.35	13
17	330	0.14	10	67.01	65.28	66.14	20
18	330	0.14	16	32.99	48.30	39.92	24
19	400	0.08	4	94.85	90.57	92.68	5
20	400	0.08	10	63.92	73.96	68.76	19
21	400	0.08	16	20.62	45.66	30.68	25
22	400	0.11	4	91.75	87.92	89.82	7
23	400	0.11	10	45.36	62.26	53.14	22
24	400	0.11	16	10.31	9.43	9.86	26
25	400	0.14	4	90.72	76.98	83.57	10
26	400	0.14	10	43.30	56.60	49.51	23
27	400	0.14	16	0	0	0	27

At first the evaluation of individual desirability (di) has been carried out for each response (y_i) based on function *Smaller–the–Better* given in Eq. (13), which is used for minimization and *Larger–the–Better* as given in Eq. (14) used for maximization of response. Then, individual desirability (di) of each response characteristics are fluxed into the composite desirability (Dc) by using Eq. (15). Tables 7 and 8 show the evaluated individual desirability and composite desirability values for each of the experimental trials.

Table 8

Evaluated Individual desirability and composite desirability with their rank for *lifespan T*

Runs	Factors		Composite desirability (Dc) (%)	Rank
	Vc m/min	f mm/rev		
1	280	0.08	100	1
2	280	0.11	94.12	2
3	280	0.14	85.29	3
4	330	0.08	50	4
5	330	0.11	29.41	5
6	330	0.14	14.71	6
7	400	0.08	14.71	7
8	400	0.11	2.94	8
9	400	0.14	0.00	9

The composite desirability (Dc) has computed and the ranks are assigned to them in ascending order and it is found that the higher composite desirability value ($Dc = 99.81\%$ for *VB* and *Ra*; $Dc = 100\%$ for *T*) obtained for the 1st trial of the experiment and its corresponding cutting combination may be regarded as optimal combination which emphasize on being closer to the experimental results. To ensure the optimal combination of levels for various factors, Table 9 shows the response mean of average composite desirability function for each level and total mean of composite desirability is also evaluated. The maximum mean of composite desirability value of each level of the factors gives the optimum cutting combination which is (Vc_1–f_1–t_1) for *VB* and *Ra*, and (Vc_1-f_1) for *T*.

Table 9

Response table for Composite desirability (Dc)

Flank wear (VB) and Surface roughness (Ra)

Levels	Average composite desirability		
	Vc, %	f, %	t, %
1	88.72	80.16	92.67
2	75.62	72.20	73.04
3	52.06	64.04	50.70
Max-Min	36.66	16.12	41.97
Optimum Level	Vc_1	f_1	t_1

Total mean of composite desirability = 72.13 %

Lifespan (T)

Levels	Average composite desirability	
	Vc, %	f, %
1	93.14	54.9
2	31.37	42.16
3	5.88	33.33
Max-Min	87.25	21.57
Optimum Level	Vc_1	f_1

Total mean of composite desirability = 43.46%

The ANOVA results of composite desirability (Dc) depicts that the cutting time (t) is the most significant parameter flowed by cutting speed (Vc) and feed rat (f) on VB and Ra. Moreover, the cutting speed is the largest factor influencing tool lifespan (T), its contribution is (92.27%) as represented in the given Table 10.

Table 10
ANOVA for composite desirability (Dc)

Source	DF	SS	MS	F-value	P-value	Cont%	Remarques
Flank wear (VB) and Surface roughness (Ra)							
Vc	1	0,517	0,517	31,262	0.000	29,37	Significant
f	1	0,186	0,186	11,222	0,003	10,54	Significant
t	1	0,678	0,678	40,945	0.000	38,47	Significant
Error	23	0,381	0,017				
Total	26	1.761					
Lifespan (T)							
Vc	1	1.194	1.194	237.402	0.0000	92.27	Significant
f	1	0.070	0.070	13.877	0.0098	5.39	Significant
Error	6	0.030	0.005				
Total	8	1.294					

The contour plot of composite desirability (Dc) has been plotted between two most significant parameter by keeping the third term constant at first level ($f = 0.08$ mm/rev) in case of flank wear and surface roughness as shown in Fig 12 and between the parameters (Vc and f) in case of tool lifespan as shown in Fig 13 . The multiple response characteristic found maximum where Vc ranges from $280 - 345$ m/min with t from $4 - 6.2$ min for VB and Ra, and Vc from $280 - 297$ m/min with f from $0.08 - 0.14$ mm/rev for T. This enclosed regions indicate the optimum zone for all responses where VB and Ra will be minimized and T will get maximized.

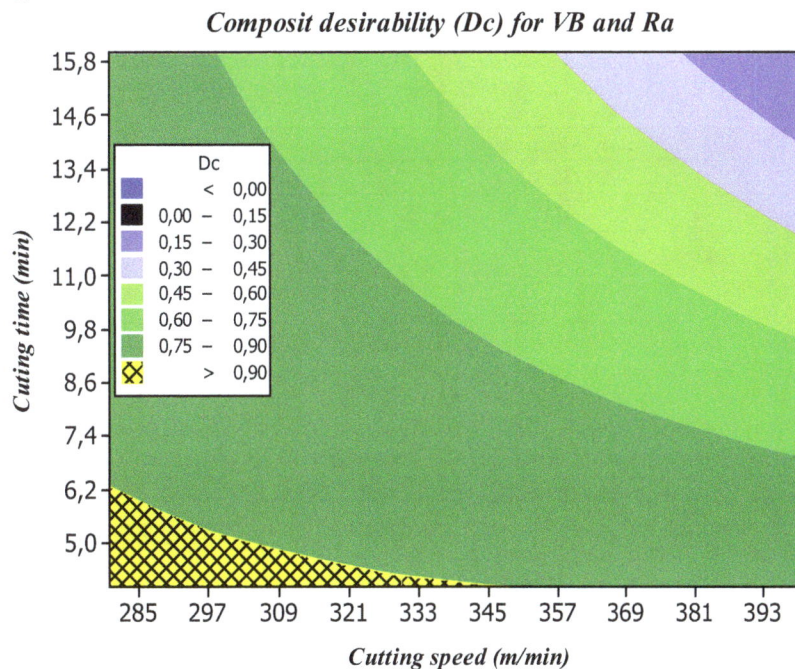

Fig. 12. Contour plot of composite desirability (Dc) for VB and Ra

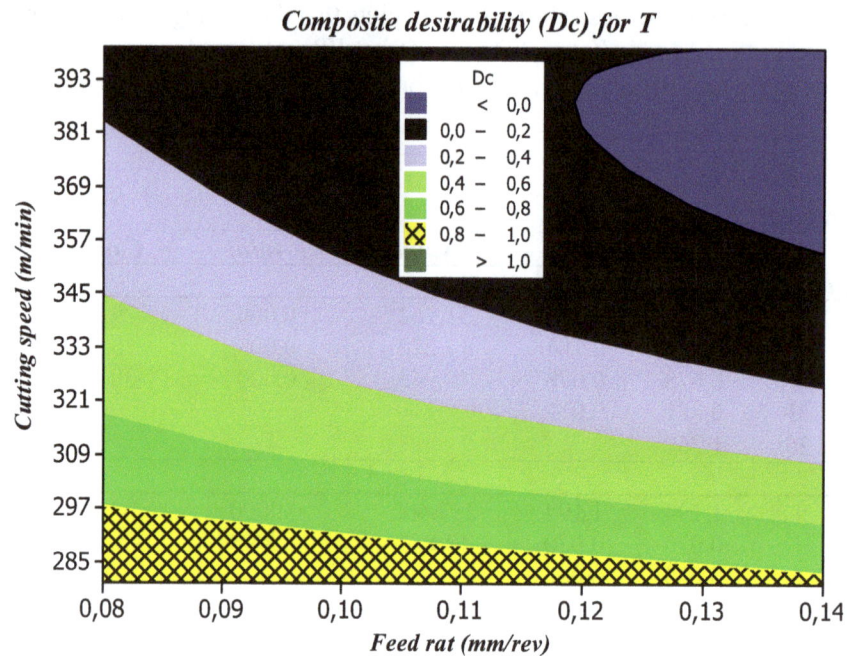

Fig. 13. Contour plot of composite desirability (*Dc*) for *T*

4.2. Confirmation of results

The predicted results based on the optimum level (*Vc₁-f₁-t₁* for *VB*, *Ra* and *Vc₁-f₁* for *T*) are evaluated and comparisons between predicted values and associated experimental values have drawn in the terms of percentage error. The error percentage is within permissible limits, Table 11 shows the maximum percentage of error is 6.38%.

Table 11
Confirmation results

Responses	Predicted $(Vc_1\text{-}f_1\text{-}t_1)$	Experimental $(Vc_1\text{-}f_1\text{-}t_1)$	Error (%)
VB (mm)	0.0235	0.025	6.38
Ra (µm)	0.586	0.56	4.44
	$(Vc_1\text{-}f_1)$	$(Vc_1\text{-}f_1)$	
T (min)	45.72	44	3.76

5. Conclusion

In this paper, the application of RSM for the turning of AISI 304 stainless steel with CVD coated carbide tool was presented. Mathematical models of flank wear (*VB*), surface roughness (*Ra*) and lifespan (*T*) evolutions according to the influence of machining parameters were investigated and optimal cutting parameters are determined through DFA optimization. The following conclusions were drawn:

1. The flank wear of CVD coated carbide tool increased with cutting speed and cutting time. The present study shows that a higher tool wear rate is noted at cutting speed 400 m/min and cutting time of 16 min.

2. The flank wear is influenced principally by the cutting time, cutting speed and the interaction effect of cutting speed/cutting time with a contribution of 46.18%, 33.6% and 16.21%, respectively.

3. The cutting time has a greater influence on the surface roughness (31.96%) followed by feed rate (23.25%), cutting speed (19.84%) and the interaction cutting speed/cutting time (17.15%).

4. Cutting speed influences lifespan (T) of GC2015 more significantly than the feed rate.

5. The tool life of the coated carbide GC2015 is 44 min.

6. The statistical models deduced define the degree of influence of each cutting regime element on flank wear and surface roughness. They can also be used for optimization of the cutting process.

7. Desirability functional approach (DFA) found very compatible to deal with multi-response optimization to obtain the optimal cutting conditions. The optimal cutting parameter combination for composite desirability reported as Vc_1–f_1–t_1 which corresponds to cutting speed of 280 m/min, feed rate of 0.08 mm/rev and cutting time of 4 min for VB, Ra and Vc_1-f_1 which corresponds to $Vc = 280$ m/min and $f = 0.08$ mm/rev for lifespan.

8. From the confirmation results, it can be showed that better parameter combination for lifespan with 3.76% error followed by 4.44% error for surface roughness and 6.38 % for VB respectively.

Acknowledgements

This work was achieved in the laboratories LMS (University of Guelma Algeria) in collaboration with LaMCoS (CNRS, INSA-Lyon, France). The authors would like to thank the Algerian Ministry of Higher Education and Scientific Research (MESRS) and the Delegated Ministry for Scientific Research (MDRS) for granting financial support for CNEPRU Research Project, CODE: J0301520140021 (University 08 May 1945, Guelma).

References

Aouici, H., Bouchelaghem, H., Yallese, M. A., Elbah, M., & Fnides, B. (2014). Machinability investigation in hard turning of AISI D3 cold work steel with ceramic tool using response surface methodology. *The International Journal of Advanced Manufacturing Technology, 73*(9-12), 1775-1788.

Bouzid, L., Yallese, M. A., Chaoui, K., Mabrouki, T., & Boulanouar, L. (2015). Mathematical modeling for turning on AISI 420 stainless steel using surface response methodology. *Proceedings of the Institution of Mechanical Engineers, Part B: Journal of Engineering Manufacture, 229*(1), 45-61.

Bouzid, L., Yallese, M. A., Belhadi, S., Mabrouki, T., & Boulanouar, L. (2014). RMS-based optimisation of surface roughness when turning AISI 420 stainless steel. *International Journal of Materials and Product Technology, 49*(4), 224-251.

Berkani, S., Yallese, M., Boulanouar, L., & Mabrouki, T. (2015). Statistical analysis of AISI304 austenitic stainless steel machining using Ti (C, N)/Al2O3/TiN CVD coated carbide tool. *International Journal of Industrial Engineering Computations, 6*(4), 539-552.

Benga, G. C., & Abrao, A. M. (2003). Turning of hardened 100Cr6 bearing steel with ceramic and PCBN cutting tools. *Journal of Materials Processing Technology, 143*, 237-241.

Çaydas, U. (2010). Machinability evaluation in hard turning of AISI 4340 steel with different cutting tools using statistical techniques. *Proceedings of the Institution of Mechanical Engineers, 224*(B7), 1043.

Ciftci, I. (2006). Machining of austenitic stainless steels using CVD multi-layer coated cemented carbide tools. *Tribology International, 39*(6), 565-569.

Dureja, J. S., Gupta, V. K., Sharma, V. S., & Dogra, M. (2009). Design optimization of cutting conditions and analysis of their effect on tool wear and surface roughness during hard turning of AISI-H11 steel with a coated—mixed ceramic tool. *Proceedings of the Institution of Mechanical Engineers, Part B: Journal of Engineering Manufacture, 223*(11), 1441-1453.

Elbah, M., Yallese, M. A., Aouici, H., Mabrouki, T., & Rigal, J. F. (2013). Comparative assessment of wiper and conventional ceramic tools on surface roughness in hard turning AISI 4140 steel. *Measurement, 46*(9), 3041-3056.

Fnides, B., Yallese, M. A., & Aouici, H. (2008). Comportement à l'usure des céramiques de coupe (Al2O3+ TiC et Al2O3+ SiC) en tournage des pièces trempées. *Algerian Journal of Advanced Materials, 5*, 121-124.

Garcia-Diaz, A., Hogg, G. L., & Tari, F. G. (1981). Combining simulation and optimization to solve the multimachine interference problem. *Simulation, 36*(6), 193-201.

Horng, J. T., Liu, N. M., & Chiang, K. T. (2008). Investigating the machinability evaluation of Hadfield steel in the hard turning with Al 2 O 3/TiC mixed ceramic tool based on the response surface methodology. *Journal of materials processing technology, 208*(1), 532-541.

Kramer, B. M., & Von Turkovich, B. F. (1986). A comprehensive tool wear model. *CIRP Annals-Manufacturing Technology, 35*(1), 67-70.

Kaddeche, M., Chaoui, K., & Yallese, M. A. (2012). Cutting parameters effects on the machining of two high density polyethylene pipes resins: Cutting parameters effects on HDPE machining. *Mechanics & Industry, 13*(5), 307-316.

Keblouti, O., Boulanouar, L., Azizi, M., & Athmane, M. (2017). Modeling and multi-objective optimization of surface roughness and productivity in dry turning of AISI 52100 steel using (TiCN-TiN) coating cermet tools. *International Journal of Industrial Engineering Computations, 8*(1), 71-84.

Meddour, I., Yallese, M. A., Khattabi, R., Elbah, M., & Boulanouar, L. (2015). Investigation and modeling of cutting forces and surface roughness when hard turning of AISI 52100 steel with mixed ceramic tool: cutting conditions optimization. *The International Journal of Advanced Manufacturing Technology, 77*(5-8), 1387-1399.

Muthukrishnan, N., & Davim, J. P. (2009). Optimization of machining parameters of Al/SiC-MMC with ANOVA and ANN analysis. *Journal of Materials Processing Technology, 209*(1), 225-232.

Neşeli, S., Yaldız, S., & Türkeş, E. (2011). Optimization of tool geometry parameters for turning operations based on the response surface methodology. *Measurement, 44*(3), 580-587.

Smith, D. E. (1973). An empirical investigation of optimum-seeking in the computer simulation situation. *Operations Research, 21*(2), 475-497.

Singh, D., & Rao, P. V. (2010). Flank wear prediction of ceramic tools in hard turning. *The International Journal of Advanced Manufacturing Technology, 50*(5-8), 479-493.

Singh, C. K., & Vajpayee, S. (1980). Evaluation of flank wear on cutting tools. *Wear, 62*(2), 247-254.

Uvaraja, V. C., & Natarajan, N. (2012). Optimization on friction and wear process parameters using Taguchi technique. *International Journal of Engineering Technology, 2*(4), 694-699.

Yallese, M. A., Boulanouar, L., & Chaoui, K. (2004). Usinage de l'acier 100Cr6 trempé par un outil en nitrure de bore cubique. *Mechanics & Industry, 5*(4), 355-368.

Yallese, M. A., Chaoui, K., Zeghib, N., Boulanouar, L., & Rigal, J. F. (2009). Hard machining of hardened bearing steel using cubic boron nitride tool. *Journal of Materials Processing Technology, 209*(2), 1092-1104.

Zahia, H., Athmane, Y., Lakhdar, B., & Tarek, M. (2015). On the application of response surface methodology for predicting and optimizing surface roughness and cutting forces in hard turning by PVD coated insert. *International Journal of Industrial Engineering Computations, 6*(2), 267-284.

6

Single machine batch processing problem with release dates to minimize total completion time

Pedram Beldar[a] and Antonio Costa[b*]

[a]*Department of Industrial and Mechanical Engineering, Qazvin Branch, Islamic Azad University, Qazvin, Iran*
[b]*University of Catania, DICAR, Viale Andrea Doria 6, 95125 Catania, Italy*

CHRONICLE	ABSTRACT
	A single machine batch processing problem with release dates to minimize the total completion time $(1\|r_j, batch\| \sum C_j)$ is investigated in this research. An original mixed integer linear programming (MILP) model is proposed to optimally solve the problem. Since the research problem at hand is shown to be *NP*-hard, several different meta-heuristic algorithms based on tabu search (TS) and particle swarm optimization (PSO) are used to solve the problem. To find the most performing heuristic optimization technique, a set of test cases ranging in size (small, medium, and large) are randomly generated and solved by the proposed meta-heuristic algorithms. An extended comparison analysis is carried out and the outperformance of a hybrid meta-heuristic technique properly combining PSO and genetic algorithm (PSO-GA) is statistically demonstrated.
Keywords: *Minimization of total completion time* *Batch processing* *Single machine scheduling* *Mathematical programming* *Scheduling with release dates*	

1. Introduction

Batch processing (BP) problem has been attracting many academics and practitioners because of its intensive applications in the real-world industry. In BP, a resource is able to process more than one job as a batch at the same time. BP problems can be observed in different industries such as cutting machines in textile industry, testers in semiconductors manufacturing, and ovens in metalworking. A similar problem wherein jobs must be grouped into batches and contemporarily sequences of jobs and batches must be managed to satisfy a certain objective is also ascribable to the fields denoted as scheduling with tool changes (Costa et al., 2016) and group scheduling (Costa et al., 2017).

Based on the required processing time for each batch, the traditional BP problems are classified into three categories as: (1) the processing time of a batch is equal to the sum of the processing times of the jobs assigned to the batch; (2) the processing time of a batch is equal to the maximum processing time of the jobs assigned to the batch; (3) the processing time of a batch is equal to a fixed processing time to process the jobs assigned to the batch. There is also another method of classifying the BP problems based on the capacity of batches: a) the number of jobs assigned to a batch is limited by the maximum number of jobs that can be assigned to a batch; b) the number of jobs assigned to a batch depends on a size capacity for

* Corresponding author
E-mail: costa@diim.unict.it (A. Costa)

each batch based on one of the jobs attributes such as weight or volume; c) the jobs assigned to a batch must respect both conditions explained in (a) and (b).

Most of the research performed in BP problems have been focused on category (2) as for processing time of each batch and categories (a) or (b) as for the capacities of the batches. Chandru et al. (1993) propose an exact procedure and several heuristics for solving the $1|batch|\sum C_j$ problem for categories (2) and (a). Uzsoy (1994) tackles both $1|batch|C_{max}$ and $1|batch|\sum C_j$ problems and proposes an exact procedure based on Branch and Bound (B&B) algorithm for the $1|batch|\sum C_j$ problem. Moreover, he devises several heuristics to solve those problems for categories (2) and (b) and proves that both problems are NP-hard. Another research of Uzsoy and Yang (1997) deals with the $1|batch|\sum w_j C_j$ problem for categories (2) and (a). They propose several heuristics and an exact approach based on B&B algorithm. Jolai and Dupont (1997) approach the $1|r_j,batch|C_{max}$ problem for categories (2) and (b) and propose a number of heuristics for the problem. Lee (1999) studies the $1|r_j,batch|C_{max}$ problem for categories (2) and (a). He proposes several polynomial and pseudo-polynomial time algorithms for a few particular instances and develop efficient heuristics for the general problem. Liu and Yu (2000) approach the $1|r_j,batch|C_{max}$ problem for categories (2) and (a) and propose a pseudo-polynomial algorithm for the case with a constant number of release dates and a greedy heuristic for the general case. Dupont and Dhaenens-Flipo (2002) address the $1|r_j,batch|C_{max}$ problem for categories (2) and (b) and develop an exact procedure based on B&B method for the research problem. Chang and Wang (2004) develop a heuristic algorithm for the $1|r_j, batch|\sum C_j$ problem for categories (2) and (b). Melouk et al. (2004) and Damodaran et al. (2006) develop several meta-heuristic algorithms based on simulated annealing (SA) and genetic algorithm (GA) for the $1|batch|C_{max}$ problem for categories (2) and (b). Damodaran et al. (2007) cope with the $1|batch|C_{max}$ problem for categories (2) and (c) and develop a meta-heuristic algorithm based on SA to solve the problem. Parsa et al. (2010) approach the $1|batch|C_{max}$ problem for categories (2) and (b) and propose an exact approach based on Branch and Price (B&P) algorithm. They compare their proposed algorithm with the B&B method proposed by Dupont and Dhaenens-Flipo (2002) and show that the B&P algorithm is better than the B&B algorithm. Xu et al. (2012) solve the $1|r_j,batch|C_{max}$ problem for categories (2) and (b) through a mixed integer linear programming (MILP) model. They develop an effective lower bound method, a heuristic algorithm, and an ant colony optimization (ACO) algorithm for solving the mentioned BP problem. Lee and Lee (2013) develop several heuristics for the $1|batch|C_{max}$ problem for categories (2) and (b). Jia and Leung (2014) provide an improved max-min ant system algorithm for the $1|batch|C_{max}$ problem for categories (2) and (b). Zhou et al. (2014) approach the $1|r_j,batch|C_{max}$ problem for categories (2) and (b) and propose various heuristics to solve the problem. Al-Salamah (2015) devises an artificial bee colony method to minimize the $1|batch|C_{max}$ problem for categories (2) and (b). Li et al. (2015) consider the $1|batch, d_j=d|\sum (E_j + T_j)$ problem for categories (2) and (b) and develop a hybrid GA by combining GA with a heuristic algorithm managing the batches. Parsa et al. (2016) introduce a hybrid meta-heuristic algorithm based on the max-min ant system for $1|batch|\sum C_j$ for categories for (2) and (b).

Accordingly to the literary contributions mentioned before, there exists a large amount of research dealing with the BP problem for categories (a) and (b). Conversely, BP problems for category (c) have been weakly investigated by the body of literature so far. In this research, a single machine BP problem with release dates to minimize the total completion time ($1|r_j,batch|\sum C_j$) is approached for category (c).

The processing time of each batch is as the one defined by category (2) and the capacity of batches is defined by category (c). The application of the proposed research problem can be observed in the burn-in operations in the final testing step of integrated circuits in semiconductor manufacturing (Uzsoy, 1994). The burn-in operations ensure that no faulty product is accepted. In burn-in operations, the integrated circuits are placed in an oven (a batch processing machine) at a fixed temperature for a long period of time. Each circuit (job) may have a different burn-in time (processing time). The jobs are loaded onto boards and then, the boards are put into an oven. The number of boards that can process at the same time is described as oven capacity. Thus, the boards must be split into batches. The processing time of a batch is defined by the longest processing time among all jobs in that batch. The processing time in burn-in operations is too lengthy compared to other testing operations. Thus, they form a bottleneck in the final stage. The minimization of the total completion time would ensure the increase of the throughput.

In this research, it is assumed that n jobs are assigned to a machine to be processed. The machine can process the jobs arranged into batches. All jobs are not available at the beginning of the planning horizon. The machine has a size capacity and also can process at most a limited amount of jobs at the same time. Each job has a different size. The processing time of each batch is equal to the maximum processing time of jobs assigned to the batch. The goal is to minimize the total completion time of jobs.

The rest of this paper is organized as follows. In Section 2 a mathematical model is developed for the proposed research problem. In Section 3 several meta-heuristic algorithms are proposed to heuristically solve the problem. The specifications of the test problems used to compare the performance of the proposed algorithms, also including parameter settings and solution time are explained in Section 4. In Section 5 numerical results and comparison analysis involving the proposed metaheuristic algorithms are presented. In Section 6 conclusions and future research areas are discussed.

2. The Mathematical Model

In this section, an original MILP model is developed for the proposed research problem. In this model, a batch is considered to be active whether it has at least one job and the maximum number of batches is equal to the number of jobs. Indexes, parameters, decision variables, and the whole mathematical model are as follows:

Indexes:

j Index of jobs

b Index of batches

Parameters:

n The number of jobs

B Capacity of machine

N The maximum number of jobs can be assigned to a batch

M A large number

p_j The processing time of job j, $j=1,2,...,n$

r_j The release date of job j, $j=1,2,...,n$

s_j The size of job j, $j=1,2,...,n$

Decision variables:

$$X_{jb} = \begin{cases} 1 & \text{if job } j \text{ is assigned to batch } b \\ 0 & \text{otherwise} \end{cases} \quad j=1,2,...,n \ \& \ b=1,2,...,n$$

$$y_b = \begin{cases} 1 & \text{if batch } b \text{ is active} \\ 0 & \text{otherwise} \end{cases} \qquad b = 1, 2, \dots, n$$

The model

$$\min \sum_{j=1}^{n} C_j \tag{1}$$

subject to

$$\sum_{b=1}^{n} X_{jb} = 1 \qquad \forall j = 1, 2, \dots, n \tag{2}$$

$$\sum_{j=1}^{n} X_{jb} s_j \leq B y_b \qquad \forall b = 1, 2, \dots, n \tag{3}$$

$$\sum_{j=1}^{n} X_{jb} \leq N \qquad \forall b = 1, 2, \dots, n \tag{4}$$

$$P^b \geq X_{jb} p_j \qquad \forall j = 1, 2, \dots, n; \ b = 1, 2, \dots, n \tag{5}$$

$$S^b \geq X_{jb} r_j \qquad \forall j = 1, 2, \dots, n; \ b = 1, 2, \dots, n \tag{6}$$

$$S^b \geq S^{b-1} + P^{b-1} \qquad \forall b = 2, 3, \dots, n \tag{7}$$

$$C_j \geq S^b + P^b - (1 - X_{jb}) M \qquad \forall j = 1, 2, \dots, n; \ b = 1, 2, \dots, n \tag{8}$$

$$\sum_{j=1}^{n} X_{jb} \geq y_b \qquad \forall b = 1, 2, \dots, n \tag{9}$$

$$y_b \geq y_{b+1} \qquad \forall b = 1, 2, \dots, n-1 \tag{10}$$

$$X_{jb} \in \{0,1\}, \ Y_b \in \{0,1\}, \ C_j \geq 0 \qquad \forall j = 1, 2, \dots, n; \ b = 1, 2, \dots, n \tag{11}$$

The objective, as presented by Eq. (1), is to minimize the total completion time. Constraint (2) is incorporated into the model to ensure that each job is assigned to one batch. Constraint (3) guarantees that the total size of the jobs assigned to a batch does not exceed the batch capacity. Constraint (4) assures that the number of jobs assigned to a batch cannot be greater than the number of jobs to be assigned to each batch. Constraint (5) allows the processing time of each batch is equal to or greater than the processing time of the jobs assigned to that batch. Constraint (6) ensures that the starting processing time of a batch is greater than or equal to the arrival time of all jobs assigned to that batch. According to Constraint (7) the single-machine cannot process more than one batch at a time. Constraint (8) states the completion time of a job is equal to the completion time of the batch the job is assigned to. Constraint (9) ensures at least one job is allocated to an active batch. Constraint (10) guarantees that all active batches are ordered consecutively.

Since Uzsoy (1994) proved that the $1|batch|\sum C_j$ problem for categories (2) and (b) is an NP-hard problem, the single machine batch processing problem under investigation, with release dates to minimize of the total completion time ($1|r_j, batch|\sum C_j$) for categories (2) and (c), can be classified as an NP-hard problem too. As a result, meta-heuristic algorithms are needed to solve large-sized issues.

3. Meta-heuristic algorithms

Basically, two different classes can be considered to categorize meta-heuristic algorithms: single-solution based algorithms and population based algorithms. In the first class, algorithms try to make an improvement on a single candidate solution at each iteration. Tabu search (TS), simulating annealing, variable neighborhood search are examples of algorithms belonging to this class. In the second class, algorithms try to make an improvement on multiple candidate solutions at each iteration. Genetic Algorithms (GAs), Ant Colony Optimization (ACO), particle swarm intelligence (PSO) fall under this class. In this research, due to the significant performance of TS and PSO in solving relevant BP scheduling problems (Liao and Huang (2011), Damodaran et al. (2012) and Damodaran et al. (2013)), both of them have been taken into account for solving the research problem under investigation. The following sections deal with the adopted algorithms in depth.

3.1 Tabu Search

Tabu Search (TS) was originally developed by Glover and Laguna (1999) and basically consists of a neighborhood search method that keeps track of the up-to-now search path to avoid to be trapped into local optima or to try an explorative search of the solution domain (Costa et al., 2015). The proposed TS was inspired to that of Liao and Huang (2011) but, differently from the original one, in this paper an enhanced two-level TS algorithm was devised. In the first level (inside level), the best assignment of jobs to batches is examined. In the second level (outside level), the best sequence of batches is investigated. The relationship between the aforementioned phases is that once the inside level search is carried out to assign jobs to batches, the search process is switched to the outside level. In fact, whenever the inside level search stopping criterion is met, the best obtained job assignment is considered and the search strategy changes to the outside level. The outside search stops when the related stopping criterion is met. The best obtained solution, which combines a sequence of batches and the allocation of jobs to each batch, is the final solution. The peculiarities of the proposed TS are as follows.

3.1.1 Initial Feasible Solution

In this research, an effective procedure inspired to the heuristic algorithm called DYNA (Jolai & Dupont, 1997) is proposed to generate the initial feasible solution to be handled by the inside level. Then, the best solution obtained by the inside level becomes the initial feasible solution for the subsequent phase, i.e., the outside level. The algorithm generating the initial solution works as follows:

Step 1: First, the jobs are ordered based on the Earliest Completion Time (ECT) sorting rule. In words, jobs are ordered in nondecreasing order of their expected completion time ($r_j + p_j$), with ties broken by r_j.
Step 2: The first job in the ECT list is assigned to the first batch.
Step 3: The subsequent job is scheduled according to the following criteria:
  Step 3.1: The job on the top of the list of remaining jobs is assigned separately to the existing batches having enough space.
  Step 3.2: If the job on the top of the list of remaining jobs cannot be allocated to any batch, a new batch is created and the job is assigned to it.
Step 4: Then the total completion time is calculated for all possible combinations.
Step 5: The state with the minimum total completion time among all possible states is selected.
Step 6: All jobs have been scheduled. Go to Step7. Otherwise go to Step3.
Step 7: Stop algorithm.

3.1.2 Neighborhood generation mechanism

After an initial feasible solution is generated, the neighborhood search at the inside level is performed by two moves, the former being a swap move and the latter being an insert move:

Swap move. The job in ath position of bth batch is swapped with the job in kth position of gth batch. if bth and gth batches have enough space the move is feasible. Otherwise the move is unfeasible and it is rejected from the neighborhood.

Insertion move. The job in ath position of bth batch is removed and inserted into kth position (empty position) of gth batch. if the gth batch has enough space the move is feasible. Otherwise the move is unfeasible and it is rejected from the neighborhood.

Let us consider a BP problem with $n=5$ jobs and two batches. The maximum number of items per batch is $N=3$ and $B=10$ is the machine capacity. The seed solution is [1-2-3-4-5] while the size of each job s_j as well as the batch i each job is allocated to, i.e., X_{ji}, are reported in Table 1. Six distinct moves are allowed by applying the swap operator as follows: (2,1), (2,4), (2,5), (3,1), (3,2), (3,5). Move (3,1) is not feasible as the obtained batch size (6+5) exceeds the provided maximum capacity $B=10$. As far as the insertion operator is concerned, since $N=3$, the following three different moves may be generated: (2,3,1), (2,3,4), (2,3,5). Only the last move is feasible under the batch size viewpoint. As a result, just six out of nine moves can be considered as candidates to be the next seed.

Table 1
An example of seed solution

seed	1	2	3	4	5
s_j	5	6	2	3	1
X_{j1}	0	1	1	0	0
X_{j2}	1	0	0	1	1

As for as the outside level, the neighborhood is generated according to a regular adjacent pairwise interchange method; thus, $\beta-1$ alternative moves may be produced, where β is the current amount of batches. Hence, $\beta-1$ additional moves have to be considered for selecting the next seed solution.

3.1.3 The Tabu List (TL)

TL is a list of the characteristics of forbidden moves. It prevents the cycling back to formerly visited solutions by storing the characteristics of these moves for a certain period. At each iteration, the best neighbor solution, i.e., that one with the best objective function value, is selected and the move that generated that candidate solution is compared with the moves stored in the TL. Whether the move does not exist in the TL, the corresponding solution is selected as a new seed solution that, subsequently, will be subject again to the neighborhood generation mechanism. Otherwise, the move associated to the next best solution is taken into account. If a tabu move allows a better objective function value than the best global one found so far, the tabu restriction is ignored and the solution related to that move is selected as the new seed solution. The best objective function value found so far is denoted as the aspiration criterion.

3.1.4 Stopping Criterion

The aforementioned process is repeated until the stopping criterion based on the maximum time designated for each problem is met.

3.2 The PSO

The PSO algorithm, introduced by Kennedy and Eberhart (1995), is a population-based algorithm to solve continuous optimization problems. It starts with a set of initial solutions (particles). Each particle has its own position and velocity vector. The position vector $X_i^t = \left[x_{i1}^t, x_{i2}^t, ..., x_{in}^t \right]$ represents the position of the ith particle in the tth iteration where x_{ij}^t illustrates the jth dimension of the n-dimensional position vector. The velocity vector $V_i^t = \left[v_{i1}^t, v_{i2}^t, ..., v_{in}^t \right]$ represents the velocity of the ith particle in the tth iteration where v_{ij}^t illustrates the jth dimension of the n-dimensional velocity vector. The dimension of search space is equal to the number of jobs n in this research. The successful behavior of each particle affects the behavior of other particles, so particles move toward areas with better objective function values according to the two following targets:

Pbest: The best position visited by the ith particle at the tth iteration is shown as *Pbest*, $Pbest_i^t = \left[Pbest_{i1}^t, Pbest_{i2}^t, ..., Pbest_{in}^t \right]$.

Gbest: The best position visited by all particles at the *t*th iteration is called *Gbest*, $Gbest_i^t = \left[Gbest_1^t, Gbest_2^t, ..., Gbest_n^t \right]$.

Since particles move towards better positions during the search process, the velocity of each particle changes on the basis of *Pbest* and *Gbest* vectors at each iteration. The velocity of each particle at each iteration is updated by:

$$V_i^{t+1} = wV_i^t + c_1 \times r_1 \left(Pbest_i^t - X_i^t \right) + c_2 \times r_2 \left(Gbest^t - X_i^t \right), \tag{11}$$

where w is the inertia weight which controls the impact of the previous velocity vector at the *t*th iteration while c_1 and c_2 are two constants called acceleration coefficients. Also, r_1 and r_2 are random numbers drawn from a uniform distribution $U[0, 1]$. The position of each particle varies as velocity changes. As a result, the position of the *i*th particle at the $(t + 1)$th iteration is updated as follows:

$$X_i^{t+1} = X_i^t + V_i^{t+1} \tag{12}$$

3.2.1 PSO encoding

The original PSO algorithm is used to solve the continuous optimization problems. Arabameri and Salmasi (2013), and Tadayon and Salmasi (2013) apply the smallest position value (SPV) rule that is an increasing order mechanism to transform a continuous PSO to a discrete one. In other words, the SPV method allows transforming a real-encoded solution into a permutation one, that is a sequence of jobs. For instance, assume that $X_i^t = [2.3, -1.5, -2.03, 1.16]$ is the position vector of the *i*th particle at the *t*th iteration whose dimension is equal to four (i.e., four jobs). Since -2.03 is the smallest value in X_i^t, the third member of the position vector, that is J_3, is located in the first position of the sequence. The second smallest value is -1.5, so J_2 will be the second digit of the sequence, and so on. Following the same fashion, the final permutation sequence will be: J_3-J_2-J_4-J_1.

3.2.2 Initial solutions

In this research, three members (particles) of the initial population are created by the following dispatching rules: shortest processing time (SPT) (Baker & Trietsch, 2009), earliest release date (ERD) (Baker and Trietsch, 2009), and ECT. The rest of the members are randomly generated.

3.2.3 Updating Particles

Based on Poli et al. (2007), if a multiplier χ is inserted into Eq. (11), the convergence speed as well as the effectiveness of PSO algorithm may be improved. The appropriate value of χ is calculated through the following Equations:

$$\varphi = c_1 + c_2, \ \varphi > 4 \tag{13}$$

$$\chi = \frac{2}{(\varphi - 2) + \sqrt{\varphi^2 - 4\varphi}} \tag{14}$$

To satisfy Eq. (13), the value of 2.05 is considered for both c_1 and c_2. Therefore, according to Eq. (14), χ is equal to 7298. As a result, the new expression for the velocity vector of each particle is:

$$V_i^{t+1} = \chi \left[wV_i^t + c_1 r_1 \left(Pbest_i^t - X_i^t \right) + c_2 r_2 \left(Gbest^t - X_i^t \right) \right] \tag{15}$$

Eq. (12) is regularly used to update the position of each particle. As mentioned in section 3.2.1, the SPV rule is employed to assess the quality of the solution related to each particle and to update the values of $Pbest_i$ and *Gbest*, if applicable.

3.2.4 Improving the performance of PSO

Pros and cons characterize the particle swarm optimization algorithm (Coello Coello et al., 2007). The main advantage of PSO consists of its ability in exploitation while it may give weak result in exploring the solution space. In order to boost the performance of PSO, a series of hybrid PSO algorithms, which combine PSO with other performing search techniques, have been developed by literature so far (Sha & Hsu, 2006; Xia & Wu, 2006; Chen et al., 2013; Gao et al., 2014; Javidrad & Nazari, 2017). Notably, Arabameri and Salmasi (2013) demonstrated as the performance of PSO may be significantly improved if it is combined with a proper neighborhood search. In this research, two different hybrid methods have been considered to boost the performance of the regular PSO, the former being based on GA (PSO-GA), the latter being focused on TS (PSO-TS).

PSO-GA. The GA, just like the PSO, is a population-based stochastic algorithm that evolves through a number of initial solutions called chromosomes. At each iteration, a series of basic operators called selection, crossover and mutation are performed on the population with the aim of performing both exploration and exploitation (Michalewicz, 1996; Lin & Kang, 1999). In this research, the selection of offspring is randomly executed on the population in order to strengthen the diversification phase. An arithmetic crossover operator based on Simon (2013) was employed to produce two chromosomes called offspring by combining two parent chromosomes. In this method, the genes related to two randomly selected individuals (ith and jth particles) are exchanged and modified through Eq. (16) and Eq. (17), where α_l is an l-dimensional vector of random weights between 0 and 1. $X1_{kl}$ and $X2_{kl}$ represent the offspring particles at the kth iteration. The part of population on which the crossover operator is executed at each iteration depends on the crossover rate p_c.

$$X1_{kl} = \alpha_l X_{il} + (1-\alpha_l) X_{jl} \quad \forall l = 1,2,\ldots,n \tag{16}$$

$$X2_{kl} = \alpha_l X_{jl} + (1-\alpha_l) X_{il} \quad \forall l = 1,2,\ldots,n \tag{17}$$

The goal of mutation is to maintain the population diversity, thus avoiding any premature convergence. It makes a new offspring up from a single parent and assists the algorithm to avoid to be trapped in local optima. A single parent (ith particle) is selected randomly to perform a Gaussian mutation procedure based on Simon (2013). Then, the jth element of the selected particle (X_i) is varied according to Kuo and Han (2011), as follows:

$$X_{ij}^{new} = X_{ij} + N(0,1) \times rand[0,1] \tag{18}$$

where N(0,1) is a number from a standard normal distribution and rand[0,1] is a random number form a uniform distribution in the range [0,1]. The portion of population in which mutation procedure is applied to depends on the mutation rate (p_c). A pseudo-code of the proposed PSO-GA is reported in the following paragraph.

Pseudo-code of PSO-GA

Step 1. Initialize control parameters and create a swarm with P particles.
Do While (stopping criterion is not met)
Step 2. Update position and velocity vectors
for i= 1 to P
        for j=1 to n
$$V_i^{t+1} = \chi \left[wV_i^t + c_1 r_1 \left(Pbest_i^t - X_i^t \right) + c_2 r_2 \left(Gbest^t - X_i^t \right) \right]$$
        end for
end for
Step 3. Arithmetic crossover operator
      for k=1 to n_c
      Select two particles randomly (ith and jth particles)

for l=1 to n
$$X1_{kl} = \alpha_l X_{il} + (1-\alpha_l) X_{jl}$$
$$X2_{kl} = \alpha_l X_{jl} + (1-\alpha_l) X_{il}$$
end for
end for

Step 4. Gaussian mutation operator
for q=1 to n_m
Select one single parent randomly (ith) and change the jth element of it
$$X_{ij}^{new} = X_{ij} + N(0,1) \times rand[0,1]$$
end for

Step 5. New particles
Merge all the newly generated particles yielded by the crossover, the mutation, and PSO operators.
Then, select the best P particles in terms of objective function value.

Step 6. Update the P_{best} and G_{best} vectors
for s=1 to P
    if ($f(X_s) < f(P_{best_s})$)
      $P_{best_s} = X_s$
    end if
if ($f(P_{best_s}) < f(G_{best})$)
  $G_{best} = P_{best_s}$
end if
end for
update inertia weight
loop

PSO-TS. Actually, three different versions of PSO-TS, hereinafter denoted with the subscripts a, b and c, have been developed as explained below:

a) In the first version, whenever the G_{best} is updated at a certain iteration, it is given to TS as a seed solution and TS is performed for a finite amount of time. If a better solution is found, it is considered as the new G_{best}.

b) In this version, whenever the P_{best} is updated at a certain iteration, it is given to TS as a seed solution and TS is performed for a finite amount of time. If a better solution is found, it is considered as the new P_{best}.

c) In the last scenario, PSO and TS are hierarchically applied to the problem at hand. The regular PSO is executed in the first phase until a switching criterion is satisfied. Subsequently, the best solution found by the PSO is handled by the TS algorithm until a time-based stopping criterion is met.

It is worthy to point out that the aforementioned versions of hybrid PSO are powered by the same two-levels tabu search described in Section 3.1.

3.2.5 Calculating the objective function

Based on empirical studies, an appropriate assignment of jobs to the batches has a considerable effect on the value of the objective function. Therefore, the proposed heuristic algorithm as in Section 3.1.1 was employed to calculate the total completion time of each solution.

4. Computational experiments

In order to compare the performances of the proposed meta-heuristic algorithms, a set of test problems have been generated, randomly. The whole set of test problems are solved by the five proposed algorithms, namely TS, PSO-GA, PSO_a, PSO_b, and PSO_c, and subsequently the performances of these algorithms have been compared. The test problems specifications, parameter setting, the adopted solution time as well as the obtained numerical results are dealt with the subsequent sub-sections.

4.1 Test problems

To evaluate the performance of the proposed algorithms, a wide range of test problems has randomly been generated on the basis of the following five different factors: the number of jobs (n), the maximum number of jobs in a batch (N), the maximum capacity of the machine (B), the size of jobs (s_j) and the processing time of jobs (p_j). Three different classes of problems (small, medium, large) have been provided for each factor, as depicted in Table 2. Hence, the total amount of scenario problems to be investigated is $3^5 = 243$.

Table 2
Settings about the test problems

Factor/Class	Small	Medium	Large
n	$U[5, 20]$	$U[21, 50]$	$U[51, 100]$
N	3	5	7
B	10	15	20
s_j	$U[1, 5]$	$U[1, 10]$	$U[4, 10]$
p_j	$U[1, 10]$	$U[1, 20]$	$U[1, 50]$

Two replicates have been considered for each class; thus, a total amount of 486 test problems have been solved by means of each algorithm. As far as the release dates are concerned, they are drawn from the uniform discrete distribution $U[0,(n/N) \times \max_j(p_j)]$ and then rounded to the closer integer value; $\max(p_j)$ is the maximum value of processing time among the n jobs.

4.2 Setting of control parameters

Both effectiveness and efficiency of metaheuristic algorithms can be enhanced by giving appropriate values to their control parameters. As for TS algorithm the most influencing parameter to be chosen is the TL size. In this paper an empirical formula for choosing the TL size has been conceived making use of a trial-and-error approach involving an extended number of instances. Table 3 shows how the TL size should be set for both the first and the second level, on the basis of the expected neighborhood size.

Table 3
The different values of TL size for different number of jobs

From	To	TL size (first level)	TL size (second level)
5	10	0.4*NS1	
11	15	0.4*NS1	0.3* NS2
16	20	0.3*NS1	
21	25	0.3*NS1	
26	30	0.3*NS1+1	
31	35	0.2*NS1+3	
36	40	0.3*NS1	0.2* NS2
41	45	0.2*NS1-1	
46	50	0.2*NS1	
51	60	0.2*NS1-3	
61	70	0.2*NS1	
71	80	0.2*NS1	0.1* NS2
81	90	0.2*NS1+3	
91	100	0.2*NS1+2	

In fact, especially for the first level of TS, the neighborhood size (NS1) is not a priori known as it depends on how the jobs are allocated to the batches, conforming to the other parameters, i.e., N, B, s_j. On the other hand, as concerns the second level of TS, the neighborhood size (NS2) is equal to β-1, where β is the current number of batches. To calculate both $NS1$ and $NS2$ the heuristic described in Section 3.3.1 has been employed. Whether the TL size in Table 3 does not yield any integer number, the obtained result must be properly rounded down. The control parameters pertaining to the other proposed algorithms (i.e., PSO_a, PSO_b, PSO_c, PSO-GA) have been tuned through the response surface methodology (RSM) method. RSM is a collection of statistical and mathematical techniques used for developing, improving, and optimizing processes. In RSM, there are various input variables (control parameters) that can potentially influence some performance measures or quality characteristics that are called response (the objective function of the algorithm). Usually, in the RSM approach, it is convenient to transform the original variables into coded variables $x_1, x_2, ..., x_l$, which are usually defined to be dimensionless with mean zero and the same spread or standard deviation (Raymond et al., 2009). Eq. (190 shows how to transform an original variable to a coded one, where X_i and x_i represent the actual variable and the coded variable, respectively. Thus, the response function can be written as in Eq. (20).

$$x_i = \frac{X_i - \left(\frac{X_{High} + X_{Low}}{2} \right)}{\left(\frac{X_{High} - X_{Low}}{2} \right)} , \tag{19}$$

$$y = \left(x_1, x_2, ..., x_l \right), \tag{20}$$

where l represents the number of input variables. The goal is to gain the most suitable level of the algorithm parameters to optimize the response value. The main hypothesis is that the independent variables are continuous and controllable by experiments with negligible errors. Finding a proper approximation for the true functional relationship between independent variables and the response surface is required (Kwak, 2005). Raymond et al. (2009) proposed a second-order model, as follows:

$$y = \beta_0 + \sum_{j=1}^{l} \beta_j x_j + \sum_{j=1}^{l} \beta_{jj} x_j^2 + \sum \sum_{i<j} \beta_{ij} x_i x_j + \varepsilon \tag{21}$$

where y is the predicted response, β_0 is the model constant, β_j is the linear coefficient, β_{jj} is the quadratic coefficient, and β_{ij} is the interaction coefficient.

Shi and Eberhart (1999) stated that a large value of inertia weight w simplifies probing new positions (global search), while a small value of inertia weight simplifies a local search. As a result, a suitable balancing between local and global search can be achieved by adaptively reducing the inertia weight linearly according to Eq. (22), where t indicates the time of the current iteration and T refers to the total time considered for each test problem, respectively. Also, w_{start} and w_{end} are the starting and the finishing values for the inertia weight, respectively.

$$w = w_{start} - \left(w_{start} - w_{end} \right) \frac{t}{T} \tag{22}$$

Basically, NP, w_{start}, and w_{end}, p_c, and p_m are the control parameters considered as the input variables for PSO-GA. The input variables of the other PSO-based algorithms (namely PSO_a, PSO_b, and PSO_c) are NP, w_{start}, w_{end} and TL size. According to Eq. (19), the PSO-GA parameters are denoted by $X_1(NP)$, $X_2(w_{start})$, $X_3(w_{end})$, $X_4(p_c)$, $X_5(p_m)$, while the other control parameters are denoted by $X_1(NP)$, $X_2(w_{start})$, $X_3(w_{end})$, X_4(TL size). Each control parameter has been varied at three levels, conforming to low (-1), center (0) and high level (+1), as depicted in Table 4 and Table 5. As concerns the NP parameter, three scenarios depending on the problem size have been taken into account. The Box-Behnken experimental design was employed to handle a family of efficient three-level designs fitting the second-order response surfaces (Raymond et al., 2009). The number of experiments is equal to $2l(l-1)+n_c$, where n_c is the number of the central points. As a result, 48 and 32 experiments have been performed for tuning PSO-GA and the different PSO-based algorithms, respectively.

Table 4

The levels of the values of each parameter for PSO-GA

Parameter/Level	Low (-1)	Center (0)	High (+1)
NP (small size)	10	20	30
NP (medium size)	40	50	60
NP (large size)	70	80	90
w_{start}	0.1	0.3	0.5
w_{end}	0.7	0.9	1.1
p_c	0.7	0.8	0.9
p_m	0.1	0.2	0.3

Table 5

The levels of the values of each parameter for PSO_a, PSO_b, and PSO_c

Parameter/Level	Low (-1)	Center (0)	High (+1)
NP (small size)	10	20	30
NP (medium size)	40	50	60
NP (large size)	70	80	90
w_{start}	0.2	0.4	0.6
w_{end}	0.9	1.1	1.3
TL size	0.1*NS	0.25* NS	0.4* NS

Tables 6, 7, 8, and 9 show the optimal values of control parameters obtained by means of the RSM-based calibration. Due to representation purposes, Figs. (1-3) illustrate the 3D surface plots involving the different PSO-GA control parameters.

Table 6

Tuned value of PSO-GA's parameters

Parameter	Small size problem	Medium size problem	Large size problem
NP	10	40	70
w_{start}	0.5	0.5	0.1
w_{end}	1.1	0.7	0.7
p_c	0.7	0.9	0.9
p_m	0.3	0.3	0.3

Table 7

Tuned value of PSO_a's parameters

Parameter	Small size problem	Medium size problem	Large size problem
NP	30	40	70
w_{start}	0.2	0.3535	0.6
w_{end}	1.3	0.9	0.9808
TL size	0.4*NS	0.1*NS	0.4*NS

Table 8

Tuned value of PSO_b's parameters

Parameter	Small size problem	Medium size problem	Large size problem
NP	10	60	70
w_{start}	0.2	0.2	0.6
w_{end}	0.9	1.3	1.3
TL size	0.1*NM	0.1*NM	0.4*NM

Table 9
Tuned value of PSO$_C$'s parameters

Parameter	Small size problem	Medium size problem	Large size problem
NP	10	60	70
w_{start}	0.2	0.2	0.6
w_{end}	0.9	1.3	0.9484
TL size	0.4*NM	0.4*NM	0.4*NM

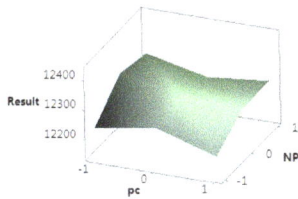

Fig. 1. 3D surface plot involving *NP* and p_c for PSO-GA

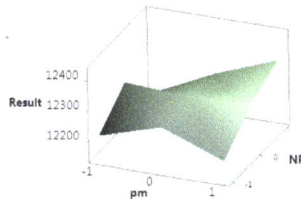

Fig. 2. 3D surface plot involving *NP* and p_m for PSO-GA

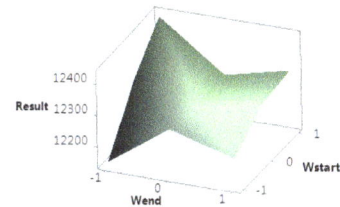

Fig. 3. 3D surface plot involving w_{start} and w_{end} for PSO-GA

4.3. Computational time

The exit criterion of the proposed metaheuristic algorithms is based on the computational time (CT) and, conforming to Gohari and Salmasi (2015), it has been parametrized as a function of the number of jobs (*n*), as follows:

$$CT = \delta \times n \ , \delta = \frac{CT_{max}}{n_{up}} \tag{23}$$

where CT_{max} and n_{up} respectively are the maximum allowed computational time and the upper value of the range related to the number of jobs; as for example, in the range [5, 20], n_{up} is equal to 20. CT_{max} has been set to 30, 90, and 180 seconds for small-, medium-, and large-sized problems, respectively. Hence, the values of δ for small-, medium-, and large-sized problems are equal to 1.5, 1.8 and 1.8, respectively. All the proposed algorithms were coded in C++ programming language. The whole set of experiments was performed on a PC with 2.8 GHz CPU and 2 GB RAM.

5. Numerical results and comparison analysis

In order to compare the different metaheuristics, a Relative Percentage Deviation (RPD) performance indicator as in Eq. (24) has been taken into account. It is worth pointing out that, for most small-size problems, the RPDs have been computed on the basis of the global optima obtained by solving the mathematical problem. Whereas, for both medium and larger-sized issues, each RPD was computed by exploiting the relative local optimum value among the different competing algorithms.

$$RPD = \frac{algorithm\ solution - optimal\ solution}{optimal\ solution} \times 100 \tag{24}$$

As for the smaller-sized class of problems, the global optima have been achieved by ILOG CPLEX (version 12.2). Actually, the MILP model was able to optimally solve only 74 smaller-sized test problems out of 162; thus, the RPD values of the remaining 88 test problems have been computed making use of the local optima instead of the global ones. Table 10 shows the average values of RPDs over the two replicates per each scenario problem, related to the smaller-sized issues.

Table 10
Small-sized problems: average RPD results over the two replicates

Test number	PSOa	PSOb	PSOc	PSO-GA	TS	Test number	PSOa	PSOb	PSOc	PSO-GA	TS
1	0	0	0	0	0	42	0	0	0	0	0.135
2	0	0	0	0	1.335	43	0.179	0	0	0	5.43
3	0	0	0	0	0	44	0	0	0	0	1.512
4	0.359	0	0	0	1.142	45	0.936	0	0	0	0.163
5	0	0	0	0	0.259	46	0.877	0	0	0	4.325
6	0	0	0	0	0.399	47	0	0	0	0	5.263
7	1.525	0	0	0	1.987	48	0.118	0	0	0	0.711
8	1.008	0	0	0	0.144	49	0	0	0	0	0.73
9	0	0	0	0	1.193	50	0	0	0	0	0.681
10	0	0	0	0	1.899	51	0.088	0	0	0	0
11	3.201	0	0	0	1.829	52	0	0	0	0	2.757
12	0	0	0	0	0	53	0.074	0	0	0	0.074
13	1.967	0	0	0	5.622	54	1.756	0	0	0	1.313
14	0	0	0	0	0.811	55	0	0	0	0	2.156
15	0.323	0.323	0.323	0	0.223	56	0.093	0	0	0	0.876
16	0.313	0	0	0.122	2.001	57	0.644	0	0	0	2.551
17	0	0	0	0	0.077	58	0	0	0	0	1.21
18	1.507	0	0	0	1.096	59	0	0	0	0	3.165
19	0	0	0	0	3.193	60	1.813	0	0	0	2.926
20	0.612	0	0	0	0.194	61	0	0	0	0	1.401
21	0	0	0	0	0	62	0	0	0	0	0
22	0	0	0	0	0.882	63	0	0	0	0	3.249
23	0.054	0	0	0	1.463	64	0	0	0	0	0
24	3.593	0	0	0.115	0.075	65	0	0	0	0	1.41
25	0	0	0	0	0.594	66	0	0	0	0	0
26	0.389	0	0	0	2.693	67	0	0	0	0	4.126
27	0	0	0	0	0	68	0	0	0	0	3.108
28	0	0	0	0	3.013	69	0.515	0	0.258	0	0
29	0.85	0	0	0	0.899	70	0.806	0	0	0	2.959
30	0.436	0	0	0	0	71	0.212	0	0	0	2.797
31	0.042	0	0	0	1.19	72	0	0	0	0	0.209
32	0	0	0	0	0.963	73	0	0	0	0	2.283
33	0.606	0.606	0	0	0	74	0	0	0	0	0.246
34	0.73	0	0	0	0	75	1.014	0	0	0	0
35	0	0	0	0	4.874	76	0.742	0	0	0	2.014
36	0	0	0	0	1.145	77	0	0	0	0	0
37	0.746	0	0	0	1.874	78	0	0	0	0	2.959
38	0	0	0	0	3.813	79	0	0	0	0	1.962
39	0	0	0	0	0	80	0.152	0	0	0	1.339
40	0.281	0	0	0	2.423	81	0	0	0	0	2.257
41	0	0	0	0	1.056						
						AVE	0.353	0.011	0.007	0.003	1.465
						ST.DEV	0.68	0.076	0.046	0.018	1.458

As the reader can notice, in the last two rows both grand averages (ave) and standard deviations (st.dev) highlight the effectiveness of PSOc and PSO-GA over the other metaheuristics. In addition, the small values of RPDs confirm that each algorithm has been properly calibrated and designed for the problem at hand, though TS is strongly less performing than the other competitors. Table 11 reports the RPD medians for each algorithm, with respect to the three different classes of problems. PSO-GA and PSOc achieve the best results, regardless of the specific class of problem. Notably, for small and medium sized issues, medians related to PSO-GA and PSOc are comparably close to zero. Conversely, a slight advantage for PSO-GA emerges for solving the case of larger sized issues. The weakness of TS comes to light again, especially when the size of the problems increases. Similarly, the performances of PSOa and PSOb drastically deteriorate with the problem size though they are able to assure a median equal to zero for the smaller sized issues.

Table 11

RPD medians for each algorithm and for each class of problems

Problem size	PSOa	PSOb	PSOc	PSO-GA	TS
Lower	0.000	0.000	0.000	0.000	0.512
Middle	0.682	0.283	0.061	0.000	2.006
Larger	1.109	8.590	0.293	0.074	1.703

To statistically infer about the whole set of results, MINITAB 16 commercial package has been implemented. Since the normality test has not been satisfied over the obtained RPD results, a Kruskal-Wallis non-parametric test on the medians (Corder & Foreman, 2014; Costa et al., 2015) was assumed to be the most appropriate statistical method to compare the proposed algorithms. Table 12 represents the output of the aforementioned non-parametric test. The results show that there was a significant statistical difference among the performance of the proposed meta-heuristic algorithms (the p-value is equal to 0.0000).

Table 12

Kruskal-Wallis Test on RPD values of competing algorithms

ALGO	N	Median	Ave Rank	Z
PSO-GA	486	0.000000000	812.0	-14.18
PSOa	486	0.430030102	1298.5	2.92
PSOb	486	0.323933278	1377.6	5.69
PSOc	486	0.000000000	962.6	-8.89
TS	486	1.597411647	1626.9	14.45
Overall	2430			1215.5
H=423.77	DF=4	P = 0.000		
H=472.14	DF=4	P = 0.000		(adjusted for ties)

The medians as well as the Z rank reveal as PSO_c and PSO-GA outperform the other competitors. The Box plot diagram at 95 percent confidence level reported in Fig. 4 confirms that there was a statistical difference among the different metaheuristics.

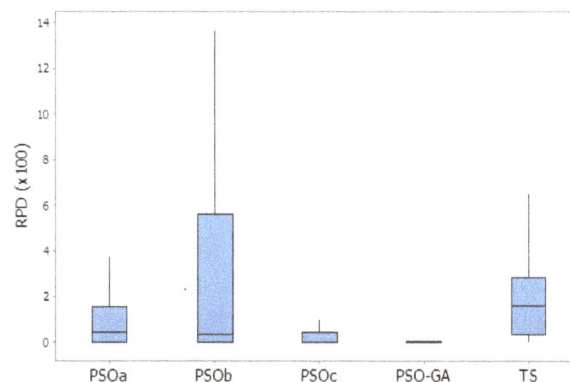

Fig. 4. Comparison of meta-heuristics: Boxplot

Since findings from the previous statistical analysis suggest PSO-GA and PSOc are the most promising algorithms, similarly being done by Costa et al. (2015), a post-hoc Mann-Whitney non-parametric pairwise test (Mann and Whitney, 1947) has been carried out. Table 13 demonstrates that there was a statistically significant difference (the p-value is equal to 0.0000) between PSO_c and PSO-GA. Therefore, PSO-GA outperforms PSO_c for solving the batch processing problem at hand.

Table 13
Comparison between PSO-GA and PSO-TS: Test of Mann Whitney

ALGO	N	Median
PSO-GA	486	0.000000000
PSOc	486	0.000000000
Point estimate for η1 - η2 is -0.00000		
95.0 Percent CI for η1 - η2 is (-0.00001; 0.00000)		
W = 218155.5		
Test of η1 = η2 vs η1 ≠ η2 is significant at 0.0000		
The test is significant at 0.0000 (adjusted for ties)		

6. Conclusions and Future Research

In this research, we have approached a single machine batch processing problem with release dates to minimize the total completion time ($1|r_j, batch| \sum C_j$). A mathematical model has been proposed to solve the research problem optimally. The proposed research problem is known to be *NP*-hard; hence, several meta-heuristic algorithms based on TS and PSO with two different approaches have been developed to heuristically solve the problem. Since the normal probability plot of residuals were not normally distributed, a non-parametric test (Kruskal-Wallis) was employed to compare the performance of the proposed algorithms. The results show that there was a significant statistical difference among the meta-heuristic algorithms. Medians and Z rank demonstrate that PSO$_a$ and PSO-GA assure the most promising performances compared to the other metaheuristics. Therefore, in order to find the algorithm with the best efficiency between PSO$_c$ and PSO-GA, a post-hoc pairwise comparison based on a non-parametric Mann-Whitney U test was employed. The result of Mann-Whitney test indicates that PSO-GA had a better performance than PSO$_c$. Since the proposed research problem was investigated in a single machine environment, the obtained results in this research can be applied in other environments such as parallel machines and flow shop. A lower bounding method can be provided to evaluate the performance of the proposed meta-heuristic algorithms for the future research.

References

Al-Salamah, M. (2015). Constrained binary artificial bee colony to minimize the makespan for single machine batch processing with non-identical job sizes. *Applied Soft Computing*, 29, 379 -85.

Arabameri, S., & Salmasi, N. (2013). Minimization of weighted earliness and tardiness for no-wait sequence-dependent setup times flow-shop scheduling problem. *Computers & Industrial Engineering*, 64(4), 902-916.

Baker, R.K., & Trietsch, D. (2009). *Principles of Sequencing and Scheduling*; New Jersey: John Wiley & Sons.

Chandru,V., Lee, C.Y., & Uzsoy, R. (1993). Minimizing total completion time on a batch processing machine with job families. *Operations Research Letters, 13*(2), 61-65.

Chang, P.C., & Wang, H.M. (2004). A heuristic for a batch processing machine scheduled to minimize total completion time with non- identical job sizes. *The International Journal of Advanced Manufacturing Technology, 24*(7), 615-620.

Chen, Y.Y., Cheng, C.Y., Wang, L.C., & Chen, T.L. (2013). A hybrid approach based on the variable neighborhood search and particle swarm optimization for parallel machine scheduling problems: a case study for solar cell industry. *International Journal of Production Economics, 141*(1), 66-78.

Coello Coello, C.A., Lamont, G.B., & Van Veldhuizen, D.A. (2007). *Alternative Meta-heuristics*: Boston, MA: Springer US.

Corder, G.W., & Foreman, D.I. (2014). *Nonparametric Statistics: A Step-by-Step Approach*. New Jersey: John Wiley & Sons.

Costa, A., Alfieri, A., Matta, A., & Fichera, S. (2015). A parallel tabu search for solving the primal buffer allocation problem in serial production systems. *Computers & Operations Research*, *64*, 97-112.

Costa, A., Cappadonna, F.A., & Fichera, S. (2016). Minimizing the total completion time on a parallel machine system with tool changes. *Computers & Industrial Engineering*, *91*, 290-301.

Costa, A., Cappadonna, F.A., & Fichera, S. (2017). A hybrid genetic algorithm for minimizing makespan in a flow-shop sequence-dependent group scheduling problem. *Journal of Intelligent Manufacturing*, *8*(6), 1269-1283.

Damodaran, P., Manjeshwar, P.K., & Srihari, K. (2006). Minimizing makespan on a batch-processing machine with non-identical job sizes using genetic algorithms. *International Journal of Production Economics*, *103*(2), 882-891.

Damodaran, P., Srihari, K., & Lam, S.S. (2007). Scheduling a capacitated batch-processing machine to minimize makespan. *Robotics and Computer-Integrated* Manufacturing, *23*(2), 208-2016.

Damodaran, P., Diyadawagamage, D.A., Ghrayeb, O., & Velez-Gallego, M.C. (2012). A particle swarm optimization algorithm for minimizing makespan of non-identical parallel batch processing machines. *The International Journal of Advanced Manufacturing Technology*, *58*(9), 1131-1140.

Dupont, L., & Dhaenens-Flipo, C. (2002). Minimizing the makespan on a batch machine with non-identical job sizes: an exact procedure. *Computers & Operations Research*, *29*(7), 807-819.

Gao, H., Kwong, S., Fan, B., & Wang, R. (2014). A Hybrid Particle-Swarm Tabu Search Algorithm for Solving Job Shop Scheduling Problems, *IEEE Transactions on Industrial Informatics*, *10*(4), 2044-2054.

Glover, F., & Laguna, M. (1999). *Tabu Search*. Boston, MA: Springer US.

Gohari, S., & Salmasi, N. (2015). Flexible flowline scheduling problem with constraints for the beginning and terminating time of processing of jobs at stages. *International Journal of Computer Integrated Manufacturing*, *28*(10), 1092-1105.

Javidrad F., Nazari M.. (2017). A new hybrid particle swarm and simulated annealing stochastic optimization method. *Applied Soft Computing*, *60*, 634-654.

Jia, Zh., & Leung, J.Y.T. (2014). An improved meta-heuristic for makespan minimization of a single batch machine with non-identical job sizes. *Computers & Operations Research*, *46*, 49-58.

Jolai, F., & Dupont, L. (1997). Minimizing mean flow times criteria on a single batch processing machine with non-identical jobs sizes. *International journal of production economics*, *55*, 273-280.

Kennedy, J., & Eberhart, R. (1995). Particle swarm optimization. *In: Neural Networks, 1995 Proceedings., IEEE International Conference on*, *4*, 1942-1948.

Kuo, R.J., & Han, Y.S. (2011). A hybrid of genetic algorithm and particle swarm optimization for solving bi-level linear programming problem – A case study on supply chain model. *Applied Mathematical Modelling*, *35*(8), 3905 – 3917.

Kwak, J.S. (2005). Application of taguchi and response surface methodologies for geometric error in surface grinding process. *International Journal of Machine Tools and Manufacture*, *45*(3), 327-334.

Lee, C.Y. (1999). Minimizing makespan on a single batch processing machine with dynamic job arrivals. *International Journal of Production Research*, *37*(1), 219-236.

Lee, Y.H., & Lee, Y.H. (2013). Minimizing makespan heuristics for scheduling a single batch machine processing machine with non-identical job sizes. *International Journal of Production Research*, *51*(12), 3488-3500.

Li, Z., Chen, H., Xu, R., & Li, X. (2015). Earliness–tardiness minimization on scheduling a batch processing machine with non-identical job sizes. *Computers & Industrial Engineering*, *87*, 590-599.

Liao, L.M., & Huang, C.J. (2011). Tabu search heuristic for two-machine flowshop with batch processing machines. *Computers & Industrial Engineering*, *60*(3), 426-432.

Lin, H., & Kang L. (1999). Balance between exploration and exploitation in genetic search, *Wuhan University Journal of Natural Sciences*, *4*(1), 28–32.

Liu, Z., & Yu, W. (2000). Scheduling one batch processor subject to job release dates. *Discrete Applied Mathematics*, *105*(13), 129-136.

Mann, H.B., & Whitney, D.R. (1947). On a test of whether one of two random variables is stochastically larger than the other. *he Annals of Mathematical Statistics*, *18*(1), 50-60.

Melouk, S., Damodaran, P., & Chang, P.Y. (2004). Minimizing makespan for single machine batch processing with non-identical job sizes using simulated annealing. *International Journal of Production Economics, 87*(2), 141-147.

Michalewicz, Z. (1996). *Genetic Algorithms + Data Structures = Evolution Programs*. Springer Science & Business Media.

Parsa, N.R., Karimi, B., & Kashan, A.H. (2010). A branch and price algorithm to minimize makespan on a single batch processing machine with non-identical job sizes. *Computers & Operations Research, 37*(10), 1720-1730.

Parsa, N. R., Karimi, B., & Husseini, S. M. (2016). Minimizing total flow time on a batch processing machine using a hybrid max–min ant system. *Computers & Industrial Engineering, 99*, 372-381.

Poli, R., Kennedy, J., & Blackwell, T. (2007). Particle swarm optimization. *Swarm Intelligence, 1*(1), 33-57.

Raymond, H.M., Douglas, C.M., & Christine, M. AC. (2009). *Response Surface Methodology: Process and Product Optimization Using Designed Experiments*. New Jersey: John Wiley & Sons.

Sha, D.Y., Hsu, C.-Y. (2006). A hybrid particle swarm optimization for job shop scheduling problem. *Computers & Industrial Engineering, 51*(4), 791-808.

Shi, Y., & Eberhart, R.C. (1950). Empirical study of particle swarm optimization. *In: Proceedings of the 1999 Congress on Evolutionary Computation-CEC99 (Cat. No. 99TH8406)*, 3, 1950 vol. 3.

Simon, D. (2013). *Evolutionary Optimization Algorithms*; New Jersey: John Wiley & Sons.

Tadayon, B., & Salmasi, N. (2013). A two-criteria objective function flexible flowshop scheduling problem with machine eligibility constraint. *The International Journal of Advanced Manufacturing Technology, 64*(5), 1001-1015.

Uzsoy, R. (1994). Scheduling a single batch processing machine with non-identical job sizes. *International Journal of Production Research, 32*(7), 1615-1635.

Uzsoy, R., & Yang, Y. (1997). Minimizing total weighted completion time on a single batch processing machine. *Production and Operations Management, 6*(1), 57-73.

Xia, W.-J., & Wu, Z.-M. (2006). A hybrid particle swarm optimization approach for the job-shop scheduling problem. *International Journal of Advanced Manufacturing Technology, 29*(3–4), 360–366.

Xu, R., Chen, H., & Li, X. (2012). Makespan minimization on single batch-processing machine via ant colony optimization. *Computers & Operations Research, 39*(3), 582-593.

Zhou, S., Chen, H., Xu, R., Li, X. (2014). Minimizing makespan on a single batch processing machine with dynamic job arrivals and non-identical job sizes. *International Journal of Production Research, 52*(8), 2258-2274.

7

Harvesting and transport operations to optimise biomass supply chain and industrial biorefinery processes

Robert Matindi[a]*, Mahmoud Masoud[b], Phil Hobson[a], Geoff Kent[a] and Shi Qiang Liu[c]

[a]*School of Chemistry, Physics and Mechanical Engineering Science and Engineering Faculty, Brisbane Qld 4001 Australia*
[b]*School of Mathematical Sciences, Queensland University of Technology, 2 George St, Brisbane Qld 4001 Australia*
[c]*School of Economics and Management, Fuzhou University, Fuzhou, 350108, China*

CHRONICLE	ABSTRACT
Keywords: *Bio-refinery* *Cane transport* *Cane harvesting* *Constraint programming*	In Australia, Bioenergy plays an important role in modern power systems, where many biomass resources provide greenhouse gas neutral and electricity at a variety of scales. By 2050, the Biomass energy is projected to have a 40-50 % share as an alternative source of energy. In addition to conversion of biomass, barriers and uncertainties in the production, supply may hinder biomass energy development. The sugarcane is an essential ingredient in the production of Bioenergy, across the whole spectrum ranging from the first generation to second generation, e.g., production of energy from the lignocellulosic component of the sugarcane initially regarded as waste (bagasse and cane residue). Sustainable recovery of the Lignocellulosic component of sugarcane from the field through a structured process is largely unknown and associated with high capital outlay that have stifled the growth of bioenergy sector. In this context, this paper develops a new scheduler to optimise the recovery of lignocellulosic component of sugarcane and cane, transport and harvest systems with reducing the associated costs and operational time. An Optimisation Algorithm called Limited Discrepancy Search has been adapted and integrated with the developed scheduling transport algorithms. The developed algorithms are formulated and coded by Optimization Programming Language (OPL) to obtain the optimised cane and cane residues transport schedules. Computational experiments demonstrate that high-quality solutions are obtainable for industry-scale instances. To provide insightful decisions, sensitivity analysis is conducted in terms of different scenarios and criteria.

1. Introduction

1.1 Backgroud

The key technical and financial issues affecting the viability of bioenergy projects relate to low energy density and the dispersed nature of biomass feedstock (Hobson, 2009 p2357). Most biomass conversion technologies, particularly those associated with second generation bioproducts production ideally from the lignocellulosic component of the biomass in question, have high associated capital cost and thus the financial viability of these second generation technologies for power and bioproducts productions is dependent on achieving sufficient economies of scale and high utilisation associated with abundant feedstock supply (Meyer et al., 2012 p2359).

* Corresponding author
E-mail: robert.matindi@qut.edu.au (R. Matindi)

One way of overcoming the inherent challenges of biomass (low density and spatial dispersion) is through the energy densification processes, among them raking and baling of cane residue, and retrofitting it conveyance (transport) to that of the cane transport system.
This in essence will take the advantage of the existing sugarcane transport infrastructure that is well established and understood, with minimal or no change in the existing set up (sidings, bins, trailers, locos and trucks).

With this innovative approach, two objectives will be accomplished, which are standardisation and the optimisation of equipment use. For instance, baling as residues recovery method will increase the density and transform the biomass into uniform units (bales).

Studies indicate that standardization and optimization lead to a reduction in residue recovery and transportation costs (Hobson & Wright, 2002 p2358). Approximately 67 % of the sugarcane trash from the field can be recovered when baling (Hassuani, 2005 p2355). This, combined with approximately 20 % of the trash that is not separated from the sugarcane in the harvester, results in 87 % of the trash reaching the mill (Hess et al., 2007 p2356) detailed the logistical data and cost per dry tonne for the baling unit operation, including capital, maintenance, ownership unit costs required for rectangular bales (1.2 m × 1.2 m × 2.4 m), and from their analysis, 30-40 % of biomass cost can be saved by simply densifying energy content into a standardised unit relative to its loose state.

Parallel approaches similar to cane supply chain, can be drawn and adopted for use in cane residue recovery (lignocellulosic supply chain for bioenergy production) in terms of unit operations, and potential areas for improvement and or optimisation. This in effect will entail invoking methodologies, approaches and strategies that had been successfully applied before albeit in conventional cane supply chain. One of such approaches, for instance, is in scheduling and management. The scheduling and management of cane supply has been shown to improve on operational time with minimal no capital outlay. Schedules when well executed can greatly improve operational production time without necessarily changing the existing investment portfolio (Masoud et al., 2011 p2348; Thorburn et al., 2006).

1.2 Literature review

The key aspects that distinguishes biomass supply chain from the well understood petrochemical supply chain and further analyse the commonality that can be leveraged upon in order to lower the production costs of either bio-products (platform chemicals), or bioenergy outputs, these tools and processes include planning and scheduling frameworks established for logistics activities, aimed at coordinating the use of resource and end to end optimisation of biomass-bio-products supply chain (Chen, 2012 p436; Yue, 2014 p970;).

Some of the distinct features that characterize it are uncertainties of different magnitude ranging from variability in climatic conditions to quality of recovered biomass. Due to the uncertainty aspect plants and in effect biomass are susceptible to adverse weather condition, which consequently has an effect on the overall moisture content (wet weather condition in terms of growing).

From an economic perspective, uncertainty have a direct bearing on total capital cost required per unit of recovered biomass. High moisture content biomass, for instance and based on technology at play (for example bio-gasification) will inevitably necessitate extra unit operation steps of pretreatment to have the biomass into the required state (Drying and ash removal process). This will be invoked in order to meet the strict requirements of a gasifier and other strict environmental regulations in a given jurisdiction. This extra pre-treatment steps increases the cost of recoverable biomass see (Faaij et al., 1997 p387; Henrich et al.,2009 p28; Hobson et al.,1998 p9; Hobson et al., 2003 p60).The other challenge associated with biomass supply chain is that their economic parameters are largely uncertain, due to the wide dispersal of biomass across geographical regions, thus making their collection and transport difficult coupled with intrinsic characteristic of biomass and that is their low energy density. These attributes does impact on the overall cost of the recovery process of biomass (Alex et al., 2012 p68; Hobson et al., 2006; Juffs et al., 2006; Meyer et al., 2012).

Other aspects of biomass supply chain that affect the cost of both the final product and the delivered raw biomass is the size and location of the biorefinery. Previous studies have analysed and developed models to study the facility location and biomass supply chain optimisation, this includes (Elms et al., 2010 p547) who studied the scheduling process of multiple feedstock for a biodiesel facility in order to optimise biodiesel production and lower greenhouse gases .

In the (Alex et al., 2012 p68) study a tool was developed and using Mixed Integer Linear program technique determined the optimal potential locations and size of bio refineries in a 9 state region in Mideastern USA.

In this study , it was deduced that the key parameters that affect the optimal sizing and location of a biorefinery are biomass supply source, the transportation distance and cost to the biorefinery ,the costs of biomass, the cost of the biofuel produced (Ethanol sale price), and biorefinery costs and capacity .In their analysis a total of 65 locations were identified as optimal with a total of 4.7 Billion Gallon per year (BGY) of ethanol, $2.99 per gallon assuming the volumetric ethanol excise tax credit.

The other feature of biomass supply chain is that biomass are low energy products , and in order to increase the energy density of biomass, pre-processing of the biomass may be an imperative step based on the transportation distance envisaged.

The location and configuration of such pre-processing facility has a bearing on the eventual cost of such biomass. Bowling et al. (2011) developed a superstructure for determining the optimal location of such a facility, that allows for simultaneous selection of a preferred configuration (centralized or distributed with maximizes profit in the entire value chain.

The modelling process entailed configuration of optimal flow rates that ensure maximum returns in terms of profit, Bowling et al. (2011) study found out that the key determinant on whether to density the biomass or not was the transportation distance and the capital cost of the processing facility based on the technology used for densification.

In the realms of management, biomass supply chain can be beamed into three rays; the strategic, tactical and operational levels. Decisions undertaken in each of those levels have an effect on cost and proper operation of the system .Some of the key strategic decision in biomass supply chain has to do with the storage of biomass in terms of capacity planning, truck scheduling among others.

Scheduling systems and strategies can minimize travel time and consequently the total costs as exemplified in the following studies (Ravula et al., 2008 p314; Mafakheri et al.,2014 p116).
Application of mathematical modelling techniques to bioenergy supply chain have facilitated the understanding, assessment and performance of the same.

The use of deterministic and stochastic mathematical models to optimize forest biomass supply has been studied and analysed. One facet of supply chain that applies have employed those class of techniques is in the scheduling operation.

A study by (Shabani et al., 2013 p299) on truck scheduling and optimization with an objective of minimizing the weighted sum of transportation costs and total operation time found that , these model can accurately capture the bottleneck in a system and produce a practical solution.

They further used simulated annealing solution technique to solve the truck scheduling problem, and it was found that total cost and working time reduced by 18 % and 15 % respectively.

Simulation technique are other sets of modelling tools that have been employed to study biomass supply chain .Simulation technique known to mimic real life scenarios and thus are ideal in study of logic flow and interactions that are potentially hard to represent mathematically.

Whilst simulation approaches may provide important insights into the operation of bioenergy supply chains, the fundamental structures are not fully understood because biomass supply chains are still evolving when compared to petroleum industry supply chain, and thus for economic ,environmental and energy efficiency improvement to be realized, end to end perspective need to be tapped into (Dunnett et al., 2007 p419).

Siting of the biorefinery is another characteristic of biomass supply chain. Mafakheri et al., (2014) in their analysis found that the choice of a biomass location depended on various factors key among them the type and characteristic of the biomass material.

Identification of the optimal location for both storage and biorefinery refinery facility may help in reducing the cost of the overall supply chain operations, previous studies have shown that co-location of storage facility to the biorefinery or near the processing facility reduces the overall cost of the biomass (Mafakheri et al.,2014 p116).

Models that have been employed to schedule and plan the storage capacity, and location have been analysed and enumerated in previous studies see (Nilsson et al., 2001 p247; Rentizelas et al., 2009 p887; Sokhansanj et al., 2010 p75; Sokhansanj et al., 2002 p347; Vadas et al., 2013 p133; Van Dyken et al., 2010 p1338; van Vliet et al., 2009 p855; Wang et al., 2014 p32).

In (Dunnett et al., 2007 p419) study, a network optimisation framework and integrated tool were developed .These systems linked the upstream and downstream activities ranging from storage scheduling, energy conversion schedules the process of integrating the system resulted in 5-20 % reduction in total cost of biomass supply chain (Mafakheri et al., 2014 p116) thus underpinning the importance of modelling tools that can be leveraged on to improve on process efficiency.

A critical synthesis of biomass-to-energy production shows gaps on some areas in research that are yet to be fully explored. According to Iakovou et al. (2010), the bottlenecks hindering development of bioenergy system are basically the cost of logistics operations. Logistic costs can be disaggregated into component costs that chain together to form the bioenergy supply chain.

Besides the logistic costs, biomass supply chain system suffers from quality and quantity differences, they vary considerably depending on the technology at play, the demands requirement of energy among other factors.

These challenges are further compounded by the uncertainty in their supply, the infancy state of their development, and the overall bioenergy environment. These challenges have a direct bearing on the cost of bioenergy product and thus proper planning and employment of innovative tools to can reduce the negative impact.

1.3 Innovation and contribution

While some research and development has been conducted in individual local areas for separate steps in supply chain from the grower to the mill, vertical integration or regional adoption based on priority has not been done for cane trash recovery. The marginal and notional gains accrued from such a process are hardly reported or explored. The next section, therefore, explores a potential transport problem that will inevitably be experienced if existing sugarcane infrastructure was to be adopted *prima facie* for recovery of cane residue.

To conclude this section on routing and scheduling problems, few points are worth highlighting:

1. The size of the optimization problem in all the types of routing and scheduling problem increases much quicker than the rate at which the size of the actual problem increases. For example, under unlimited train/truck capacity, if there are three nodes (sidings), there is just one feasible route (path) according to the rule of number of routes (paths) $(n - 1)!/2$ routes or trips for n nodes (sidings). As a result, the problem in reality will be more complicated under limited train/truck capacity. The impact of this property is that even if exact solution techniques exist, many are rendered useless in practical situations because of the excessive time requirements.
2. All problems deal with discrete quantities, which are difficult to deal with using traditional techniques.
3. From practical standpoints, the optimal is not necessarily one which has to be obtained; solutions which are good (i.e., close to the optimal) are equally important provided they can be obtained quickly. Therefore, unsurprisingly, heuristic techniques (which typically guarantee near-optimal solutions) continue to play an important role as viable solution techniques for these problems.

2. The Transport Rail/Route Scheduling Problem

There are two major classes of cane transport problems, namely the truck routing problem and the train railing problem. The sheer number of different possible paths and various different constraints representing several resource limitations make the development of such a transport system difficult. The road/rail scheduling problem refers to all problems where optimal closed loop paths which touch different points of interest are to be determined. There may be one or more trains or trucks. Generally the points of interest are referred to as nodes (siding or conjunction); further, the start and end nodes of a path are the same and often referred to as the depot (mill). Train/truck scheduling approach has been used to optimize train/truck transport systems. The problem is that given a set of routes (or rail tracks), one needs to develop schedules for trains or trucks arrivals and departures at all the sidings of the network. A good or efficient schedule is one which minimizes the total operating times of trains or trucks with minimal waiting time of harvesters and mill. The harvesters are waiting for empty bins at sidings and mill for full bins under a set of resource and service related constraints. The total waiting time of harvesters and mill have two components: (i) the total initial waiting time (IWT) of harvesters and mill, which is the sum of the waiting times of all the harvesters and mill at their point of origin (time 0 of starting the system), and (ii) the total transfer time (TT), which is the sum of the transfer times of all the transferring empty and full bins between harvesters and mill. The resource and service related constraints are:

a. **Limited fleet size**: only a fixed number of trucks or locomotives are available for operating on the different routes.

b. **Limited Trucks or locomotives capacity**: each truck or locomotive has a finite capacity

c. **Shunting time bounds**: Trucks or locomotives cannot stop for a very short or a very long time at a siding/pad.

d. **Policy headway**: on a given route, a minimum frequency needs to be maintained.

e. **Maximum transfer time (cane age)**: no full bins should have to wait too long for a transfer (less than 16 hours after harvesting is the recommended standard for the cut cane to retain its integrity).

Under the previous limitations, the historical data show that sugarcane transportation between harvesters and mills satisfy delays in delivering the cane as shown in Fig. 1(a). Delivering the crop 16 hours or more after harvest affects the crop quality and reduces the sugar rates. Optimizing the delivery and collection times to siding and mill will reduce the delays.

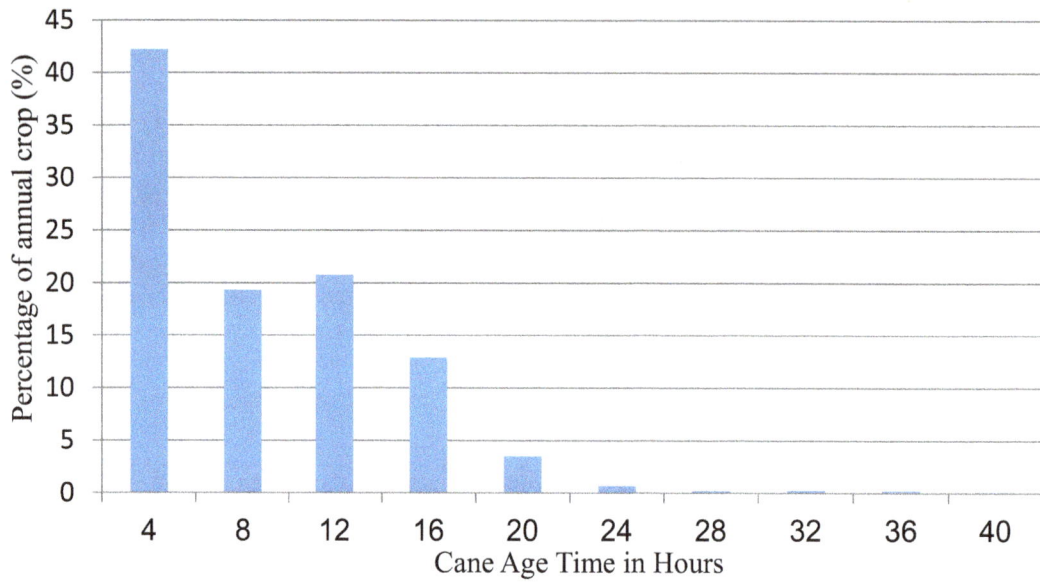

Fig. 1(a). Delay before cane is picked up from siding (hours) to mill

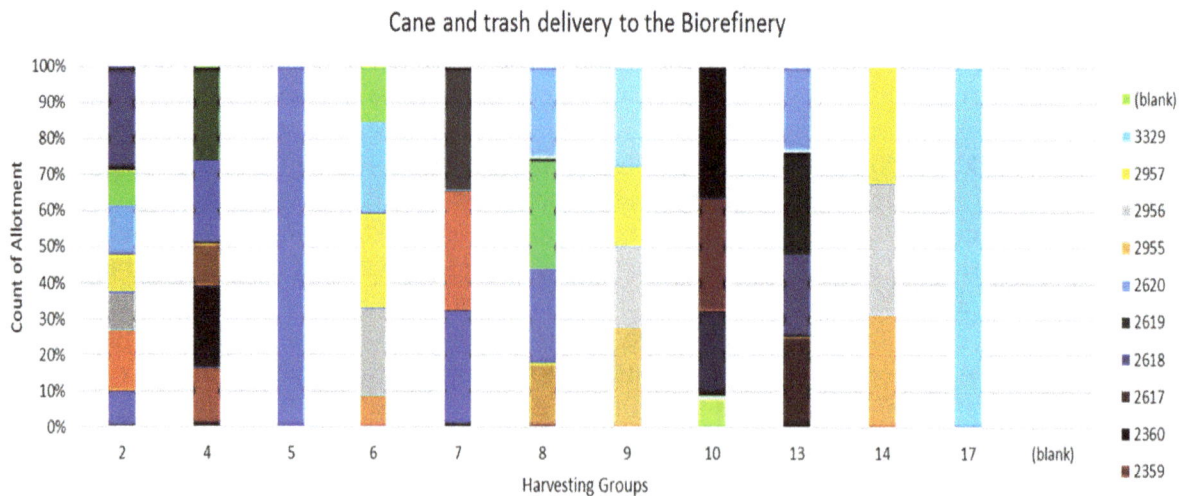

Fig. 1(b). A simulation of the flow of biomass from various harvesting groups to the factory / /biorefinery

Fig. 1(b), shows the flow of various allotments from the field to the biorefinery from various harvesting groups (biomass sources), the above flow underscores the need for proper management and scheduling of biomass through various allotments, to among other things avoid oversupply of biomass beyond the biorefinery capacity, or undersupply which may result to suboptimal use of the biorefinery. Some of the features related to the train/truck scheduling problem which the proposed methodology designed to solve the problem must be capable of handling are:

a. Arrival time of a truck/ train at one siding/pad is dependent on the arrival time of the truck/ train at the previous stop.

b. Arrival times of trains/trucks at a siding are generally not exactly as per the schedule. Arrival times are generally randomly distributed around the scheduled arrival times. Since arrival times are not exactly as per schedule, departure times are also not exactly as per schedule.

c. If demand for a route is very high during a particular period, then the queue developed for that route at the siding may not be cleared entirely by the next truck in the route due to limited train/truck capacity. In such cases, the formation and the dissipation of the queue must be tracked so that realistic values for the waiting times and transfer times can be obtained.

d. The arrival patterns of full bins at sidings/pads may vary widely. Sidings/pads which primarily have cane and trash in effect may see a surge in bins arrivals just before the arrival time of train/truck, consequently leading to stoppage of harvesting process until sufficient bins have been towed away and empty ones loaded. Synchronizing such operations real time makes that scheduling aspect of the operation NP difficult.

Therefore, given the combinatorial nature of the problem, the number of variables (especially the integer ones) and the number of constraints increase at a fast rate with the increase in number of routes and fleet size. Further, given the restricted ability of traditional optimization methods to handle MINLP problems, it is seen that even extremely small problems (for example, three routes and ten trains/trucks in each route/rail) cannot be solved within a reasonable time frame using traditional methods. As in the case of transport scheduling problem, most of the earlier work on transport scheduling with transfer considerations (e.g. Bookbinder & Desilets, p2360; Rapp & Gehner, 1967 p2361) rely on heuristics and user intervention at various stages of the solution process. Further, that MP formulations make two important simplifying assumptions of unlimited train/truck capacity or unlimited number of truck/train, or deterministic arrival times.

In Australia, transport systems play a vital role in the raw sugar production process and, to a large extent, bioenergy production by transporting the sugarcane crop between farms and mills. Most of the cane transport system uses a specific schedule of runs in order to meet the requirements of various industry stakeholders, ideally the mill and the harvesting contractor (Masoud et al., 2011 p2348; Masoud et al., 2012 p2349). Some of the methods used to transport cane entails rail only, road only, or a mixture of the two modes (rail and road) (Pernase & Pekol, 2012 p2362).

In 2013, 87 per cent of sugarcane was transported to mills by cane railway, a direct road transport network is the next most significant way to transport 8 per cent of sugar cane, and a further 5 per cent was transported by a combination of road and cane (Masoud et al., 2015 p2363;(Mitchell, 2015 p2364))). More cane lands are being established in remote locations and due to the high capital cost of establishing a railway, the amount of road transport to mills is growing, which various studies can be optimised to match up with a biorefinery facility (Nguyen, 1996 p2367). Road systems between harvesting areas and pads use small trucks (Infield Haul-out unit) with one bin fleet size (8 tonnes or 5 tonnes), while the road system (factory road system) from pads and mill use big truck with 2 bins (B-Double) with fleet size of 20 tonnes per bin (Higgins, 2006 p2368) see Fig. 2.

The sugarcane and cane residue transport system is very complex due to the need to effectively satisfy the requirements of several harvesters at different locations and keep a continuous supply of cane for mill processing, given limited resources of train /truck, trailer and bin fleet. Scheduling of sugarcane using a rail systems in Australia is complicated by the fact that cane railway networks have single tracks with few dedicated passing loops, requiring the passing of trains to be addressed. There are many branches where trains may wait to allow other trains to pass. Some sidings may also act as passing loops when not in use. The challenge is further compounded by the simultaneous recovery of cane residues which, in effect, means a near doubling of the numbers of bins to be railed on an already constrained system in order to fulfil the daily requirements of a biorefinery. Such a problem will require a seamless scheduling system. Scheduling problems in single track railway systems have proven difficult to solve and there have been many studies to improve system performance.

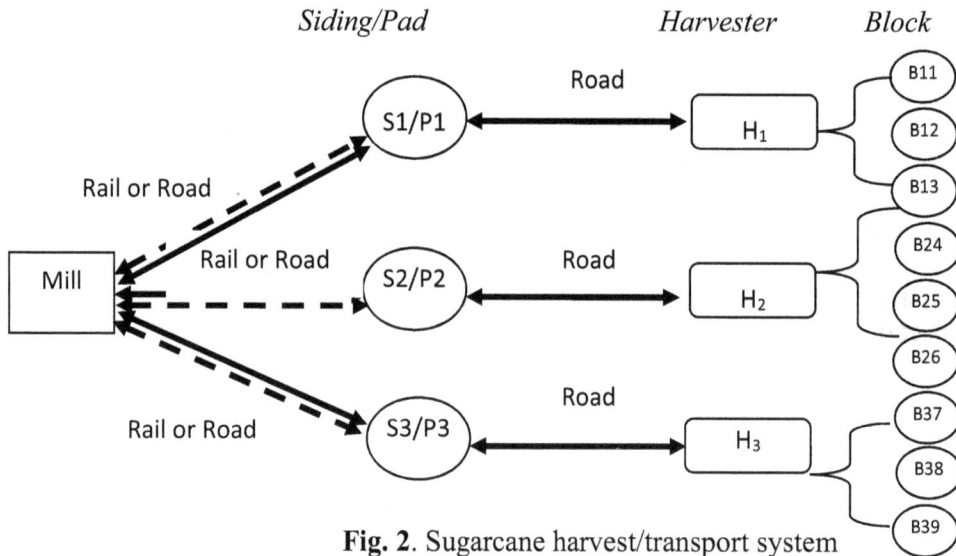

Fig. 2. Sugarcane harvest/transport system

3. Solution Approach

Many strategies are followed to find the solution. Search trees have been explored in past studies (Goloboff, 1999 p2369 (Van Hentenryck, 2000 p2370) (Jain, 1999 p2371))). Where most of them uses search strategies as prime methods when using parallel branch and bound technique to solve combinatorial optimisation problems. The main advantage of these methods is in obtaining the optimal solution for many NP-hard problems but on the other hand, using some of these techniques individually is time consuming coupled with expensive resource requirements in terms of memory footprint that is required to solve large scale problems.

The framework of the proposed methodology therefore depends on integrating optimisation methods such as Limited Discrepancy Search (LDS) with transport scheduling algorithms to obtain a good solution but not necessarily optimal. The LDS will be used to develop the proposed main paths of locomotives or trucks without transport constraints and then use the transport methods to construct the practical constraints that are valid for real cases studies. Fig. 3 shows the main proposed solution approach for the scheduling transport problem. ExtendSIM software (Simulation software) was used to run un-scheduled approach to the transport problem, while the ILOG-OPL software was used to in development of scheduling the delivery and collection processes using scheduling algorithms as stipulated in Fig. 3(a).

a. Adapted Limited Discrepancy Search

Limited Discrepancy Search (LDS) method depends on building the search tree by a good heuristics (Harvey, 1995 p2365). The first leaf which is visited is likely to be a solution. If this leaf is not a solution, it is likely the number of mistakes along the path from the root to this leaf is a small number. Then, the next leaf nodes will be visited which have paths from the root that differ only in one choice from the initial path. This process continues by visiting the leaves which have a higher discrepancy from the initial path. For a discrepancy $d > 0$, *LDS* visits leaves with discrepancy less than d, so leaves are visited many times. This can be avoided by keeping track of the remaining depth to be searched. Let the discrepancy of a node v be c, and the length of a path from v to a leaf be 1, then we consider descendants which is the discrepancy between $d - 1$ and d. This search strategy is called improved limited discrepancy search. It is noted that, *DFS* and *LDS* are complete search strategies that are not redundant where these techniques have to visit all paths from the root to a leaf exactly once.

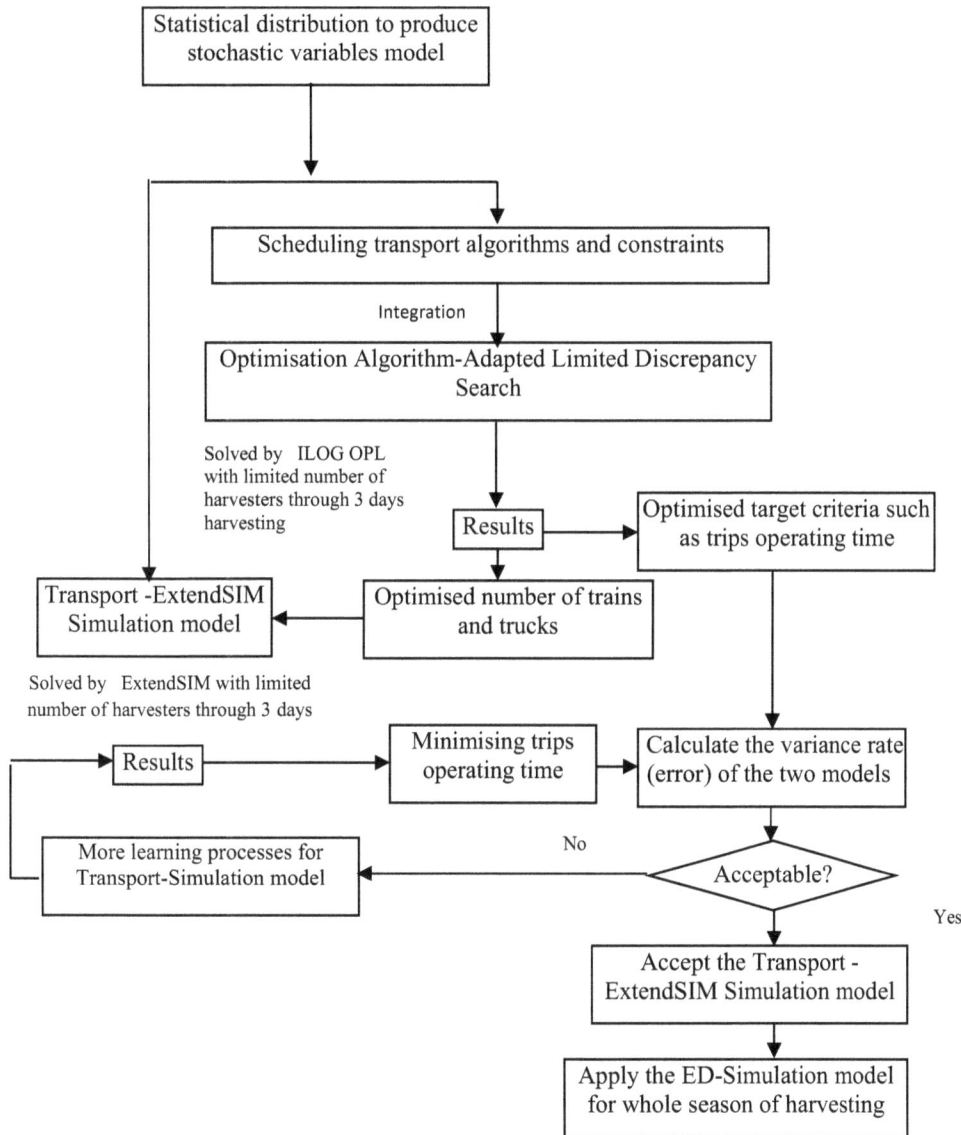

Fig. 3(a). The framework of the integrated Optimisation-Simulation approach for the transport system

The *LDS* has been adapted in this paper to be used to optimise the delivery and collection at all nodes (mill, siding and biorefinery) through a transport network. Formally, let γ_0 be a node with ordered descendants $\gamma_1, \gamma_2 ... \gamma_k$. The discrepancy of γ_j is the discrepancy of $\gamma_0 + j - 1$ for $j = 1,2,..,k$ and the discrepancy of the root node is 0. The main steps of *LDS* are listed as follows:

Step 1: Let the level of discrepancy be $d = 0$ starts at the root node.
Step 2: Proceeds by descending to its first descendant γ_1, which its discrepancy is not higher than d.
Step 3: This process continues until we reach a leaf.
Step 4: Backtrack to the parent of the leaf and the descent to its next existing descendant, which has a discrepancy that is not higher than k.
Step 5: This process continues until we are back at the root node and when all its descendants that have a discrepancy that is not higher than k have been visited.
Step 6: Set $d = d + 1$ and repeat this process until we are back at the root node and all its descendants have been visited.

3.2 Transport Scheduling Algorithms

The cane transport system includes two main stages; transporting cane from harvester's pad to nearest siding by trucks (Infield transport system) and transporting the cane from siding to mill by trains or roads (Factory transport System). Siding area has a limited capacity; as a result full bins must be collected immediately from siding by trains or trucks before the siding capacity get exhausted in terms of having bins fully filled by the harvester (with harvested cane), or having zero empty bins at the siding thus stopping the harvesting process. Delivery and collection times at siding must be optimised where collecting time for full bins from siding should be after the deliveries of empty bins to the siding by trucks/trains. This process should work continuously to avoid any blocking issues for siding or interruption in harvester or mill work or the biorefinery. Fig. 3 shows rail and road system to transport cane from harvester area (siding/pad) to mill within the siding area. In this scenario, two harvesters have been assigned to one pad, where according to the industry practice one pad can serve one or more harvesters sequentially.

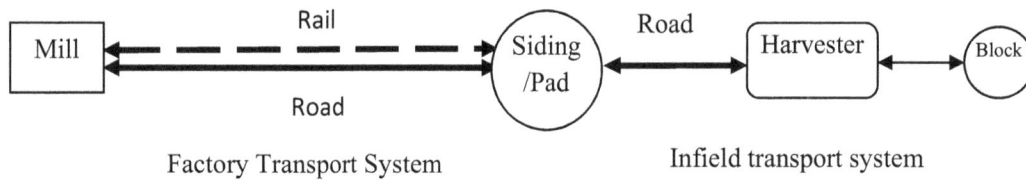

Fig. 3(b). Integrating Rail and Road systems with assigning two harvesters to one pad

3.2.1 Factory Rail Scheduling Algorithm (RSA)

The rail systems play an important role in transporting harvested sugarcane from siding/pad points to mill. In Figure 4, the rail network has been used to deliver empty bins to siding/pad and full bins to mill, and on the another side, collecting empty bins from mill and full bins from siding/pad. The algorithm *RSA* integrates the infield transport system with factory rail system. The rail transport system uses the trains with 120 bins for each, while the infield transport system uses truck with one bin fleet size (8 tonnes or 5 tonnes).

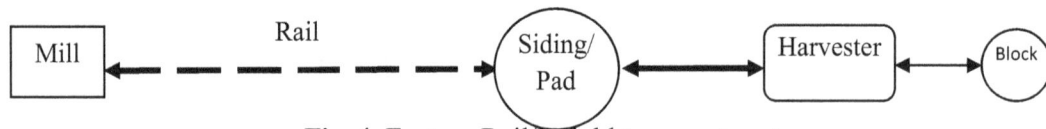

Fig. 4. Factory Rail/Infield transport systems

The Rail Scheduling algorithm (RSA) has been developed to optimise the deliver and collection times through the rail network at siding/pad points and Mill. The main procedures of RSA algorithm has been detailed below:

Select Train r; $r \in R$
Set the train capacity e_r and f_r
Select run n; $n \in N_r$
   Select siding s; $s \in S$ // step 3
   Set the siding capacity; y_s
   Set the siding Allotment; A_s
   Sum delivers to siding s by train r at run n; α_s //at the beginning of the system $\alpha_s = 0$
   Sum Collects from siding s by train r at run n; B_s //at the beginning of the system $B_s = 0$
If $o' \leq o$ then   (o & o' are two operations are implemented on same siding during same run   //
outbound direction
       If $\{sum(o'$ in operations$)\ \alpha_{rnos}\} \leq A_s$ then

$$\alpha_{rno's} \leq e_r$$
$$B_{rno's} = 0 \qquad\qquad // \textit{ no full bins in the outbound direction}$$
$$t_{rlos} \geq \left(\alpha_{rnos} / \text{Hrate}_s\right) + \text{Hstart}_s$$
$$t_{r+1nos} \geq t_{rnos} + \left(\alpha_{rnos} / \text{Hrate}_s\right) + \text{Hstart}_s$$
$$\alpha_{rno's} + B_{rno's} \leq y_s$$
$$\alpha_s = \alpha_s + \alpha_{rno's}$$
$$t_{rns} \geq tt_{kus}$$

  go to step 3
 Else
 End if

      Else   *// inbound direction*

           If $B_s < A_s$ then

$$B_{rno's} \leq C_{Kc}$$
$$\alpha_{rno's} = 0 \quad // \textit{ no empty bins in the inbound direction}$$
$$t_{rlos} \geq \left(\beta_{rnos} / \text{Hrate}_s\right) + \text{Hstart}_s$$
$$t_{r+1nos} \geq t_{rnos} + \left(\beta_{rnos} / \text{Hrate}_s\right) + \text{Hstart}_s$$
$$t_{rns} \geq tt_{kus}$$
$$\alpha_{rno's} + B_{rno's} \leq C_s$$
$$B_{kr} = B_{kr} - B_{kro'su}$$
$$B_s = B_s + B_{rno's}$$

  go to step 3
 Else
 End if
    End if;
Else
  $\alpha_{rno's} = 0$
$B_{rno's} = 0$
go to step 3
End if;
End.

3.2.2 Infield Road Scheduling (IFRS) Algorithm

The Infield and Factory Road Scheduling (IFRS) Algorithm integrates factory road scheduling system and infield transport system as shown in Figure 5. Factory road transport system from pads and mill uses big truck using 2 bins (B-Double) with fleet size 20 tonnes per bin, while the infield transport system between harvesting areas and pads uses small trucks with one bin fleet size (8 tonnes or 5 tonnes).

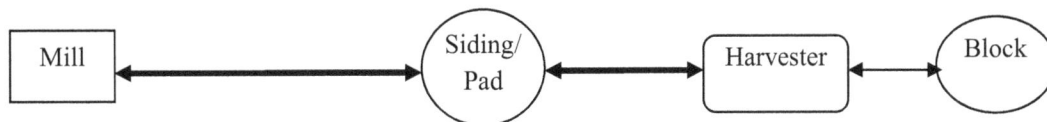

Fig. 5. Factory Road/Infield transport systems

Select Truck k; k $\in$ K
Set the Truck capacity e_k and f_k
Select run u; u $\in$ U_k
 Select siding s; s $\in$ S   // step 3
 Select pad p; p $\in$ P   // step 4
 Set the siding capacity; y_s
 Set the siding Allotment; A_s
Delivers full to siding s by truck k at run u; z_{kus} //at the beginning of the system $z_{kus}=0$
Collect empty from siding s by truck k at run u; l_{kus}; //at the beginning of the system l_{kus}

$$If\ z_{kus} \leq A_s\ then$$
$$z_{kus} \leq e_r$$
$$z_{kus} + l_{kus} \leq y_s$$
$$z_s = z_s + z_{kus}$$
$$tt_{k1s} \geq \left(z_{kus} / \text{Hrate}_s\right) + \text{Hstart}_s$$
$$tt_{k+1us} \geq tt_{kus} + \left(z_{kus} / \text{Hrate}_s\right) + \text{Hstart}_s$$
$$t_{rns} \geq tt_{kus}$$

 go to step 3
Else
End if

      Else   // inbound direction
$$If\ l_s < A_s\ then$$
$$l_{kus} \leq C_{Kc}$$
$$z_{kus} + l_{kus} \leq C_s$$
$$l_s = l_s + l_{kus}$$
$$tt_{k1s} \geq \left(l_{kus} / \text{Hrate}_s\right) + \text{Hstart}_s$$
$$tt_{k+1us} \geq tt_{kus} + \left(l_{kus} / \text{Hrate}_s\right) + \text{Hstart}_s$$
$$t_{rns} \geq tt_{kus}$$

go to step 3
 Else
End if
   End if;
Else
 $\alpha_{rno\ s} = 0$
$B_{rno\ s} = 0$
go to step 3
End if;
End.

3.2.3 Harvester Allocation Optimisation (HAO) algorithm

In the sugarcane harvesting system, each harvester serves several blocks or farms to satisfy the daily allotment requirement. The harvested crop transported is transported to a siding r for eventual transportation by truck or train to the mill and biorefinery. *The HAO algorithm* was developed to build *Harvester/block/siding assignment* for all harvesters in the system the idea of neighbourhood change. Active siding (with harvester) neighbourhoods are considered to create new paths for the harvesting period through the sugarcane rail network, see Fig. 6.

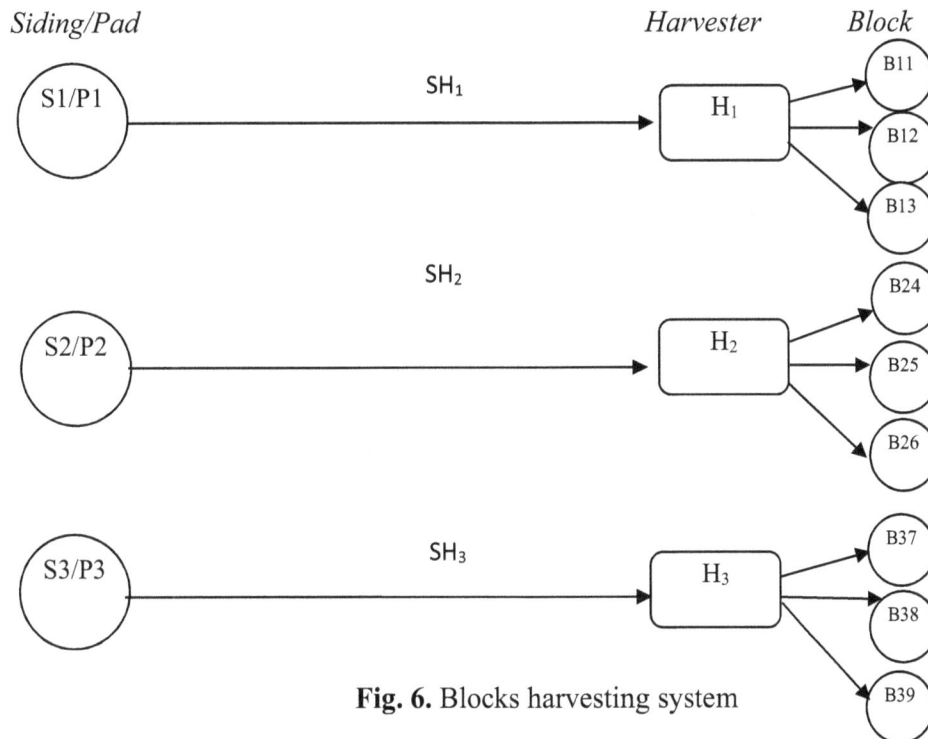

Fig. 6. Blocks harvesting system

This algorithm includes the following main steps:

Begin
Step 1 *Construct the harvesters list (1, 2, 3... H) For the current harvesting season*
Step 2 *Construct the sidings list (1, 2... S) with its allotment A_s*
Step 3 *Set the average distance list between the sidings and harvesters*
Step 4 *Assigning each harvester for a siding using shortest distance from the harvester and pad siding.*
Step 5 *Set the priority list of harvesters to be served using two main criteria*
      5.1 Cane age
      5.2 Daily allotment
Step 6 While total deliveries and collections <= total daily allotment
      6.1 Assigning each harvester in the priority list to closest siding
      6.2 Completing allotment (delivers and collections) of each harvester
      6.3 Removing the selected sidings and harvesters form the list
      *6.4 **Updating** the siding and harvester lists*
End.

3.2.4 Harvester Siding/Pad Blocking (HPB) Algorithm

Harvester Pad/Siding blocking algorithm has been developed to assign several harvesters to specific siding/pad points according to shortest distance. This can reduce the total operating time and accordingly the associated costs. Fig. 7 shows a small sector of assigning several harvesters with two siding/pad points considering start time and finish time of each harvester to avoid conflict times or overlapping times.

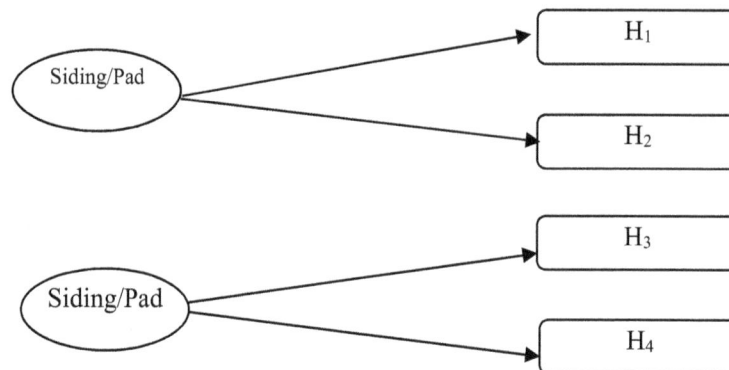

Fig. 7. Harvester Pad/Siding system

The main procedures have been detailed below

Begin

Step 1 *Construct the harvesters list (1, 2, 3, H) for the current harvesting season*
Step 2 *Construct the sidings list (1, 2... S) With its allotment A_s*
Step 3 *Set the distance list between the sidings and harvesters*
Step 4 *Assigning each harvester for a siding using shortest distance from the harvester and pad siding.*
Step 5 *Set the priority list of harvesters to be served using two main criteria*
        3.1 Cane age
        3.2 Daily allotment
Step 6 While total delivers and collections <= total daily allotment
        6.1 Assigning each harvester in the priority list to closest siding
        6.2 Completing allotment (delivers and collections) of each harvester
        6.3 Removing the selected sidings and harvesters form the list
        6.4 Updating the siding and harvester lists
End

4. Experimental Results

4.1 A Rail Transport Case Study

The proposed algorithms have been tested in a real case study that has been detailed in Appendix A, where Table A1 shows the rail section number and the distance between all network sections. Additionally, each rail section that has been used as a delivery and collection point is called a siding, where the empty and full bins are stored separately in each siding. This siding has two types of capacities, empty capacity (maximum number of empty bins that can be stored in this siding) and full capacity (maximum number of full bins that can be stored in this siding). Table A2 shows the total number of locomotives that has been used to produce the optimised locomotive scheduler, where each locomotive is described by two types of capacities (with full and empty bins) and average speed per hour.

Scheduler for three days has been produced including daily allotment for each harvester. The supply chain of empty and full bins of cane between harvesting points and mill for three days has been figured out in Tables 1, 2 and 3, and Figures 11, 12, and 13. Table 1 shows the delays of collection full bins from several sidings within day 1.

Table 1
Delays of crop collections from several sidings in day 1

Day Index	Siding Index	Allotment in bins	Harvesting Rate	Target Time	Finish Time	Delays in hours
1	85	124	13.433	13.231	22.11667	11.11667
1	55	96	12.581	12.63056	19.65	5.716667
1	62	70	11.194	12.25335	17.28333	5.033333
1	61	108	12.222	12.83652	17.7	1.616667
1	67	144	14.754	13.76007	18.11667	7.633333
1	99	112	10.312	14.86113	18.83333	6.083333
1	91	104	12.281	12.46837	16.16667	3.7
1	42	134	12.712	14.04122	16.56667	3.466667
1	19	126	12.264	16.27397	17.9	1.133333
1	17	140	13.208	15.59964	17.2	0
1	32	96	11.875	12.58421	13.56667	0
1	34	118	12.667	13.31554	13.41667	0
1	66	30	14.754	6.033347	6.016667	0
1	86	52	13.433	7.871064	7.85	0
1	16	107	12.459	12.08817	12.06667	0
1	20	24	12.264	7.956947	7.933333	0
1	11	82	11.972	11.34932	11.31667	0
1	49	32	4.561	10.01601	9.983334	0
1	82	84	10.5	11	10.96667	0
1	45	84	10	13.4	13.31667	0

Figure 8 shows number of empty bins (red graph) and full bins (green graph) at mill in the first day of harvesting, where a specific allotment is required to be transported from harvesting points to mill. From Figure 8, we note that there is no shortage in the number of empty and full bins in the first day, where number of bins at mill is above zero. The system started with 670 full bins and 730 empty bins, where the full bins at the beginning of the system means the number of empty bins at harvesters that has been distributed previous night as inventory. The total number of bins that have been used in the system equal 1400 bins. The number of bins that has been used at the rest of day is less than the initial bins number.

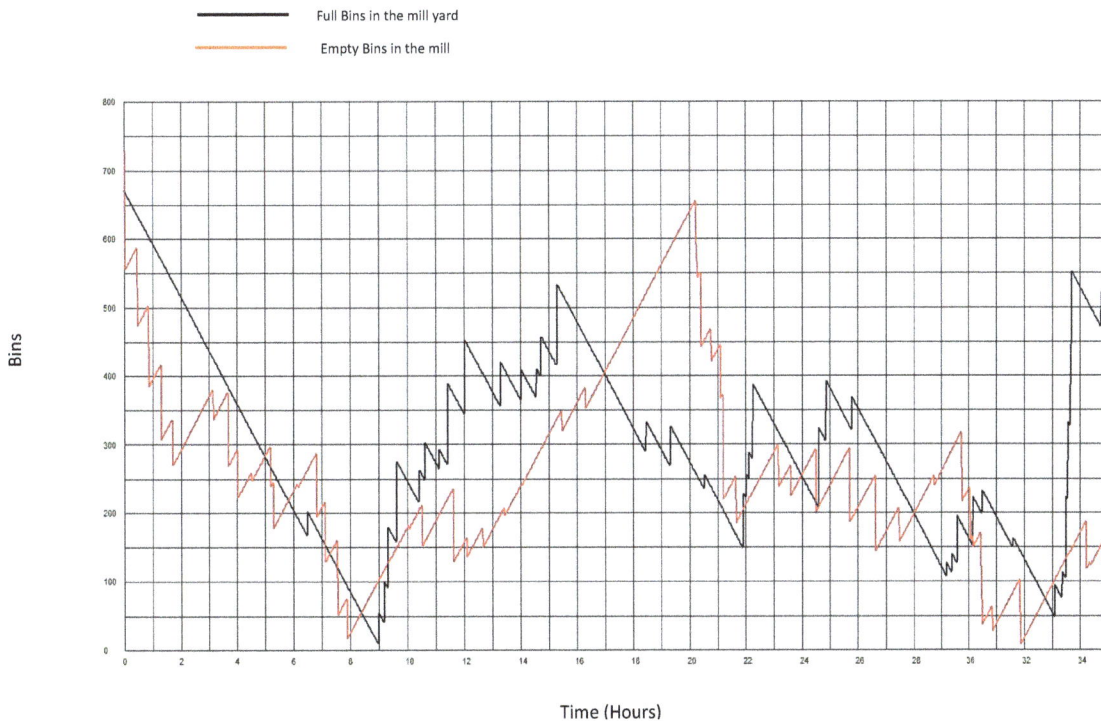

Full Bins in the mill yard

Empty Bins in the mill

Bins

Time (Hours)

Fig. 8. Empty and full bins at Mill in the first day

Table 2 shows the collection delays for the second day where the maximum delays was 10.366 hours for siding 67. Fig. 9 shows the number of empty and full bins at mill that has been used in the second day.

Table 2
Delays of crop collections from several sidings in day 2

Day Index	Siding Index	Allotment in bins	Harvesting Rate	Target Time	Finish Time	Delays in hours
2	86	52	13.433	31.87107	39.46667	7.6
2	42	134	12.712	38.04122	47.2	14.26667
2	67	144	14.754	37.76006	43.9	10.36667
2	82	84	10.5	35	38.6	3.6
2	34	118	12.667	37.31554	38.93333	0
2	55	96	12.581	36.63055	37.96667	2.1
2	17	140	13.208	39.59964	40.86666	0
2	66	30	14.754	30.03335	31.3	1.266667
2	32	96	11.875	36.58421	37.7	0
2	62	70	11.194	36.25335	37.33333	1.1
2	61	108	12.222	36.83652	37.81667	0.983333
2	19	126	12.264	40.27397	40.81667	0.05
2	45	84	10	37.4	37.4	0.016667
2	91	104	12.281	36.46837	36.45	0
2	16	107	12.459	36.08817	36.06667	0
2	20	24	12.264	31.95695	31.93333	0
2	99	112	10.312	38.86113	38.83333	0
2	11	82	11.972	35.34932	35.31667	0
2	49	32	4.561	34.01601	33.98333	0

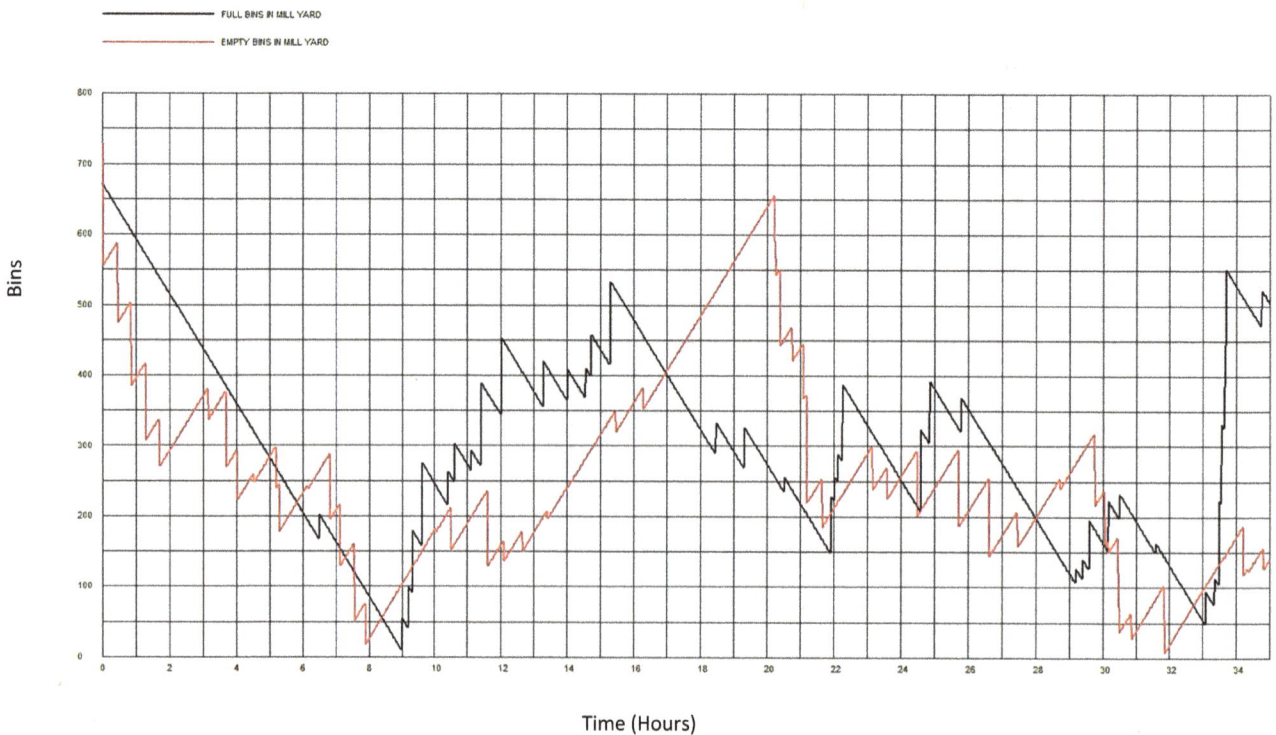

Fig. 9. Empty and full bins at Mill in the second day

Table 3 shows the delays of collections during the third day where the maximum delays is 6.11 for siding 45. Fig. 10 shows the empty and full bins at mill that has been used to transport the third day's allotment of sugarcane.

Table 3
Delays of crop collections from several sidings in day 3

Day Index	Siding Index	Allotment in bins	Harvesting Rate	Target Time	Finish Time	Delays in hours
3	45	100	10	63	69.1	6.116667
3	49	82	4.561	68.97852	75.06667	6.1
3	55	38	12.581	56.02043	61.66667	5.683333
3	54	60	13.44467	57.46273	62.9	5.45
3	66	31	14.754	54.60112	56.8	2.2
3	61	93	12.222	59.60923	61.33333	0.333333
3	19	92	12.264	61.50163	62.4	0
3	17	153	13.208	64.58389	65.3	0
3	29	38	12.66667	55.5	55.48333	0
3	2	16	12.459	52.78421	52.76667	0
3	11	101	11.972	60.93635	60.91667	0
3	91	85	12.281	58.92126	58.9	0
3	34	116	12.667	61.15765	61.13334	0
3	86	31	13.433	54.80775	54.78333	0
3	32	48	11.875	56.5421	56.51667	0
3	100	110	15.323	58.17875	58.15	0
3	16	93	12.459	58.96449	58.93333	0
3	42	118	12.712	60.78257	60.75	0

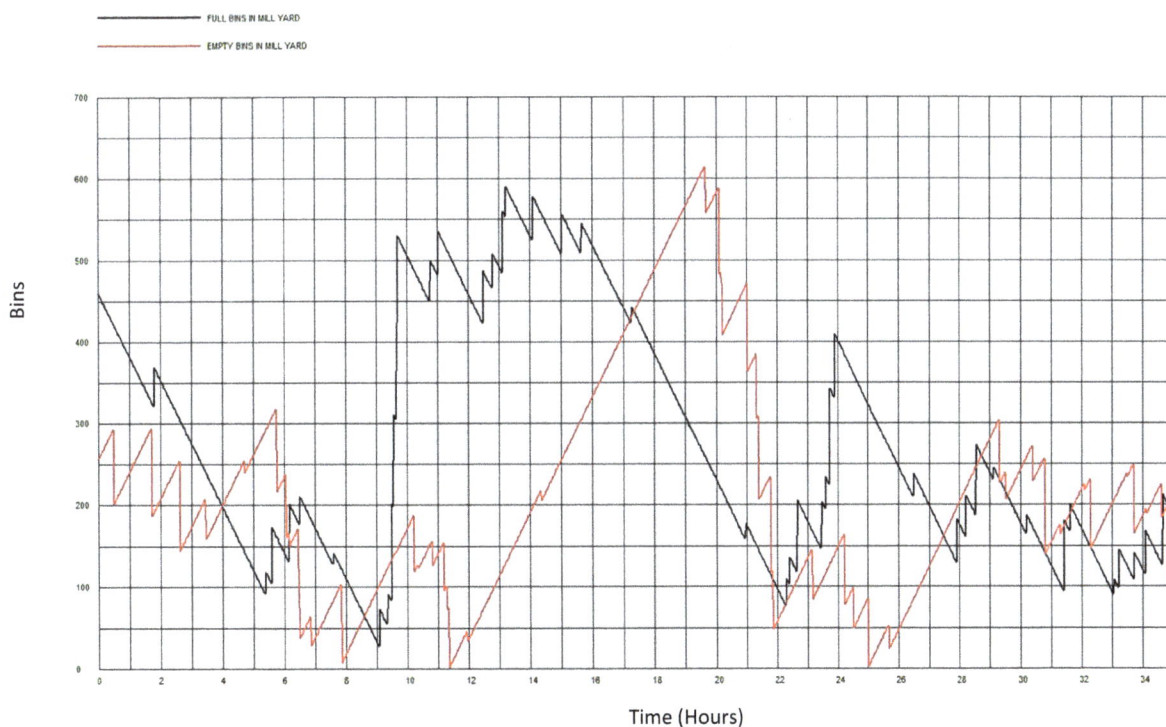

Fig. 10. Empty and full bins at Mill in the Third day

The efficient performance of the transport system can be measured by keeping the supply chain of empty and full bins of cane between harvesting points and mills without any interruptions and delays. Additionally, the stocks rate of empty and full bins at mill is considered where the transport system works through limited number of resources from bins and Locomotives. This prevent any work interruptions of mill and keep empty bins supply for harvesters. A continuous supply chain of bins between harvesting points and mill has been satisfied using the proposed methodology as shown previously in Figs. (10-12).

Trains scheduling is produced for three days using 6 trains with 100 rail sections. Fig. 11 shows the train scheduling with makespan, 4915 minutes, total number of runs sums up to 134 trips and the total operating time is 219.5 h. The deliveries and collections number of full and empty bins have been shown in Fig. 12, where green number is full collected bins and red number is empty delivered bins. Operating times inbound and outbound directions have been detailed in Fig. 13 with showing the waiting time at each section.

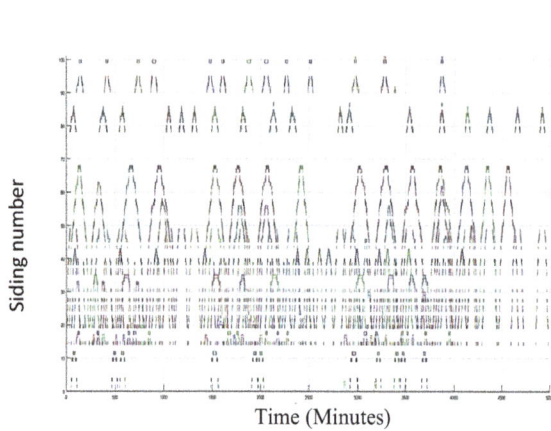

Fig. 11. Gantt chart of 6 trains scheduling with 27km/h

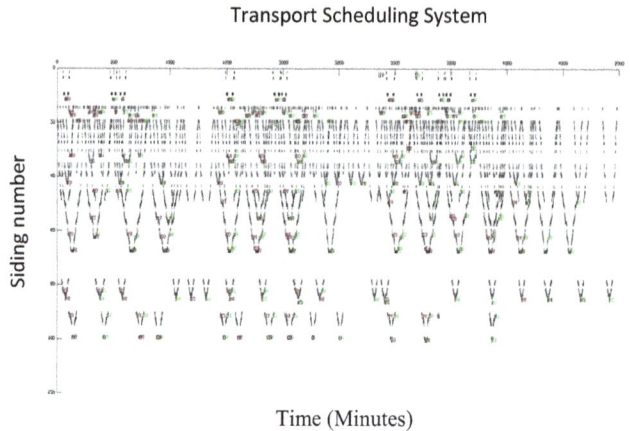

Fig. 12. Delivery and collections at marked sidings within several trips

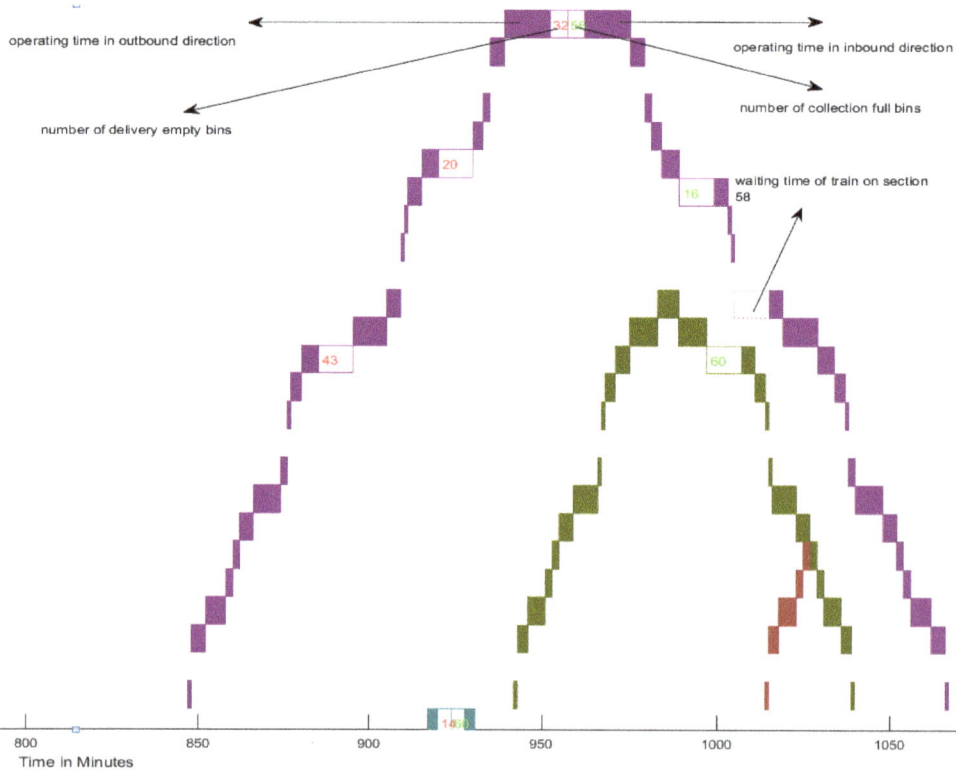

Fig. 13. Close up of train scheduling showing delivery and collection bins

Fig. 14 concludes the total operating time for the three days and the number of trips that have been used by each locomotive to satisfy the total allotments of all harvesters. In Figure 15, the trips distribution for each locomotive has been shown and concluded.

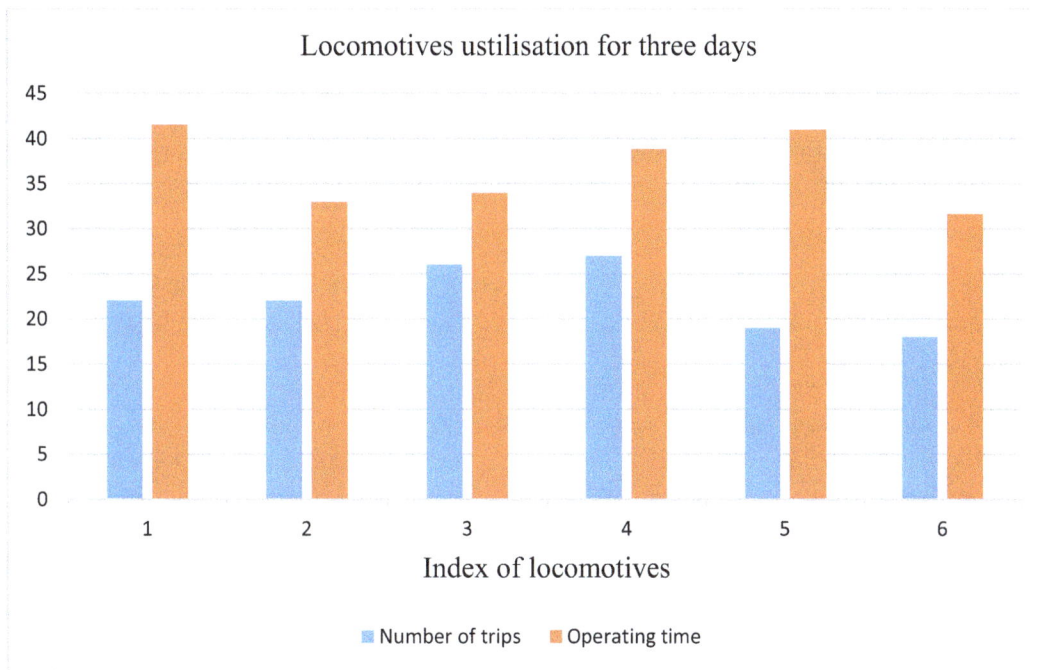

Fig. 14. Operating time for the three days schedule using 6 locomotives

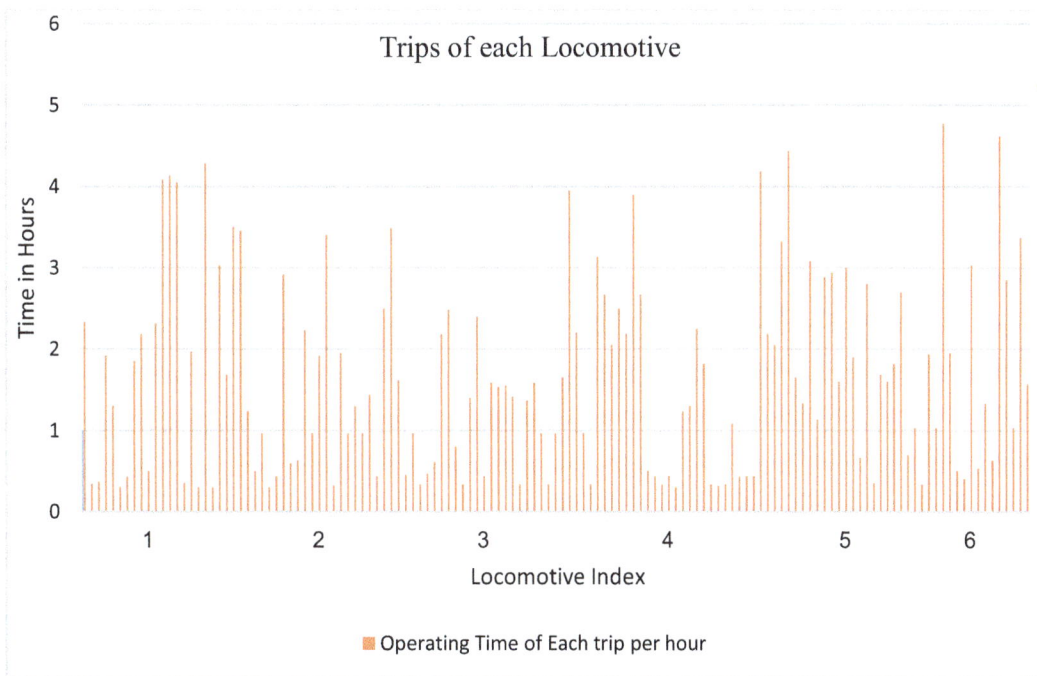

Fig. 15. Trips distribution for the 6 locomotives during three days

The proposed methodology has been applied for the whole season of sugarcane harvesting and delivery of both cane and trash from harvesting areas to mill. In this system the total operating time has greatly improved. Figure 16(a) shows the total operating time and the total deliveries without applying the proposed scheduling methodology, while Figure 16b shows the values with applying the scheduling approach. The total cane delivered captioned in blue, while the processed cane is captured is orange .From Figure 16 (b), the total cane delivered is 893086 tonnes and delivered in 5286 Hours without applying scheduling approach, while the total cane delivered and trash in figure 16(c) 1025470 tonnes and delivered in 3504 Hours using the proposed scheduling approach which is a great improvement in terms of total operation time resource utilisation.

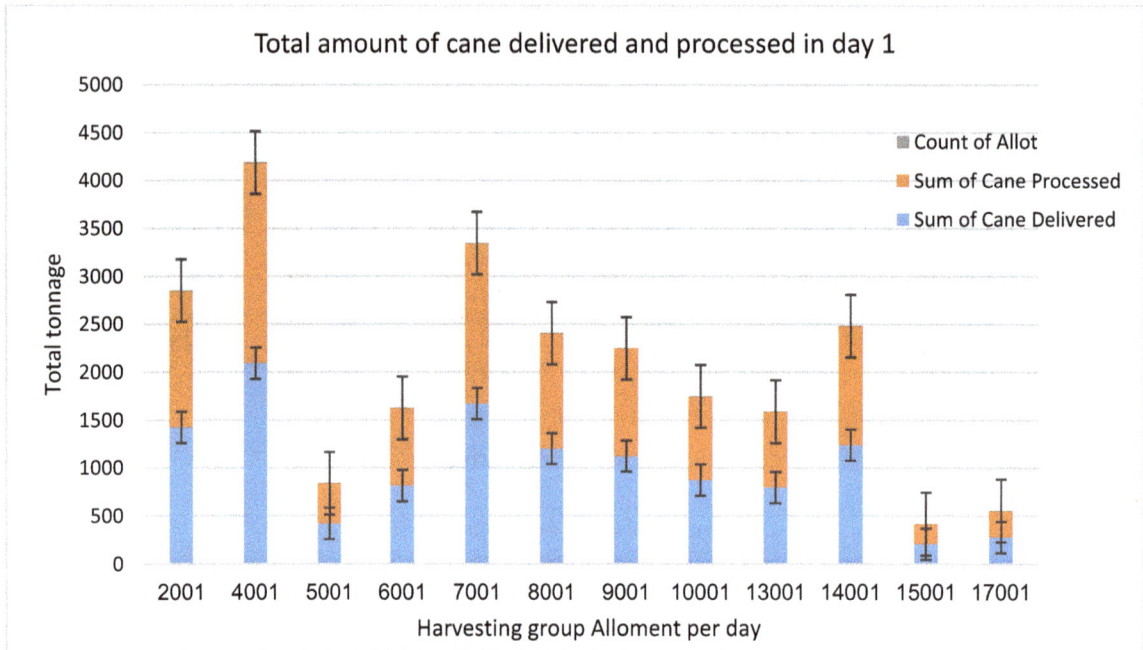

Fig. 16 (a). Total amount of cane delivered and processed in day number 1 per harvesting group

Further, Figure 16 (a) above shows a graph of activities in a sugar factory for one day operation, which from the simulation a total 12,136.59 tonnes of cane were delivered and processed. The sugar factory is assumed to operates at a rate of $600\ tonnes/hour$ (crushing rate), continuously for $24\ hours$. Therefore, assuming an efficiency of $90\ \%$ for the factory, the total amount of tonnage required per day (24 hours) continuous crushing

$$Factory\ requirement\ = 600 tonnes * 24 hours * \frac{90}{100}\ (efficiency)$$

Loss of crushing efficiency (ρ_{loss}) therefore is equal to

$$\rho_{loss} = \left\{ \frac{600 tonnes * 24 hours * \frac{90}{100}\ (eff) - 600 tonnes * 24 hours * \frac{90}{100}\ (eff)}{\left[600 tonnes * 24 hours * \frac{90}{100}\ (efficiency) \right]} \right\} * 100$$

$$\rho_{loss} = 90\ \% - 84.28188\% = 5.71812\ \%\ Loss$$

$$\rho_{loss} = 5.71812\ \%$$, which can be corrected, by proper scheduling of transport and harvesting system as shown in Fig. 16(c).

Factory cane accumulation

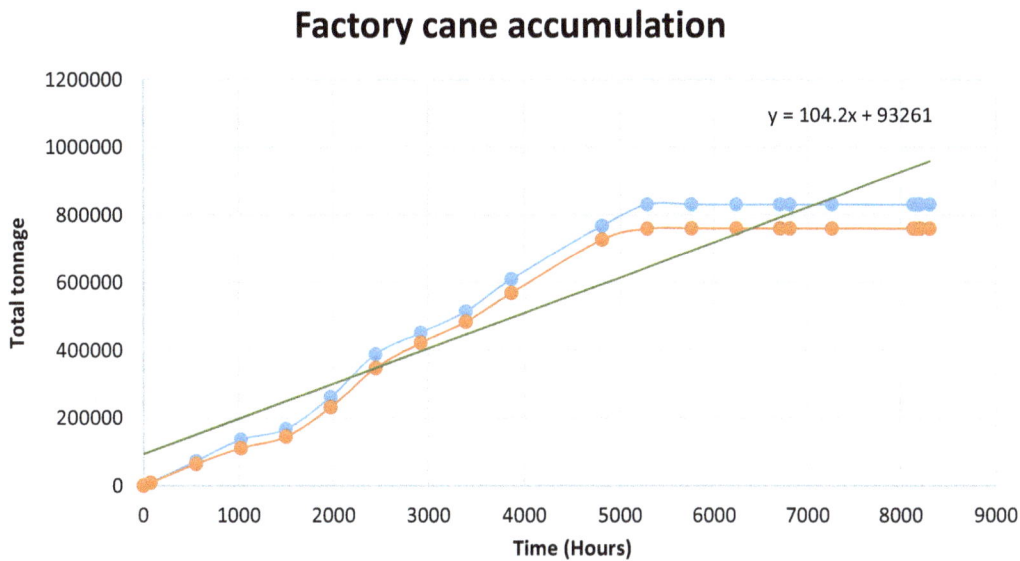

$y = 104.2x + 93261$

Fig. 16 (b). The operating time and total deliveries without scheduling theory

Fig. 16 (a) depicts a system without scheduling, where at the start of the crushing period (day 1), a total of 94,000 tonnes were delivered to the factory per hour, which point to the fact that most of the time the factory would run out of cane, and would require longer period to finish the operation of harvesting, transport and processing of cane to produce bagasse and other product .This however, is in sharp contrast to the expectation of an ideal system, where a total of 14400 *tonnes* is required to guarantee continuous crushing.

Fig. 16(c), on the other hand from the line of best fit, at the start of the season, approximately 535224 *tonnes* are delivered per hour (Y- Intercept) on the graph, thus slightly meeting factory requirements, which in terms of efficiency is approximately 89.204 % , which is within the margin of error of an efficient system (90 %), for this case .Scheduling subsequently has improved the operational time of the whole system, and ensured maximum utilisation of the resources thus saving on time and cost

.

Factory cane accumulation

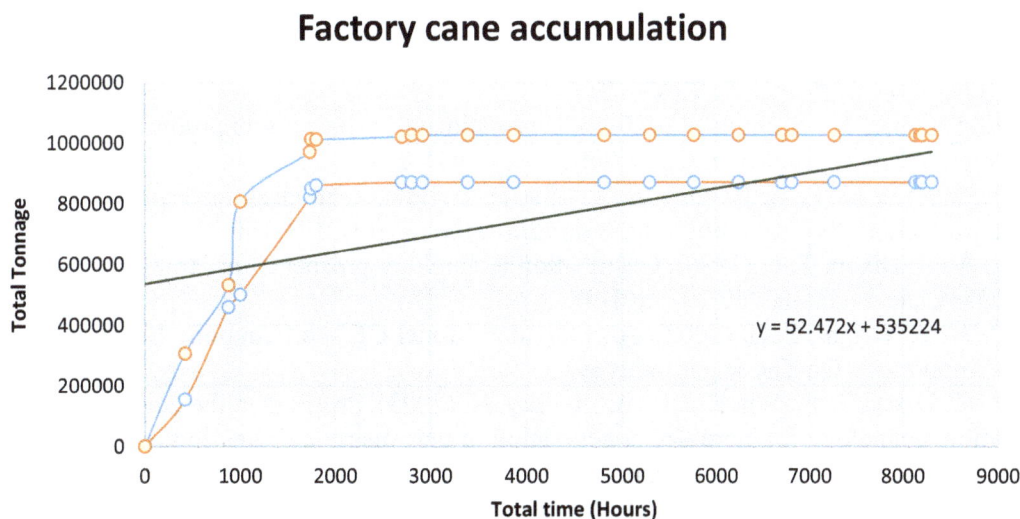

$y = 52.472x + 535224$

Fig. 16 (c). The operating time and total deliveries with scheduling theory

5. Conclusion

Cane and cane residue transport systems using both rail and road networks play a vital role in the Australian bio-energy supply chain. Delivery and collection times have a high impact on optimising bio-energy production and consequently the associated costs. Constraint programming approach has been introduced with adapted Limited Discrepancy Search (LDS) to optimise the delivery and collection times through the transport network, where the cane and cane residue has been taken from harvesting points to siding or pad points and then transported to mill. The surplus bagasse (a by-product of the crushed cane) is stored at the mill, and is transported later to bio-refinery stations to produce bioenergy.

The constraint programming approach focused on satisfying the proposed model constraints were involved in harvesting operations, transport operations and milling operations. The LDS has been used as search tree that provides a solution with each node (delivery or collection points). The proposed daily trips have been constructed to serve the harvesting points and mills. This numerical case study (based on historical data) indicates that the modelled system has the potential of improving the operational time and in effect save on cost of operation and improve of equipment utilisation.

References

Asikainen, A., Liiri, H., Peltola, S., Karjalainen, T., & Laitila, J. (2008). Forest energy potential in Europe (EU27). Finnish Forest Research Institute, Vantaa.

Bookbinder, J. H., & Desilets, A. (1992). Transfer optimization in a transit network. *Transportation Science, 26*(2), 106-118.

Chen, C. W., & Fan, Y. (2012). Bioethanol supply chain system planning under supply and demand uncertainties. *Transportation Research Part E: Logistics and Transportation Review, 48*(1), 150-164.

De Meyer, A., Cattrysse, D., Rasinmäki, J., & Van Orshoven, J. (2014). Methods to optimise the design and management of biomass-for-bioenergy supply chains: A review. *Renewable and Sustainable Energy Reviews, 31*, 657-670.

Faaij, A., Van Ree, R., Waldheim, L., Olsson, E., Oudhuis, A., Van Wijk, A., ... & Turkenburg, W. (1997). Gasification of biomass wastes and residues for electricity production. *Biomass and Bioenergy, 12*(6), 387-407.

Harvey, W. D., & Ginsberg, M. L. (1995, August). Limited discrepancy search. In IJCAI (1) (pp. 607-615).

Hassuani, S. J., Da Silva, J. E. A. R., & Neves, J. L. M. (2005). Sugarcane trash recovery alternatives for power generation. *In Proc. ISSCT* (Vol. 25).

Henrich, E., Dahmen, N., & Dinjus, E. (2009). Cost estimate for biosynfuel production via biosyncrude gasification. *Biofuels, Bioproducts and Biorefining, 3*(1), 28-41.

Hess, J. R., Wright, C. T., & Kenney, K. L. (2007). Cellulosic biomass feedstocks and logistics for ethanol production. *Biofuels, Bioproducts and Biorefining, 1*(3), 181-190.

Hobson, P. A., & Dixon, T. F. (1998). Gasification technology-prospects for large-scale. In Proc. Aust. Soc. Sugar Cane Technol (Vol. 20, pp. 1-9).

Hobson, P. A., & Wright, P. G. (2002). An Extended Model Of The Economic Impact Of Extraneous Matter Components On The Sugar Industry.

Hobson, P. A., Dixon, T. F., Wheeler, C., & Lindsay, N. (2003). Progress in the development of bagasse gasification technology for increased cogeneration in the Australian sugar industry. In Processdings-Australian society of sugar cane technology (pp. 60-60). PK Editorial Services; 1999.

Hobson, P. A., Edye, L. A., Lavarack, B., & Rainey, T. J. (2006). Analysis of bagasse and trash utilization options-SRDC Technical Report 2/2006.

Iakovou, E., Karagiannidis, A., Vlachos, D., Toka, A., & Malamakis, A. (2010). Waste biomass-to-energy supply chain management: a critical synthesis. *Waste Management, 30*(10), 1860-1870.

Mafakheri, F., & Nasiri, F. (2014). Modeling of biomass-to-energy supply chain operations: applications, challenges and research directions. *Energy Policy, 67*, 116-126.

Masoud, M., Kozan, E., & Kent, G. (2011). A job-shop scheduling approach for optimising sugarcane rail operations. *Flexible Services and Manufacturing Journal, 23*(2), 181-206.

Masoud, M. M. A. A., Kozan, E., & Kent, G. A. (2012). A new approach to automatically producing schedules for cane railways. *In Proceedings of the 34th Conference of the Australian Society of Sugar Cane Technologists* (Vol. 34). Australian Society of Sugar Cane Technologists.

Masoud, M. M. A. A., Kozan, E., & Kent, G. (2015, November). A near Optimal Cane Rail Scheduler under Limited and Unlimited Capacity Constraints. In *Proceedings of 21st International Congress on Modelling and Simulation* (pp. 1738-1744).

Meyer, J. C., Hobson, P. A., & Schultmann, F. (2012). The potential for centralised second generation hydrocarbons and ethanol production in the Australian sugar industry. In *Proceedings of the Australian Society of Sugar Cane Technologists* (Vol. 34).

Nilsson, D., & Hansson, P. A. (2001). Influence of various machinery combinations, fuel proportions and storage capacities on costs for co-handling of straw and reed canary grass to district heating plants. *Biomass and Bioenergy, 20*(4), 247-260.

Pernase, S., & Pekol, A. (2012, September). The Transportation of Sugar Cane in Queensland's Sugar Industry. *In Australasian Transport Research Forum 2012 Proceedings (pp. 26-28).*

Rapp, M. H., & Gehner, C. D. (1967). Transfer optimization in an interactive graphic system for transit planning (No. Intrm Rpt.).

Ravula, P. P., Grisso, R. D., & Cundiff, J. S. (2008). Cotton logistics as a model for a biomass transportation system. *Biomass and Bioenergy, 32*(4), 314-325.

Rentizelas, A. A., Tolis, A. J., & Tatsiopoulos, I. P. (2009). Logistics issues of biomass: the storage problem and the multi-biomass supply chain. *Renewable and Sustainable Energy Reviews, 13*(4), 887-894.

Shabani, N., Akhtari, S., & Sowlati, T. (2013). Value chain optimization of forest biomass for bioenergy production: a review. *Renewable and Sustainable Energy Reviews, 23*, 299-311.

Sokhansanj, S., Turhollow, A., Cushman, J., & Cundiff, J. (2002). Engineering aspects of collecting corn stover for bioenergy. *Biomass and Bioenergy, 23*(5), 347-355.

Sokhansanj, S., Mani, S., Tagore, S., & Turhollow, A. F. (2010). Techno-economic analysis of using corn stover to supply heat and power to a corn ethanol plant–Part 1: Cost of feedstock supply logistics. *Biomass and Bioenergy, 34*(1), 75-81.

Thorburn, P. J., Archer, A. A., Hobson, P. A., Higgins, A. J., Sandell, G. R., Prestwidge, D. B., ... & Juffs, R. (2006). Value chain analyses of whole crop harvesting to maximise co-generation. In *PROCEEDINGS-AUSTRALIAN SOCIETY OF SUGAR CANE TECHNOLOGISTS* (Vol. 2006, p. 37). PK Editorial Services; 1999.

Vadas, P. A., & Digman, M. F. (2013). Production costs of potential corn stover harvest and storage systems. *Biomass and Bioenergy, 54*, 133-139.

Van Dyken, S., Bakken, B. H., & Skjelbred, H. I. (2010). Linear mixed-integer models for biomass supply chains with transport, storage and processing. *Energy, 35*(3), 1338-1350.

Van Vliet, O. P., Faaij, A. P., & Turkenburg, W. C. (2009). Fischer–Tropsch diesel production in a well-to-wheel perspective: a carbon, energy flow and cost analysis. *Energy Conversion and Management, 50*(4), 855-876.

Wang, J. J., Yang, K., Xu, Z. L., Fu, C., Li, L., & Zhou, Z. K. (2014). Combined methodology of optimization and life cycle inventory for a biomass gasification based BCHP system. *Biomass and Bioenergy, 67*, 32-45.

Yue, D., You, F., & Snyder, S. W. (2014). Biomass-to-bioenergy and biofuel supply chain optimization: overview, key issues and challenges. *Computers & Chemical Engineering, 66*, 36-56.

Appendix A

Table A1
Rail Transport Network

Segment	Section	Distance	Capacity Empty	Capacity Full	Segment	Section	Distance	Capacity	Capacity
1, 0	1	0.48	0	0	33, 0	58	1.48	66	68
2, 0	2	0.32	60	60	34, 0	59	0.07	58	64
3, 0	3	1.59	0	0	34, 0	60	0.33	44	54
4, 0	4	0.98	80	82	34, 0	61	1.21	60	74
4, 0	5	1.82	0	0	34, 0	62	1.48	50	62
5, 0	6	0.19	68	68	34, 0	63	1	0	0
6, 0	7	0.93	38	38	35 0	64	0.5	0	0
6, 0	8	1.6	110	110	35, 1	65	0.5	0	0
7, 0	9	4.22	0	0	36, 0	66	1.07	62	70
8, 0	10	0.27	72	68	36, 0	67	3.81	56	63
9, 0	11	2.03	80	80	36, 0	68	2.89	0	0
10, 0	12	1.68	58	70	37, 0	69	3.23	82	90
10, 0	13	2.4	52	52	38, 0	70	0.98	56	66
11, 0	14	0.71	0	0	38, 0	71	1.06	208	208
12, 0	15	0.66	0	0	38, 0	72	1.16	62	68
13, 0	16	0.36	66	66	38, 0	73	1.23	50	58
14, 0	17	1.53	110	110	38, 0	74	2.54	80	82
14, 0	18	3.13	66	60	38, 0	75	0.24	88	88
15, 0	19	1.11	64	66	38, 0	76	2.77	212	212
15 0	20	0.44	28	34	38, 0	77	2.71	80	96
15, 0	21	1	0	0	39, 0	78	1.43	58	68
16, 0	22	0.5	0	0	39, 0	79	1.73	0	0
16, 1	23	0.5	0	0	40, 0	80	0.22	62	62
17, 0	24	0.19	74	74	41, 0	81	1.57	26	36
18, 0	25	0.5	0	0	41, 0	82	1.13	50	56
18, 1	26	0.5	0	0	41, 0	83	2.49	50	56
19, 0	27	1.07	0	0	41, 0	84	0.83	0	0
20, 0	28	0.32	50	50	42, 0	85	0.31	46	46
20, 0	29	0.01	48	48	43, 0	86	0.71	62	64
21, 0	30	0.13	0	0	43, 0	87	2.61	0	0
22, 0	31	0.33	34	34	44, 0	88	0.22	34	34
22, 0	32	1.4	70	80	45, 0	89	0.17	40	40
22, 0	33	1.06	66	70	46, 0	90	0.76	52	66
22, 0	34	1.01	72	82	46, 0	91	0.14	52	66
23, 0	35	0.5	0	0	46, 0	92	1.75	60	86
24, 0	36	0.5	0	0	46, 0	93	1.39	54	66
24, 1	37	0.5	0	0	46, 0	94	0.73	0	0
25, 0	38	1	0	0	47, 0	95	1.46	56	62
26, 0	39	1.9	0	0	47, 0	96	0.15	0	0
27, 0	40	1.2	94	94	47, 0	97	0.39	62	62
27, 0	41	1.24	64	68	47, 0	98	0.37	52	52
27, 0	42	0.99	60	60	48, 0	99	1.5	46	54
28, 0	43	0.19	126	126	48, 0	100	0.43	76	76
28, 1	44	0.19	0	0					
29, 0	45	1.22	60	64					
29, 0	46	1.94	60	60					
29, 0	47	0.68	62	78					
29, 0	48	0.67	44	44					
30, 0	49	1.3	48	54					
30, 0	50	2.42	86	88					
31, 0	51	0.5	0	0					
31, 1	52	0.5	0	0					
32, 0	53	0	86	88					
32, 0	54	0.95	62	52					
32, 0	55	1.41	52	60					
32, 0	56	2.97	64	56					
32, 0	57	1.1	0	0					

Table A2
Locomotives capacity and speed

Locomotive Number	Locomotive Full Capacity	Average Speed (Km/h)	Locomotive Empty Capacity
1	120	22	120
2	120	22	120
3	120	22	120
4	110	22	120
5	90	18	110
6	72	18	110

Parameters optimization of fabric finishing system of a textile industry using teaching–learning-based optimization algorithm

Rajiv Kumar[a*], P.C. Tewari[b] and Dinesh Khanduja[b]

[a]Reseach Scholar, Department of Mechanical Engineering, National Institute of Technology, Kurukshetra, Haryana, India -136119
[b]Professor, Department of Mechanical Engineering, National Institute of Technology, Kurukshetra, Haryana, India -136119

CHRONICLE	ABSTRACT
	In the present work, a recently developed advanced optimization algorithm named as teaching–learning-based optimization (TLBO) is used for the parameters optimization of fabric finishing system of a textile industry. Fabric Finishing System has four main subsystems, arranged in hybrid configuration. For performance modeling and analysis of availability, a performance evaluating model of fabric finishing system has been developed with the help of mathematical formulation based on Markov-Birth-Death process using Probabilistic Approach. Then, the overall performance of the concerned system has first analyzed and then, optimized by using teaching–learning-based optimization (TLBO). The results of optimization using the proposed algorithm are validated by comparing with those obtained by using the genetic algorithm (GA) on the same system. Improvement in the results is obtained by the proposed algorithm. The results of effect of variation of the algorithm parameters on fitness values of the objective function are reported.
Keywords: *Performance modeling* *TLBO* *Markov process* *Genetic algorithm* *Probabilistic Approach*	

1. Introduction

Availability is a performance criterion for repairable systems that contains both the reliability and maintainability features of a system. Any industrial system comprises of subsystems arranged in series, parallel or hybrid configuration of the subsystems. The Textile Industry comprises of large complex engineering systems arranged in fusion configurations. Some of the important systems of a Textile Industry are Yarn Manufacturing, Yarn and Fiber Dyeing, Fabric Weaving, Sewing Thread, Fabric Dyeing, Fabric Finishing. The important system of a Textile Industry, upon which the quality of products mainly depends, is the Fabric Finishing System. In the present work an attempt has been made to analyze the performance and optimize the availability parameters of fabric finishing system. It has four main subsystems, arranged in hybrid configuration. In the process of fabric finishing, fabric from storage are fed into a stenter machine to impart various chemical finishes and make them set on the fabric. Finishing chemicals are fixed in curing machine. Seuding machine is used to impart peach finish to the fabric.

* Corresponding author
E-mail: rajivpansra@gmail.com (R. Kumar)

Firstly cloth is passed through water in the padding mangle and then squeezed under pressure so as to remove excess of the liquor. The fabric is wetted in order to remove creases from the fabric. Then fabric is dried over Vertical Dryers consisting of five to eight vertical steam heated cylinders depending upon the type of Seuding Machine. From here fabric moves the main Seuding Section. Twenty four small rollers are covered with emery paper and these rollers are mounted on the Drum. Twelve rollers are known to be energy pile rollers and are rotating in the direction of the fabric. Twelve remaining rollers are counter energy pile rollers and are rotating against the direction of the fabric. Sanforizing or shrinkage is the final step of finishing before the fabric is forwarded to the folding department. The function of this machine is to impart pre-determined shrinkage to the fabric so that there is no further shrinkage in fabric during washing. The schematic flow diagram of fabric finishing system is shown in Fig. 2.

The available literature shows the many approaches have been used to analyze the system performance in terms of availability. These are Reliability Block Diagram (RBD), Monte Carlo simulation, Markov approach, failure mode and effect analysis, Fault tree analysis and petri nets. Cafaro et al. (1986) explained the use of Markov models for evaluating the availability and reliability of a system when the transition rates of each component depend on the state of the system. The Markov approach was also discussed by Fu et al. (1986, 1987). Chung (1987) presented a mathematical model of a repairable parallel system with standby units involving human error and common cause failures. Laplace transforms of state probabilities and steady state availabilities of the system were evaluated. Kumar et al. (1988, 1989, 1990, 1991, 1992) derived the expressions for steady state availabilities of feeding, washing, screening, paper production and crystallization system in paper and sugar industries. Failure and repair rates were taken to be constant. Coit and Smith (1996) developed a problem specific Genetic Algorithm (GA) to analyze series-parallel systems and to determine the optimal design configuration. Singh and Mahajan (1999) examined the reliability and availability of a utensils manufacturing plant assuming constant failure and repair rates for various machines. Lai et al. (2002) studied the availability of distributed software/hardware systems. A Markov model was developed and equations were derived to obtain the steady state availability. Tewari et al. (2003) framed out a decision support system for refining system with the help of mathematical modeling using probabilistic approach. Gupta et al. (2005) proposed a method to compute reliability and long-run availability of the main parts of the serial processes system. Mathematical formulation of the model was carried out using mnemonic rule and the differential equations were solved by Runge-Kutta method. Gupta et al. (2009) assessed the reliability and availability of a critical ash handling unit of a steam thermal power plant using the concept of performance modeling and analysis. Mathematical formulation for reliability of ash handling unit of plant has been carried out using probability theory and Markov birth-death process. Khanduja et al. (2009, 2010) dealt with the mathematical modeling and performance optimization for the screening unit and paper making system in a paper plant using GA. Garg et al. (2010) dealt with availability optimization for screw plant using genetic algorithm (GA). Gupta (2011) demonstrated a mathematical model of a repairable spinning solution preparation system, a part of an acrylic yarn manufacturing plant with an effort to improve its availability. The proposed derived methodology relied on Markov Modeling. Garg and Sharma (2012) presented a technique for analyzing the behavior of an industrial unit. The synthesis unit of a urea plant situated in northern part of India has been considered to demonstrate the proposed approach. Goyal and Gupta (2012) developed a mathematical model of a complex bubble gum production system with an attempt to improve its availability. The methodology for determining the availability of the system was based on Markov modeling. The mathematical model was established using probability considerations and supplementary variable technique. Wang et al. (2012) dealt with two availability systems with warm standby units and different imperfect coverage. The time-to-failure and the time-to-repair of the active and standby units are assumed to be exponentially and generally distributed, respectively. Supplementary variable technique has been used to develop the steady-state availability for two systems. Modgil et al. (2013) dealt a performance model based on Markov process for shoe upper manufacturing unit and find out the time dependent system availability with long term availability of the system. Levitin et al. (2013) proposed a recursive and exact method for reliability evaluation of phased-mission systems with failures originating from some system elements that can propagate causing the common cause failures of groups of elements. The number of spare parts required for an item can be effectively estimated based on its reliability performance. Ravinder Kumar (2014) developed a mathematical model based on Markov birth-death process for a boiler air circulation system of a thermal power plant. The differential equations associated with the model have been solved recursively in order to find out the system's steady state availability. They focused mainly on coherent systems and series connection of k-out-of-n stand by subsystems with exponentially distributed component lifetimes. Sabouhi et al. (2016) discussed the Reliability modeling and Availability analysis of combined cycle power plants (CCPP). Kumar et al. (2017) dealt the performance analysis and optimization for Carbonated Soft Drink Glass Bottle (CSDGB) filling system of a beverage plant using Particle Swarm Optimization (PSO) approach. Çekyay and Özekici (2015) presented the Reliability, MTTF and Steady-State Availability Analysis of systems with exponential failure.

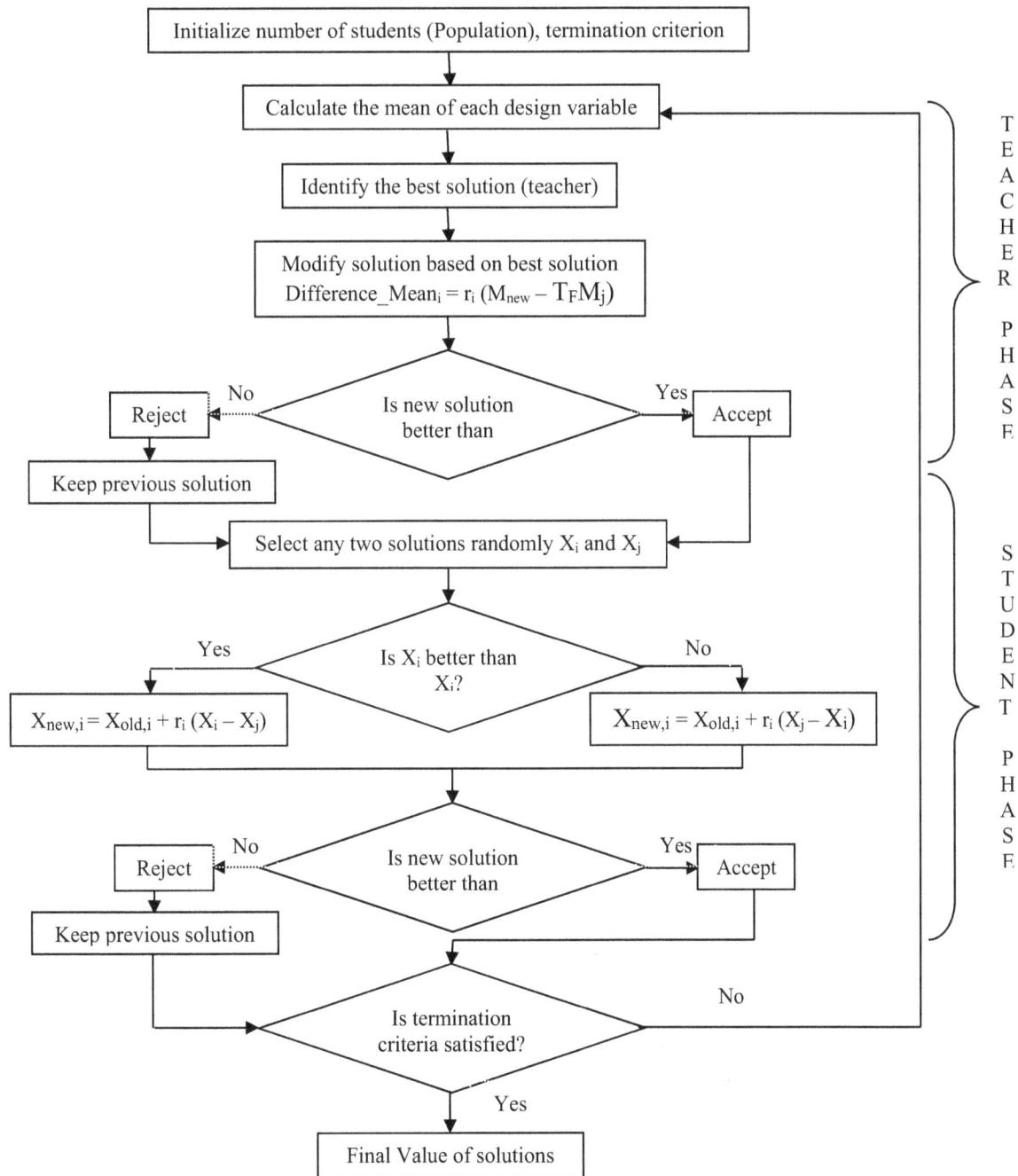

Fig. 1. Flowchart of TLBO algorithm (Rao et al., 2012)

1.1 Teaching–learning-based optimization algorithm

Teaching–learning-based optimization algorithm is a teaching–learning process inspired algorithm recently proposed by Rao et al. (2011, 2012) and Rao and Patel (2012) based on the effect of influence of a teacher on the output of learners in a class. In this algorithm, a group of learners are considered as population and different subjects offered to the learners are considered as different design parameters and a learner's result is analogous to the 'fitness' value of the optimization problem. The best solution in the entire population is considered as the teacher. The design parameters are actually the parameters involved in the objective function of the given optimization problem and the best solution is the best value of the objective function. The working of TLBO algorithm is divided into two parts, 'Teacher phase' and 'Learner phase'. Working of both these phases is described in detail by Rao et al. (2011, 2012). The same explanation of teacher phase and learner phase is referred here for the working of TLBO algorithm. Fig.1. represents the flowchart of TLBO algorithm (Rao et al. (2012)). The TLBO algorithm has been already tested on several constrained and unconstrained benchmark functions and proved better than the other advanced optimization techniques by Rao and Patel (2012). It is also proving better in various field of engineering such as those reported by Niknam et al. (2012) in the field of electrical engineering, Togan (2012) in the field of civil engineering. Similarly, Krishnanand et al. (2011) used it for the problems related to economic load dispatch, Rao and Kalyankar (2012,2013) used it for various fields related to manufacturing processes such as machining processes, modern machining processes, laser beam welding process, etc. and Rao and Patel (2013) used it to attempt multi-objective mathematical models in the field of thermal engineering. In the literature, it is observed that, the TLBO algorithm is not yet used in the field of optimization of mathematical models of textile industry. Hence the same is now used for the parameters optimization of a system of textile industry under consideration. In this work, efforts are carried out to prove the importance of advanced optimization techniques in the field of parameters optimization of fabric finishing system so that the maintenance personnel can achieve their objectives along with satisfying various constraints and limits of the respective system.

2. System description

Fabric Finishing System of a Textile Industry consists of four subsystems with the following description:

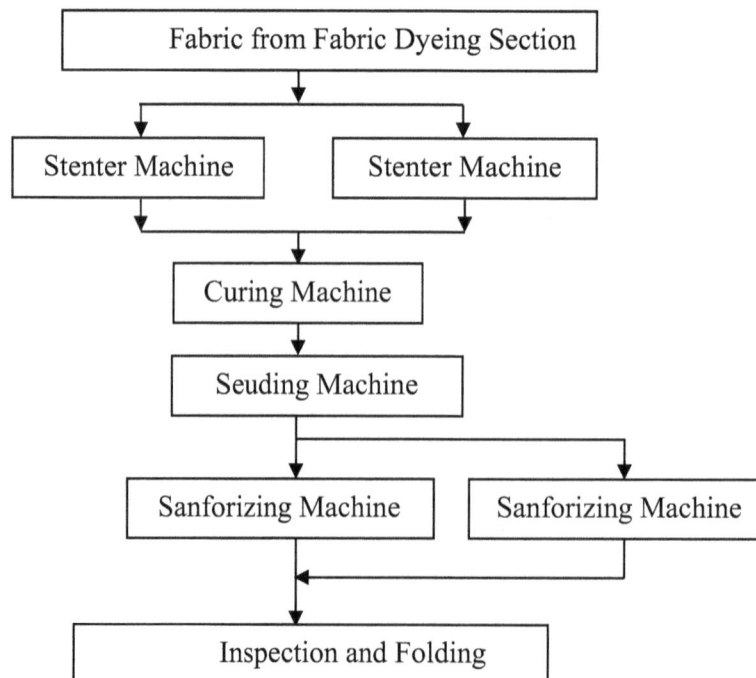

Fig. 2. Schematic Flow Diagram of Fabric Finishing System of a Textile Industry

Stenter Machine (A): The main function of Stenter Machine is to impart various chemical finishes and made them set on the fabric. There are two units of Stenter Machines working in parallel. Failure of any one reduces the capacity of the system. Complete failure of the system occurs when both the machines fail.

Curing Machine (B): Finishing chemicals are fixed here i.e. resin finishes (crosslinks). It consists of four chambers and uses thermic oil to heat these chambers. Failure of anyone chamber causes failure of this subsystem.

Seuding Machine (C): Seuding machine is used to impart peach finish to the fabric. It consists of padding mangle, vertical dryer, emery paper covered twenty four small rollers mounted on the Drum, energy pile roller and counter energy pile roller. This subsystem fails due to failure of some component.

Sanforizing Machine (D): Sanforizing is used to impart pre-determined shrinkage to the fabric so that there is no further shrinkage in fabric during washing. It consists of guided rollers, a rubber belt and steam heated cylinders. The subsystem consists of two units of Sanforizing Machines. The standby unit operates only upon the failure of first one. Complete failure of system occurs when standby unit also fails.

2.1 Assumptions

 i. Failure/repair rates are constant over time and statistically independent.
 ii. A repaired unit is as good as new, performance wise for a specified duration.
 iii. Sufficient repair facilities are provided, i.e. no waiting time to start the repairs.
 iv. Standby units (if any) are of the same nature and capacity as the active units.
 v. System failure /repair follow exponential distribution.
 vi. Service includes repair and /or replacement.
 vii. System may work at a reduced capacity/ efficiency.
 viii. There are no simultaneous failures among the system.

2.2 Notations

The notions associated with the Transition Diagram (Fig. 3.) are as follows:

A, B, C, D : Subsystems in good operating state.

$\bar{A}$: indicates that A is working in reduced capacity.

D* : One unit of subsystem D is in failed state and the system is working in full capacity with standby unit.

a, b, c, d : indicate the failed states of A, B, C, D.

λ_i : Mean constant failure rates from A, B, C, D to the states a, b, c, d.

μ_i : Mean constant repair rates from states a, b, c, d to the states A, B, C, D.

$P_0(t)$: Probability of full working capacity (without standby unit).

$P_2(t)$: Probability of full working capacity (with standby unit).

$P_1(t), P_3(t)$: Probability of reduced working capacity.

$P_4(t) - P_{15}(t)$: Probability of failed states.

(') :Derivatives w.r.t. 't'.

○ : System working at full capacity.

◇ : System working at reduced capacity.

▭ : System in failed state.

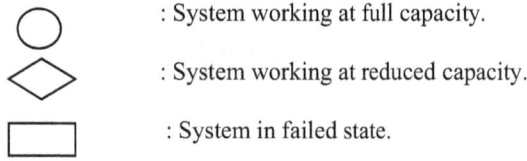

Based on above assumptions and notations the state transition diagram of fabric finishing system of a textile industry has been developed as shown in Fig. 3.

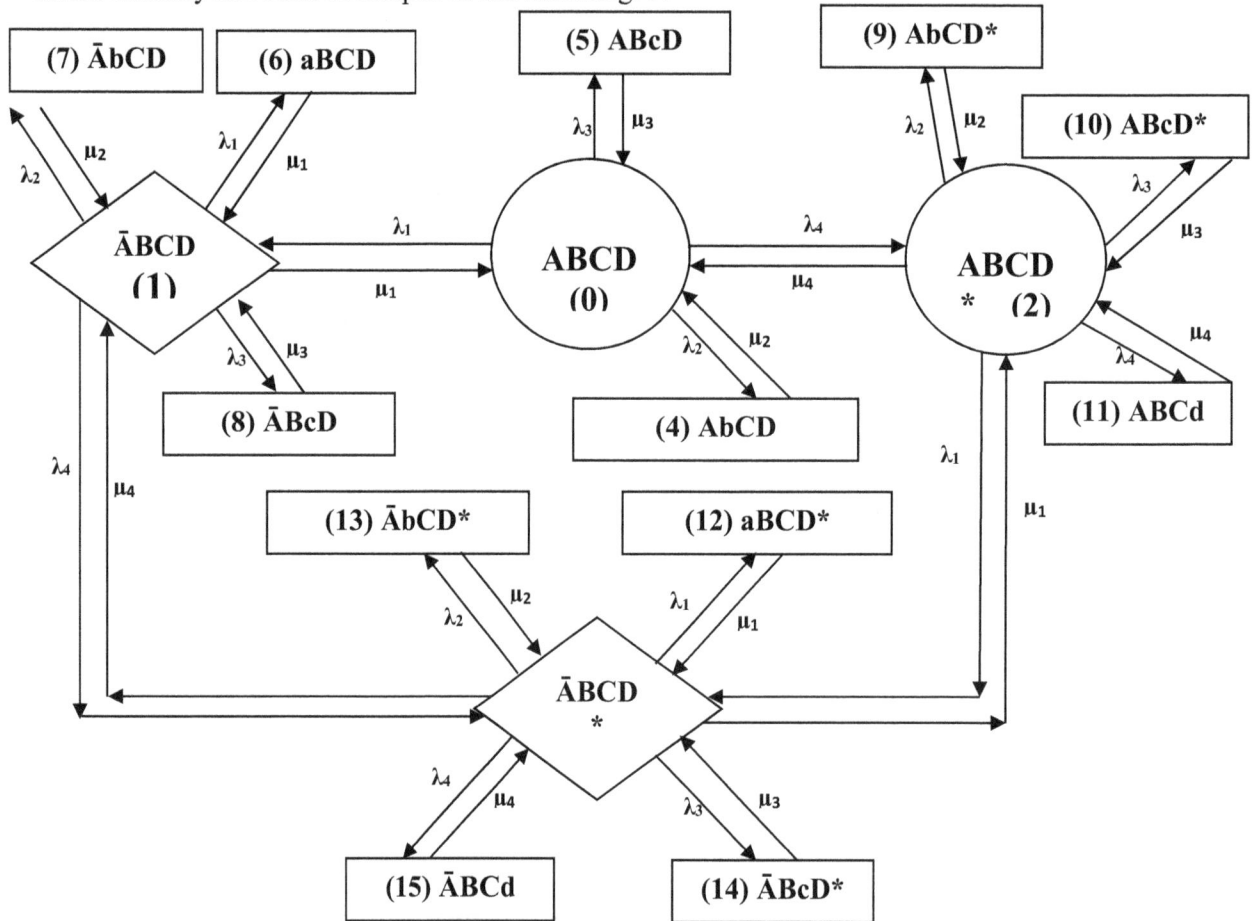

Fig. 3. Transition Diagram of Fabric Finishing System of a Textile Industry

3. Performance modeling

The mathematical modeling of the system based on Markov birth-death process is carried out using various probabilistic considerations. The first order Chapman-Kolmogorov differential equations associated with the state transition diagram shown in fig.3 are developed by using mnemonic rule as stated by Khanduja et al. (2012).Various probability considerations generate the following sets of differential equations:

$$P_0{}'(t) + K_0 P_0(t) = \sum_{i=1}^{4} \mu_i \left(P_1(t) + P_4(t) + P_5(t) + P_2(t) \right) \tag{1}$$

$$P_1{}'(t) + K_1 P_1(t) = \sum_{i=1}^{3} \mu_i P_{5+i}(t) + \mu_4 P_3(t) + \lambda_1 P_0(t) \tag{2}$$

$$P_2{'}(t) + K_2 P_2(t) = \sum_{i=2}^{4} \mu_i P_{7+i}(t) + \mu_1 P_3(t) + \lambda_4 P_0(t) \tag{3}$$

$$P_3{'}(t) + K_3 P_3(t) = \sum_{i=1}^{4} \mu_i P_{11+i}(t) + \lambda_4 P_1(t) \tag{4}$$

$$P_4{'}(t) + \mu_2 P_4(t) = \lambda_2 P_0(t) \tag{5}$$

$$P_5{'}(t) + \mu_3 P_5(t) = \lambda_3 P_0(t) \tag{6}$$

$$P_6{'}(t) + \mu_1 P_6(t) = \lambda_1 P_1(t) \tag{7}$$

$$P_7{'}(t) + \mu_2 P_7(t) = \lambda_2 P_1(t) \tag{8}$$

$$P_8{'}(t) + \mu_3 P_8(t) = \lambda_3 P_1(t) \tag{9}$$

$$P_9{'}(t) + \mu_2 P_9(t) = \lambda_2 P_2(t) \tag{10}$$

$$P_{10}{'}(t) + \mu_3 P_{10}(t) = \lambda_3 P_2(t) \tag{11}$$

$$P_{11}{'}(t) + \mu_4 P_{11}(t) = \lambda_4 P_2(t) \tag{12}$$

$$P_{12}{'}(t) + \mu_1 P_{12}(t) = \lambda_1 P_3(t) \tag{13}$$

$$P_{13}{'}(t) + \mu_2 P_{13}(t) = \lambda_2 P_3(t) \tag{14}$$

$$P_{14}{'}(t) + \mu_3 P_{14}(t) = \lambda_3 P_3(t) \tag{15}$$

$$P_{15}{'}(t) + \mu_4 P_{15}(t) = \lambda_4 P_3(t) \tag{16}$$

where

$$K_0 = (\lambda_1 + \lambda_2 + \lambda_3 + \lambda_4) \qquad K_1 = (\lambda_1 + \lambda_2 + \lambda_3 + \lambda_4 + \mu_1)$$

$$K_2 = (\lambda_1 + \lambda_2 + \lambda_3 + \lambda_4 + \mu_4) \qquad K_3 = (\lambda_1 + \lambda_2 + \lambda_3 + \lambda_4 + \mu_1 + \mu_4)$$

Since Textile Industry is process industry, it's every subsystem should be available for long period of time. So, long run availability of the system is computed by taking t→∞ and d/dt→ 0 applying on set of first order differential equations and solving them recursively we get:

$$P_1 = L_1 P_0 \qquad P_2 = L_2 P_0 \qquad P_3 = L_3 P_0 \qquad P_4 = H_2 P_0$$
$$P_8 = H_3 L_1 P_0$$

$$P_5 = H_3 P_0 \qquad P_6 = H_1 L_1 P_0 \qquad P_7 = H_2 L_1 P_0$$

$$P_9 = H_2 L_2 P_0 \qquad P_{10} = H_3 L_2 P_0 \qquad P_{11} = H_4 L_2 P_0 \qquad P_{12} = H_1 L_3 P_0$$

$$P_{13} = H_2 L_3 P_0 \qquad P_{14} = H_3 L_3 P_0 \qquad P_{15} = H_4 L_3 P_0$$

where

$$H_1 = \frac{\lambda_1}{\mu_1} \qquad H_2 = \frac{\lambda_2}{\mu_2} \qquad H_3 = \frac{\lambda_3}{\mu_3} \qquad H_4 = \frac{\lambda_4}{\mu_4}$$

$$N_1 = \mu_1 H_4 + \frac{\mu_1 \mu_1}{\mu_4} + \frac{\lambda_1 \mu_1}{\mu_4} + \mu_1 \qquad N_2 = \frac{\lambda_1 \lambda_1}{\mu_4} + \lambda_1 H_4 + \lambda_1 + \lambda_4 + \frac{\mu_1 \lambda_1}{\mu_4} - \lambda_4 \qquad L_1 = \frac{N_2}{N_1}$$

$$L_2 = \frac{\lambda_1}{\mu_4} + H_4 - \frac{\mu_1 L_1}{\mu_4} \qquad L_3 = H_4 L_1 + \frac{\mu_1 L_1}{\mu_4} - \lambda_1$$

Now using Normalizing condition, i.e., sum of all the state probabilities is equal to one, we get:

$$\sum_{i=0}^{15} P_i = 1$$

$$P_0 = [1 + L_1 + L_2 + L_3 + H_2 + H_3 + H_1L_1 + H_2L_1 + H_3L_1 + H_2L_2 + H_3L_2 + H_4L_2 + H_1L_3 + H_2L_3 + H_3L_3 + H_4L_3]^{-1}$$

Now, the Steady State Availability of the Fabric Finishing System may be obtained as the summation of all the working state probabilities, i.e.

$A_v = P_0 + P_1 + P_2 + P_3 = [1 + L_1 + L_2 + L_3]P_0$

4. Proposed advanced optimization algorithms

Two advanced optimization algorithms are considered in the present work for fabric finishing system's parameters optimization and are described in the following sections.

4.1 Teaching-learning-based optimization (TLBO)

TLBO is a teaching-learning process inspired algorithm proposed by Rao et al. (2011) based on the effect of influence of a teacher on the output of learners in a class. The algorithm mimics teaching-learning ability of teacher and learners in a classroom. Teacher and learners are the two vital components of the algorithm which describes two basic modes of the learning, through a teacher (known as teacher phase) and interacting with the other learners (known as learner phase).

The output in TLBO algorithm is considered in terms of results or grades of the learners which depend on the quality of the teacher. So, a teacher is usually considered as a highly learned person who trains learners so that they can have better results in terms of their marks or grades. Moreover, learners also learn from the interaction among themselves which also helps in improving their results.

TLBO is a population based method. In this optimization algorithm a group of learners is considered as a population; different design variables are considered as different subjects offered to the learners, and learners' results are analogous to the 'fitness' value of the optimization problem. In the entire population the best solution is considered as the teacher. The working of TLBO is divided into two parts, 'Teacher phase' and 'Learner phase'. The working of both phases is explained below.

4.1.1. Teacher phase

This is first part of the algorithm where learners learn through the teacher. During this phase a teacher tries to increase the mean result of the classroom from any value M_1 to his or her level (i.e. T_A). However, practically this is not possible and a teacher can move the mean of the classroom M_1 to any other value M_2 which is better than M_1, depending on his or her capability. Consider M_j to be the mean and T_i to be the teacher at any iteration i. Now T_i will try to improve existing mean M_j towards it so the new mean will be T_i designated as M_{new} and the difference between the existing mean and new mean is given by Rao et al. (2011),

Difference Mean$_i$ = r_i ($M_{new} - T_F M_j$)   (17)

where T_F is the teaching factor which decides the value of mean to be changed, and r_i is a random number in the range [0, 1]. The value of T_F can be either 1 or 2 which is a heuristic step and it is decided randomly with equal probability as,

TF = round [1 + rand(0, 1){2 − 1}]   (18)

Based on this Difference Mean, the existing solution is updated according to the following expression

$$X_{new,i} = X_{old,i} + \text{Difference Mean}_i \qquad\qquad (19)$$

4.1.2. Learner Phase

This is the second part of the algorithm where learners increase their knowledge by interaction among themselves. A learner interacts randomly with other learners for enhancing his or her knowledge. A learner learns new things if the other learner has more knowledge than him or her. Mathematically, the learning phenomenon of this phase is expressed below:

At any iteration i, considering two different learners X_i and X_j where $i \neq j$

$$X_{new,i} = X_{old,i} + r_i\,(X_i - X_j) \quad \text{If } f(X_i) < f(X_j) \qquad (20)$$
$$X_{new,i} = X_{old,i} + r_i\,(X_j - X_i) \quad \text{If } f(X_j) < f(X_i) \qquad (21)$$

Accept X_{new} if it gives better function value. Implementation steps of the TLBO are summarized below.

Step 1: Initialize the population (i.e. learners) and design variables of the optimization problem (i.e. number of
subjects offered to the learner) with random generation and evaluate them.

Step 2: Select the best learner as a teacher and calculate mean result of learners in each subject.

Step 3: Evaluate the difference between current mean result and best mean result according to Equation (i) by
utilizing the teaching factor (TF).

Step 4: Update the learners' knowledge with the help of teacher's knowledge according to Equation (iii).

Step 5: Update the learners' knowledge by utilizing the knowledge of some other learner according to Equations (iv)
and (v).

Step 6: Repeat the procedure from step 2 to 5 until the termination criterion is met.

5. Optimization results using TLBO algorithms

The performance behavior of the Fabric Finishing System (FFS) is highly influenced by the failure and repair parameters of each subsystem. These parameters ensure high performance of the Fabric Finishing System. The optimum value of system's performance (Availability) is 89.01%, for which the best possible combination of failure and repair rates is λ_1=0.0069, μ_1=0.036191, λ_2=0.004852, μ_2=0.040192, λ_3=0.000521, μ_3=0.007225, λ_4=0.0000643 and μ_4= 0.010099 at population size 160 as given in Table 1.

Table 1
Effect of Population Size on the Availability of FFS Using Teaching Learning Based Optimization (TLBO)

Population Size	Availability	λ_1	μ_1	λ_2	μ_2	λ_3	μ_3	λ_4	μ_4
40	0.8276	0.007147	0.034566	0.006459	0.040217	0.000344	0.007474	0.000883	0.018458
80	0.843580	0.008141	0.037917	0.005364	0.043723	0.000314	0.008529	0.000411	0.012968
120	0.845421	0.007192	0.033310	0.005741	0.040379	0.000401	0.009425	0.000604	0.023436
160	**0.890136**	**0.0069**	**0.036191**	**0.004852**	**0.040192**	**0.000521**	**0.007225**	**0.0000643**	**0.010099**
200	0.831758	0.007723	0.031681	0.004694	0.041389	0.000160	0.015491	0.000159	0.016469
240	0.823870	0.007366	0.038866	0.005645	0.042167	0.000075	0.015787	0.001120	0.022416

Availability Vs Population Size

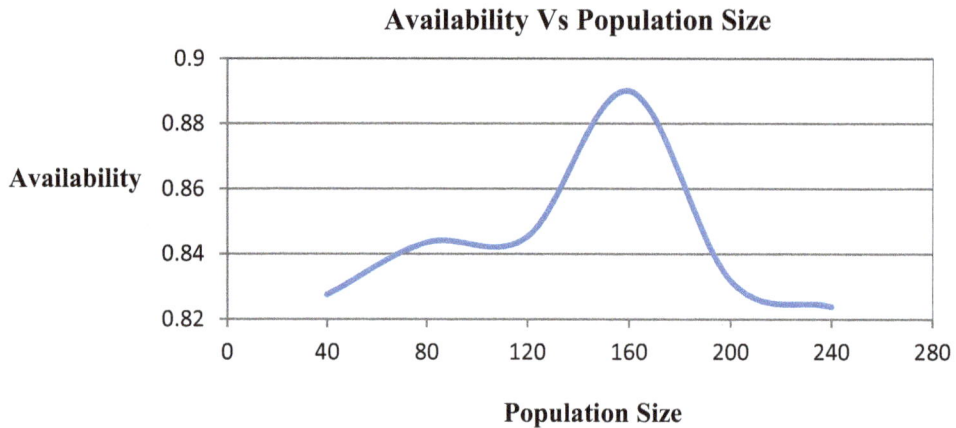

Fig. 4. Effect of Population Size on Fabric Finishing System Availability using TLBO

The effect of population size on availability of the Fabric Finishing System is shown in Fig.4. This Figure represents the optimal curve obtained by using the TLBO algorithm for availability and its comparison with the optimal curve obtained by using GA as shown in Fig. 5.

6. Performance optimization using genetic algorithm

Genetic Algorithm Technique (GAT) is hereby proposed to coordinate the failure and repair parameters of each subsystem for stable system performance, i.e., high availability. Here, the number of parameters is eight (four failure parameters and four repair parameters). The design procedure is described as follows: To use GAT for solving the given problem, the chromosomes are to be coded in real structures. Here, concatenated, multi-parameter, mapped, fixed point coding is used. Unlike, unsigned fixed-point integer coding parameters are mapped to a specified interval $[X_{min}, X_{max}]$, where X_{min} and X_{max} are the maximum and minimum values of system parameters. The maximum value of the availability function corresponds to the optimum values of system parameters. These parameters are optimized according to the performance index, i.e., desired availability level. To test the proposed method, failure and repair rates are determined simultaneously for optimal value of unit availability. Effects of population size on the availability of fabric finishing system are shown in Tables 3. To specify the computed simulation more precisely, trial sets are also chosen for GA and system parameters.

Table 2

Failure and Repair Rate Parameter Constraints

Parameter	λ_1	μ_1	λ_2	μ_2	λ_3	μ_3	λ_4	μ_4
Minimum	0.0069	0.0287	0.0052	0.0316	0.0001	0.0084	0.0005	0.0124
Maximum	0.0089	0.0387	0.0072	0.0416	0.0021	0.0184	0.0025	0.0224

The performance (availability) of the fabric finishing system is determined by the designed values of the system parameters as shown in Table 2.

Here, real-coded structures are used. The simulation is done to a maximum number of population size, which is varying from 40 to 240. The effect of population size on availability of the fabric finishing system is shown in Fig. 5.

The optimum value of system's performance (Availability) is 83.31%, for which the best possible combination of failure and repair rates is λ_1=0.006905, μ_1=0.0387, λ_2=0.005201, μ_2=0.0416, λ_3=0.000103, μ_3=0.0184, λ_4=0.000507 and μ_4= 0.0224 at population size 200 as given in Table 3.

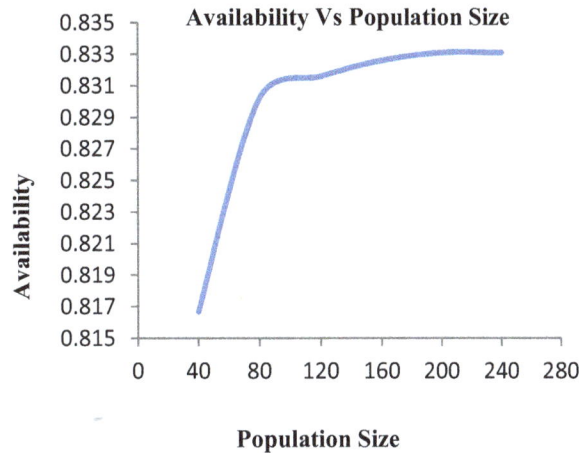

Fig. 5. Effect of Population Size on Fabric Finishing System Availability using GA

Table 3
Effect of Population Size on the Availability of FFS Using Genetic Algorithm

Population Size	Availability	λ_1	μ_1	λ_2	μ_2	λ_3	μ_3	λ_4	μ_4
40	0.8167	0.007469	0.037461	0.005242	0.041597	0.000152	0.018399	0.001141	0.0224
80	0.8302	0.006993	0.038699	0.005229	0.041599	0.000136	0.018398	0.00052	0.02239
120	0.8316	0.006919	0.038699	0.005205	0.0416	0.000131	0.018399	0.000502	0.022245
160	0.8326	0.006907	0.038698	0.005206	0.041599	0.000101	0.0184	0.00054	0.022398
200	**0.8331**	**0.006905**	**0.0387**	**0.005201**	**0.0416**	**0.000103**	**0.0184**	**0.000507**	**0.0224**
240	0.8331	0.006905	0.0387	0.005201	0.0416	0.000103	0.0184	0.000507	0.0224

Number of Generations=100; Reproduction (Crossover Fraction) = 0.80; Fitness Scaling: Rank.

7. Conclusions

The present work demonstrates the successful application of the TLBO algorithm for the single objective optimization of fabric finishing system of a textile industry. A set of optimal points is obtained by using the normalized objective function with the considered algorithm. The ability of the proposed algorithm is demonstrated by using a fabric finishing system and the performance of the proposed algorithm is compared with the performance of genetic algorithm (GA) approach. Improvement in the availability corresponding to 83.31% to 89.01% is obtained using the proposed TLBO algorithm, as compared to the GA approach showing the improvement potential of the algorithm for such industrial maintenance optimization. Unlike other population-based optimization techniques, TLBO does not require any algorithm parameters to be tuned which makes the implementation of TLBO simpler and easier.

Acknowledgment

The authors are very thankful to the management of Auro Textile Mills, Baddi, Himachal Pradesh, for granting the permission of industrial visits and also for informative discussions with engineers, maintenance personnel essential to attain optimum level of availability of the system concerned.

References

Cafaro, G., Corsi, F., & Vacca, F. (1986). Multi state markov models and structural properties of the transition rate matrix. *IEEE Transactions on Reliability, 35*(2), 192-200.

Çekyay, B., & Özekici, S. (2015). Reliability, MTTF and steady-state availability analysis of systems with exponential lifetimes. *Applied Mathematical Modelling, 39*(1), 284-296.

Chung, W.K. (1987). Reliability analysis of repairable parallel system with standby involving human error and common-cause failures. *Microelectronics Reliability, 27*(2), 269-271.

Coit, D.W., & Smith, A.E. (1996). Reliability optimization of series parallel systems using genetic algorithm. *IEEE Transactions on Reliability*, *45*(2), 254-260.

Fu, J.C. (1986). Reliability of large consecutive-K-out-of-N: F systems with k−1 step markov dependence. *IEEE Transactions on Reliability*, *35*(5), 602–606.

Fu, J.C., & Hu, B. (1987). On reliability of large consecutive-K-out-of-N: F systems with k-1 step markov dependence. *IEEE Transactions on Reliability*, *36*(1), 75-77.

Garg, D., Kumar, K., & Meenu (2010). Availability optimization for screw plant based on genetic algorithm. *International Journal of Engineering Science and Technology*, *2*(4), 658-668.

Garg, H., & Sharma, S.P. (2012). Behavior analysis of synthesis unit in fertilizer plant. *International Journal of Quality & Reliability Management*, *29*(2), 217 – 232.

Goyal, A., & Gupta, P. (2012). Performance evaluation of a multi-state repairable production system – a case study. *International Journal of Performability Engineering*, *8*(3), 330-338.

Gupta, P., Lal, A.K., Sharma, R.K., & Singh, J. (2005). Numerical Analysis of reliability and availability of the serial processes in butter-oil processing plant. *International Journal of Quality & Reliability Management*, *22*(3), 303 – 316.

Gupta, S., Tewari, P.C., & Sharma, A.K. (2009). Reliability and Availability analysis of ash handling unit of a steam thermal power plant. *South African Journal of Industrial Engineering*, *20*(1), 147-158.

Gupta, P. (2011). Markov modeling and availability analysis of a chemical production system-a case study. *Proceedings of the World Congress on Engineering*, Vol. I, WCE -2011, July 6 - 8, 2011, London, U.K.

Khanduja, R., Tewari, P.C., & Chauhan, R.S. (2009). Performance analysis of screening unit in a paper plant using genetic algorithm. *Journal of Industrial and Systems Engineering*, *3*(2), 140-151.

Khanduja, R., Tewari, P.C., Chauhan, R.S. & Kumar, D. (2010). Mathematical modeling and performance optimization for paper making system of a paper plant. *Jordan Journal of Mechanical and Industrial Engineering*, *4*(4), 487-494.

Khanduja, R., Tewari, P. C., & Chauhan, R.S. (2012). Performance modeling and optimization for the stock preparation unit of a paper plant using genetic algorithm. *International Journal of Quality Reliability and Management, 28*(6), 688-703.

Khanduja, R., Tewari, P. C. & Gupta, M. (2012). Performance enhancement for crystallization unit of sugar plant using genetic algorithm technique. *Journal of Industrial Engineering International, 28*(6), 688-703.

Krishnanand, K. R., Panigrahi, B. K., Rout, P. K., & Mohapatra, A. (2011, December). Application of multi-objective teaching-learning-based algorithm to an economic load dispatch problem with incommensurable objectives. In *International Conference on Swarm, Evolutionary, and Memetic Computing* (pp. 697-705). Springer Berlin Heidelberg.

Kumar, D., Singh, I.P., & Singh, J. (1988). Reliability analysis of the feeding system in the paper industry. *Microelectronics Reliability*, *28*(2), 213-215.

Kumar, D., Singh, J., & Pandey, P.C. (1989). Availability of a washing system in the paper industry. *Microelectronics Reliability*, *29*(5), 775-778.

Kumar, D, Singh, J., & Pandey, P.C. (1990). Cost analysis of a multi-component screening system in the paper industry. *Microelectronics Reliability*, *30*(3), 457-461.

Kumar, D., Singh, J., & Pandey, P.C. (1991). Behavioral analysis of a paper production system with different repair policies. *Microelectronics Reliability*, *31*(1), 47-51.

Kumar, D., Singh, J., & Pandey, P.C. (1992). Availability of the crystallization system in the sugar industry under common-cause failure. *IEEE Transactions on Reliability*, *41*(1), 85-91.

Kumar, R. (2014). Availability analysis of thermal power plant boiler air circulation system using Markov approach. *Decision Science Letters*, *3*(1), 65-72.

Kumar, P., & Tewari, P. (2017). Performance analysis and optimization for CSDGB filling system of a beverage plant using particle swarm optimization. *International Journal of Industrial Engineering Computations*, *8*(3), 303-314.

Lai, C. D., Xie, M., Poh, K. L., Dai, Y. S., & Yang, P. (2002). A model for availability analysis of distributed software/hardware systems. *Information and Software Technology*, *44*(6), 343-350.

Levitin, G., Xing, L., Amari, S. V., & Dai, Y. (2013). Reliability of non-repairable phased-mission systems with propagated failures. *Reliability Engineering & System Safety, 119*, 218-228.

Modgil V., Sharma, S. K., & Singh, J. (2013). Performance modeling and availability analysis of shoe upper manufacturing unit. *Int J Quality Reliability and Management, 30*(8), 816-831.

Niknam, T., Fard, A.K. & Baziar, A. (2012). Multi-objective stochastic distribution feeder reconfiguration problem considering hydrogen and thermal energy production by fuel cell power plants. *Energy, 42*, 563–573.

Niknam, T., Azizipanah-Abarghooee, R., & Narimani, M. R. (2012). An efficient scenario-based stochastic programming framework for multi-objective optimal micro-grid operation. *Applied Energy, 99*, 455-470.

Niknam, T., Azizipanah-Abarghooee, R., & Narimani, M. R. (2012). A new multi objective optimization approach based on TLBO for location of automatic voltage regulators in distribution systems. *Engineering Applications of Artificial Intelligence, 25*(8), 1577-1588.

Niknam, T., Golestaneh, F., & Sadeghi, M. S. (2012). θ-Multiobjective Teaching–Learning-Based Optimization for Dynamic Economic Emission Dispatch. *IEEE Systems Journal, 6*(2), 341-352.

Rao, R. V., Savsani, V. J., & Vakharia, D. P. (2011). Teaching–learning-based optimization: a novel method for constrained mechanical design optimization problems. *Computer-Aided Design, 43*(3), 303-315.

Rao, R. V., Savsani, V. J., & Vakharia, D. P. (2012). Teaching–learning-based optimization: an optimization method for continuous non-linear large scale problems. *Information Sciences, 183*(1), 1-15.

Rao, R., & Patel, V. (2012). An elitist teaching-learning-based optimization algorithm for solving complex constrained optimization problems. *International Journal of Industrial Engineering Computations, 3*(4), 535-560.

Venkata Rao, R., & Kalyankar, V. D. (2012). Parameter optimization of machining processes using a new optimization algorithm. *Materials and Manufacturing Processes, 27*(9), 978-985.

Rao, R.V., & Kalyankar, V.D. (2012). Multi-objective multi-parameter optimization of the industrial LBW process using a new optimization algorithm. In Proceedings of the Institution of Mechanical Engineers, Part B: *Journal of Engineering Manufacture, 226*(6), 1018–1025.

Rao, R. V., & Kalyankar, V. D. (2013). Parameter optimization of modern machining processes using teaching–learning-based optimization algorithm. *Engineering Applications of Artificial Intelligence, 26*(1), 524-531.

Rao, R. V., & Patel, V. (2013). Multi-objective optimization of heat exchangers using a modified teaching-learning-based optimization algorithm. *Applied Mathematical Modelling, 37*(3), 1147-1162.

Rao, R. V., & Patel, V. (2013). Multi-objective optimization of two stage thermoelectric cooler using a modified teaching–learning-based optimization algorithm. *Engineering Applications of Artificial Intelligence, 26*(1), 430-445.

Sabouhi, H., Abbaspour, A., Fotuhi-Firuzabad, M., & Dehghanian, P. (2016). Reliability modeling and availability analysis of combined cycle power plants. *International Journal of Electrical Power & Energy Systems, 79*, 108-119.

Singh, J., & Mahajan, P. (1999). Reliability of utensils manufacturing plant-a case study. *Opsearch, 36*(3), 260-269.

Tewari, P.C., Kumar, D., & Mehta, N.P. (2003). Decision support system of refining system of sugar plant. *Journal of Institution of Engineers (India), 84*, 41-44.

Togan, V. (2012). Design of planar steel frames using teaching–learning based optimization. *Engineering Structure, 34*, 225–232.

Wang, K.H., Yen, T.C., & Fang, Y.C. (2012). Comparison of availability between two systems with warm standby units and different imperfect coverage. *Quality Technology and Quantitative Management, 9*(3), 256-282.

9

Dynamic capacitated maximal covering location problem by considering dynamic capacity

Jafar Bagherinejad[a*] and Mahnaz Shoeib[b]

[a]Associate Professor, Department of Industrial Engineering, Alzahra University, Tehran, Iran
[b]MSc student of Industrial Engineering, Alzahra University, Tehran, Iran

CHRONICLE	ABSTRACT
	Capacitated maximal covering location problems (MCLP) have considered capacity constraint of facilities but these models have been studied in only one direction. In this paper, capacitated MCLP and dynamic MCLP are integrated with each other and dynamic capacity constraint is considered for facilities. Since MCLP is NP-hard and commercial software packages are unable to solve such problems in a rational time, Genetic algorithm (GA) and bee algorithm are proposed to solve this problem. In order to achieve better performance, these algorithms are tuned by Taguchi method. Sample problems are generated randomly. Results show that GA provides better solutions than bee algorithm in a shorter amount of time.
Keywords: *Capacitated MCLP* *Multi-period MCLP* *Dynamic capacity* *Genetic algorithm* *Bee algorithm*	

1. Introduction

Facility location problem is a special class of optimization problems whose primary goal is to locate a limited number of facilities that satisfy particular constraints (Máximo et al., 2017). Facility location problems have been studied widely during recent years due to their extensive application in real situations (Correia & Captivo, 2006). Location problems can be defined according to two factors; space (planning area) and time (time of location). Space and time issues have been taken into account in static facility location problems and dynamic facility location problems respectively (Boloori Arabani & Zanjirani Farahani, 2012). Boloori Arabani and Zanjirani Farahani (2012) classified different types of static and dynamic location problems that have been studied by the literature review. They studied multi-period facility location problem as a type of dynamic location problems. Static location problem considers only one period. If a time horizon is considered for more than one period, the location problem becomes dynamic (Canel et al., 2001). By considering a time horizon with more than one period, determining the appropriate time for facility location, specifying the best locations and better prediction of favorable and unfavorable fluctuations of demand in time horizon can be achieved; whereas single period models do not have these characteristics (Miller et al., 2006).

* Corresponding author
E-mail: jbagheri@Alzahra.ac.ir   (J. Bagherinejad)

Dynamic models can be classified into two categories: explicitly dynamic models and implicitly dynamic models. In explicitly dynamic models, in order to respond to changes in parameters over time, facilities are closed and opened in pre-specified times and locations. In implicitly dynamic models, all facilities are to be open in the beginning of time horizon and remain open throughout the time horizon. These models are considered to be dynamic because they try to consider changes in parameters such as demand changes over time (Current et al., 1998).

Due to largely capital outlaid, facility location problems are frequently long-term in nature. Facilities such as schools, hospitals and dams operate for decades (Current et al., 1998). While, some facilities such as buses may move around in order to meet the demands of the population (Datta, 2012). Decision makers should select locations and consider time of relocation of facilities which are able to response demand fluctuations over the time (Daskin et al., 1992). Therefore, in order to control probable fluctuations in the future as well as parameters fluctuations a dynamic model seems to be necessary (Boloori Arabani & Zanjirani Farahani, 2012). Theoretically, opening/closing of facilities could impose no cost (Hormozi & Khumawala, 1996). One of the objectives in facility location problem is to minimize the total cost for assigning facilities to satisfy the demand nodes (Jahantigh & Malmir, 2016).

One of the traditional location problems is covering location problem. Covering location problem seeks for a solution to cover a subset of customers by considering one or more objective (Davari et al., 2013). Although covering models are not new, due to their application in real cases especially for emergency service facilities, they have always been attractive topics for researchers (Fallah et al., 2009). Static models can be transformed to their equivalent dynamic models. In these models instead of single period, T periods are considered. Therefore, maximum covering location problem could be studeid as a multi-period and dynamic problem (Boloori Arabani & Zanjirani Farahani, 2012).

The rest of the paper is organized as follows: First, a concise literature review of covering problems and related issues are presented in Section 2. Section 3 is dedicated to the definition of the problem. The proposed solution algorithm is presented in Section 4 and numerical examples and parameter setting appear in Section 5. Moreover, results are analyzed and discussions are given in this section. Finally, to bring the paper to a close, conclusions and outlooks for potential future research are given in Section 6.

2. Literature review

Schilling et al. (1993) classified covering location problems in two categories named maximal covering location problem (MCLP) and set covering location problem (SCLP). In covering problems, a demand is said to be covered if at least one facility is located within a predefined distance of it. This predefined distance is often called coverage radius. The objective of SCLP is to cover all demand with the minimum number of facilities. Covering location problem was introduced for the first time by Hakimi (1965). The objective of that model was to determine the minimum number of polices that was necessary to cover demand nodes on a network of highways. SCLP was introduced formally by Toregas et al. (1971) and was extended slightly by Berlin and Liebman (1974). MCLP was introduced for the first time by Church and Revelle (1974). The objective of the MCLP is to locate a fixed number of facilities in such a way that the total covered demand is maximized.

The main assumption of covering location problems is that the facilities are uncapacitated (Salari, 2013). But, practically this assumption is not always valid (Pirkul & Schilling, 1991) and usually limits the applications of covering models (Current & Storbeck, 1988). Most service facilities are capacitated (Murray & Gerrard, 1998; Liao & Approach, 2008). Therefore some covering models considered a capacity constraint for facilities. Although incorporating capacity constraint in formulation of location problems is not difficult, it increases computational complexity. Therefore, most research efforts focus on improvement of solving method (Pirkul & Schilling, 1989). Current and Storbeck (1988) incorporated capacity limitation to MCLP and LSCP. There is a theoretical link between these models and capacitated plant location problem, the capacitated P-median location problem and generalized assignment problem. Small and moderately sized problems can be solved with existing solution methods. Theoretical links give insight into developing new heuristics for large sized capacitated covering problems. Pirkul and Schilling (1991) developed the model by considering workload limits on the facilities and quality of service delivered to the uncovered demand zones. Facility's workload limits the demand amount which a facility can serve. In this model, all demands are allocated to facilities regardless of whether they are in covering radius or not. The quality of service is modeled as the total distance from uncovered demand zones to the nearest facility. This model is solved by an approach based on the Lagrangian relaxation (Pirkul & Schilling, 1991). Haghani (1996) proposed a capacitated maximal covering location problem in which weighted covered demand is maximized and average distance from the uncovered demands to the located facilities is minimized. They solved the problem with two heuristic methods. The first one was based on greedy adding technique and the second one was based on Lagrangian relaxation. Correia and Captivo (2003) considered modular capacitated location problems. In this model, instead of considering only one fixed capacity level for each facility, they considered a discrete and limited set including available capacity levels. Capacity level of facility is selected from this set. This model can be applied in schools, warehouses and other public facilities. Griffin et al. (2008) proposed capacitated maximal covering location problem by considering three capacity levels for each facility. In their model, there is no composing relationship (such as that between the number of ambulances and emergency stations) between facilities' capacity levels. Yin and Mu (2012) proposed modular capacitated maximal covering location problem (MCMCLP) in two situations. In these models, it is assumed that each facility has a capacity which is related to number of vehicles assigned to that facility. Vehicles have a fixed capacity but the capacity of each facility is equal to the total capacity of vehicles assigned to that facility. In the first model, the number of vehicles is predefined but in the second model, number of vehicles as well as number of facilities is predefined. Yin and Mu (2012) stated that this is a static model and disregard dynamic factors such as daily population movement. On the other hand, since MCMCLP is NP-hard, proposing a heuristic for this problem is important (Yin & Mu, 2012). Although the papers surveyed above, considered capacitated MCLP in only one period, the concept of dynamic (multi-period) covering location problem is not new in the literature (Fazel Zarandi et al., 2013). Schilling (1980) is among the first researchers who considered dynamic maximal covering location problem. Also, other researchers have taken into account this problem. Fazel Zarandi et al. (2013) considered large scale dynamic MCLP. Dell'Olmo et al. (2014) proposed a multi period MCLP for the optimal location of intersection safety cameras on an urban traffic network. Vatsa and Jayaswal (2016) present a new formulation for a multi-period MCLP with server uncertainty, motivated by its relevance with respect to primary health centers.

In this paper, by integrating modular capacitated maximal covering location problem and multi-period maximal covering location problem, a developed model is proposed.

3. Problem definition

In the proposed model, a time horizon consisted of T periods is considered. The objective of this model is to find optimal location of q facilities in a time horizon in such a way that with locating at most p_t vehicles in period t, the maximum covering is achieved in the whole time horizon. This model can be applied in location facilities such as ambulance bases and vehicles such as ambulances. In this model, it is assumed that each vehicle has a fixed capacity (Yin & Mu, 2012) equal to maximum amount of demand that it can serve in each period. Capacity of each facility in each period is related to the number of vehicles stationed in that facility (Yin & Mu, 2012). So, facilities' capacity is changing periodically and we call it as dynamic capacity. For example, if capacity of ambulance is C and Z_{jt} is the number of ambulances located in ambulance base in location j in period t, the capacity of that ambulance base will be CZ_{jt}.

Another assumption is that potential locations are identical in all periods. In each potential location, only one facility can be located and if a facility were located in location j in period t, facility would serve in this location until the end of the time horizon. In other words, the cost/penalty of closing facilities is so high that prevents closing facilities (for instance buildings such as hospital). Since the importance of costs in public sectors is inconsequential compared to provided services (Fazel Zarandi et al., 2013), it is assumed that opening and closing of vehicles and relocation of them has no cost. Therefore, vehicles are closed at the end of each period and are relocated again in the next period (if they were available).

Since some vehicle may become unavailable in each period because of being out of use, etc. ($d_t < 0$) or some new facilities are added ($d_t > 0$), the number of vehicles are not considered to be identical in all periods. In especial situation, the number of facilities is identical in each period ($p_1 = p_2 = \cdots = p_T$). In the proposed model, each facility in each period is as a potential location for stationing of vehicles. If there is no facility in a period, there would be no potential location for stationing of vehicles. Therefore, to maintain the feasibility of the problem, the constraint on the number of vehicles in each period is considered as the maximum number of vehicles. The maximum number of vehicles is given in the beginning of each period and no limitation on the number of vehicles which can be stationed in a facility is considered. In this model, a constraint on the number of facilities is considered in the whole time horizon. If a constraint on the minimum number of new facilities which can be located in period t is not imposed, the minimum number of new facilities in period t will be zero ($m_t = 0$), if a constraint on the maximum number of new facilities which can be located in period t is not imposed, the maximum number of new facilities in period t will be q ($n_t = q$). q is the total number of facilities in the time horizon. In some periods, a constraint on the minimum or maximum number of new facilities located might be imposed in each period. In this situation, the decision maker determines the minimum number of facilities in each period in such a way that sum of minimum number of facilities in the time horizon would not exceed the total number of facilities and considers the maximum number of facilities in each period more than the minimum number of facilities in each period. It is assumed that the minimum and maximum number of new facilities in each period is certain and predefined. It is assumed that at the end of each time horizon, all facilities and vehicles are closed. So, in the beginning of each time horizon, no facilities are located in potential locations ($x_{j0} = 0$). Hereby, the proposed dynamic capacitated MCLP is presented. First, problem parameters and variables are defined.

Sets and parameters

i, I: The index and set of demand nodes
j, J: The index and set of eligible facility locations
t, T: The index and set of time periods
a_{it}: The population or demand at node i in period t
d: The Euclidean distance from demand node i to facility at j
S: The distance (or time) standard within which coverage is desired
N={j|d ≤ S}: the set of nodes that are within a distance less than S from node i
p_t: Maximum number of vehicles in period t
q: The number of facilities to be located throughout the time horizon
m_t: Minimum number of new facility in period t ($\sum_{t=1}^{T} m_t \leq q$)
n_t: Maximum number of new facility in period t ($n_t \geq m_t$)
c: capacity of each vehicle

Variables

x_{jt} : A binary variable which equals one if a facility is sited at location j in period t.
y_{it}: A binary variable which equals one if node i in period t is covered by one or more facilities stationed within a distance of S.
z_{jt}: An integer variable which indicates the number of vehicles which are located in period t and site j. ($0 \leq Z_{jt} \leq p_t$).

Then, the proposed model will be as follows:

$$MaxZ = \sum_{t=1}^{T}\sum_{i=1}^{I} a_{it}\, y_{it} \tag{1}$$

$$\sum_{j=1}^{J}\sum_{t=1}^{T}(x_{jt} - x_{jt-1}) = q \tag{2}$$

$$\sum_{j=1}^{J} Z_{jt}x_{jt} \le p_t \qquad\qquad for\ \forall t \tag{3}$$

$$m_t \le \sum_{j=1}^{j}(x_{jt} - x_{jt-1}) \le \min\{n_t, q\} \qquad\qquad t = 1 \tag{4}$$

$$m_t \le \sum_{j=1}^{j}(x_{jt} - x_{jt-1}) \le \min\{n_t,\ \left(q - \sum_{j=1}^{J}\sum_{t=1}^{t-1}(x_{jt} - x_{jt-1})\right)\} \quad t = 2 \dots T \tag{5}$$

$$Y_{it}a_{it} \le \sum_{j\in N_i} C\, X_{jt}Z_{jt} \qquad\qquad for\ \forall t, i \tag{6}$$

$$\sum_{j=1}^{J} Z_{jt}x_{jt}C \ge \sum_{i=1}^{I} a_{it}\, y_{it} \qquad\qquad for\ \forall t \tag{7}$$

$$x_{jt-1} \le x_{jt} \qquad\qquad \forall j, t \tag{8}$$

$$Z_{jt} \le P_t x_{jt} \qquad\qquad \forall j, t \tag{9}$$

$$x_{jt}, y_{it} \in \{0,1\} \qquad\qquad \forall j, i, t \tag{10}$$

$$0 \le Z_{jt} \le P_t\ and\ int \qquad\qquad \forall j, t \tag{11}$$

The objective function (1) maximizes the overall covered demand. Constraint (2) shows that q facilities are to be established in T periods. Constraint (3) specifies the maximum number of vehicles to be located in each period. Constraint (4) assures that if minimum or maximum number of new facilities in period 1 be predefined, the number of new facilities in period 1 ($t=1$) will be in the predetermined interval. Otherwise, it will be between zero and q (the constraint on the minimum and maximum number of new facilities in $t=1$). Constraint (5) specifies that if minimum or maximum number of new facilities for $t>1$ be predefined, the number of new facilities will be in the predetermined interval. Otherwise, it will be between zero and the number of available facilities (number of facilities is not located until period t). Constraint (6) specifies that the demand point i in period t is covered, if it does not exceeds the total capacity of facilities which can cover this demand point. Constraint (7) ensures that the total covered demand in each period cannot exceed total capacity of facilities in that period. Constraint (8) specifies that if a facility is located in period t, it will remain open until the end of the time horizon. Constraint (9) specifies that if a facility is located in period t in site j, the number of vehicles assigned to this facility cannot exceed p_t (the maximum number vehicles in period t). In other words, in each period vehicles can be stationed in site j if a facility is located in that site and this cannot be more than the number of vehicles predefined for each period. Constraint (10) specifies that decision variables x_{jt} and y_{it} are binary. Constraints (11) restrict the integer decision variable z_{jt}, which ranges from zero to p_t.

Linearization of the proposed model

If we have a non-linear constraint in the form of $y \le min(x_1, x_2)$, it could be linearized by definition of a binary variable (δ) and a large enough positive value ($G \ge q$) as follows (linearization of constraints (4) and (5) by Eq. (12-15)):

$$x_1 \le x_2 + G\delta \tag{12}$$

$$x_2 \le x_1 + G(1 - \delta) \tag{13}$$

$$y \le x_2 + G(1 - \delta) \tag{14}$$

$$y \le x_1 + G\delta \tag{15}$$

By consdering $z_{jt}x_{jt} = O_{jt}$, constraints (3), (6) and (7) are linearized as follows (linearization of constraints (3), (6) and (7) by Eq. (16-19)):

$$O_{jt} \le z_{jt} \tag{16}$$

$$O_{jt} \ge z_{jt} - p_t \times (1 - x_{jt}) \tag{17}$$

$$O_{jt} \le p_t \times x_{jt} \tag{18}$$

$$0 \le O_{jt} \le p_t \text{ , } int \tag{19}$$

4. Solution methods

4.1. Genetic algorithm (GA)

4.1.1. Review of GA

GA was first proposed by Holland (1975) as an evolutionary algorithm. It is based on Darwinian evolution: good traits survive and mix to form new while the bad traits are eliminated from the population (Zanjirani Farahani & Hekmatfar, 2009). Beasley and Chu (1996) seem to be the first to apply GA for covering model. The Simple GA is as follows:

1. Generate an initial population mostly in a random way.
2. Select individuals for reproduction.
3. Perform genetic operations and generate a new generation.
4. Insert offspring into population and form the new population.
5. If the predefined stopping criteria are met, stop the algorithm, otherwise, return to step 2.

The rest of this section is devoted to elaboration of the proposed GA.

4.1.2. Encoding scheme

An appropriate chromosome representation must be defined for a GA. Encoding very depends on the problem. Most previously adopted representations, such as the bit string, are linear or one-dimensional. Some real problems are naturally suitable for two-dimensional representation. In this paper, GA with multi chromosome representation is applied and a separate chromosome is considered for each variable (Matthias et al., 2013). Therefore, three chromosomes are defined which are two-dimensional. The first chromosome is binary matrix which has I rows and T columns. Each bit in this matrix represents the status (covered/uncovered) of the node i in period t. The second chromosome is a binary matrix which has J rows and T column and each bit in this matrix represents the status (located/unlocated) of facility at location j in period t. In other words, one value in the ith bit means that there is a facility at location j in period t. The third chromosome is a matrix consisted of integer number which has J rows and T column and each bit in this matrix represents the number of vehicles at location j in period t. Each bit of third chromosome is an integer number between zero and p_t in such a way that total number of vehicles in each period is not more than p_t. Therefore, an upper and lower bound are considered for the number of vehicles in each period. At first, an initial population is generated randomly for each chromosome.

4.1.3. Selection

Selection is the stage of GA in which chromosomes are selected from the population to be parented for the next generation. There are many methods in selecting the best chromosomes. In this paper, the Roulette Wheel Selection (RWS) has been used as the selection method. Each individuals of the population is allocated a section of an imaginary roulette wheel. The size of each sections is directly proportional to their fitness values, such that the fittest individual has the biggest slice of the wheel and the weakest individual has the smallest. The wheel is then spun and the individual associated with the winning section is selected.

4.1.4. Crossover

Crossover and mutation are two basic genetic operators used to make new off springs. Type and implementation of operators depend on encoding and also on a problem. In this paper, crossover in first and second chromosomes is performed using one of these two methods: one-point crossover and two-point crossover. But in the third chromosome, the arithmetic crossover is applied. Fig.1 shows different types of crossover.

4.1.4.1. One-point crossover

One-point crossover randomly select a crossover point and then copy everything before this point from the first parent and then everything after the crossover point copy from the second parent. Here chromosomes are two-dimensional and crossover point can be a bit or column. In one-point crossover (column), third column in each parent is selected randomly and exchanged. In one-point crossover (bit), two bits in each parent selected randomly and exchanged (Fig.1).

4.1.4.2. Two-point crossover

Two-point crossover randomly selects two crossover points within a chromosome then interchanges the two parent chromosomes between these points to produce two new offspring. In Fig.1, in the case of two point crossover first and forth columns in each parent are selected and they are swapped.

4.1.4.3. Arithmetic crossover

Arithmetic crossover is a linear combination of two chromosomes as follows:

$$C_i^{new} = \alpha.C_i^{old} + (1-\alpha)C_j^{old} \tag{20}$$

$$C_j^{new} = (1-\alpha)C_i^{old} + \alpha C_j^{old}, \ 0 \le \alpha \le 1 \tag{21}$$

where C^{old} is a parent, C^{new} is an offspring, and α is random matrix (between 0 and 1). This operation is performed for each bit (Köksoy & Yalcinoz, 2008). In Fig.1, random matrix is given. The number of bits are selected randomly. According to the Eq. (20) and Eq. (21) arithmetic crossover is operated.

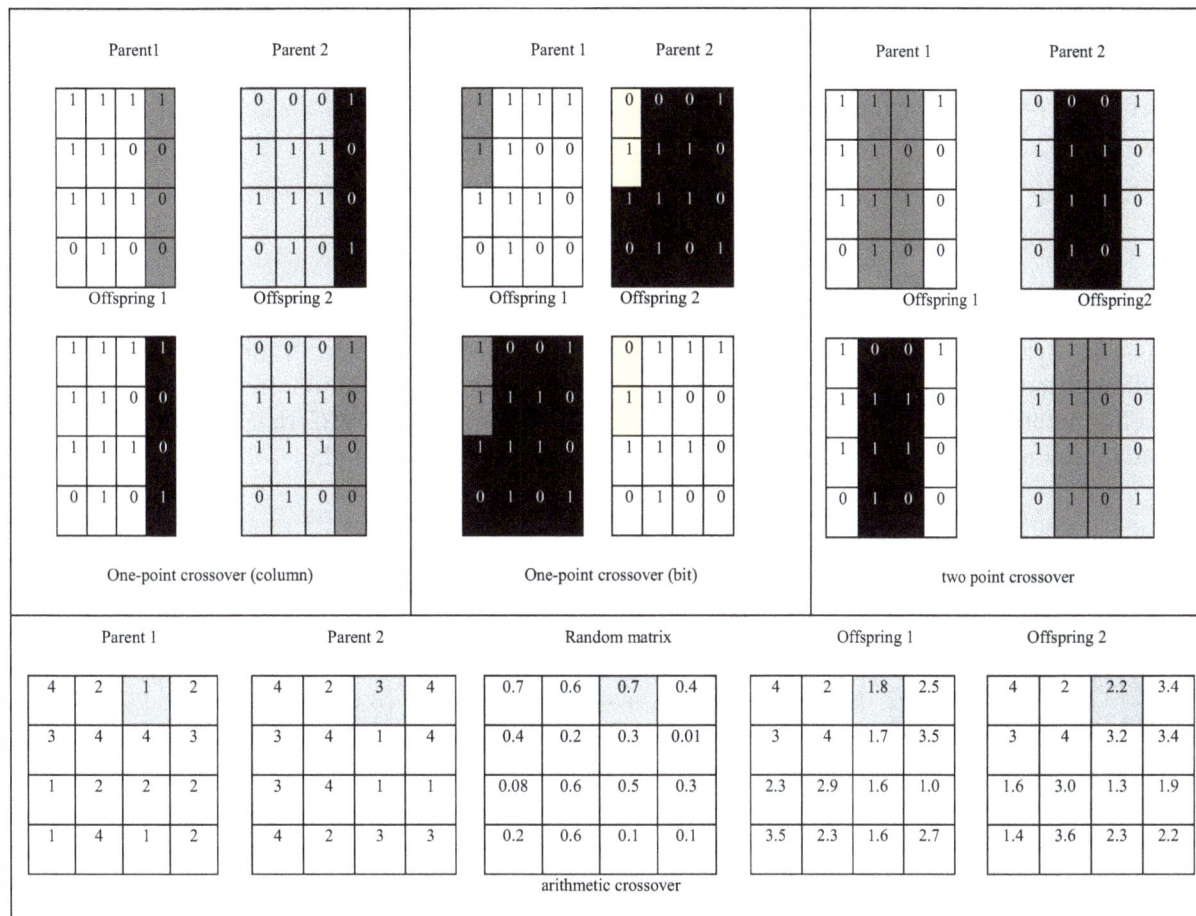

Fig. 1. Crossover

4.1.5. Mutation

Mutation is the genetic operator that randomly changes one or more of the chromosome's gene. The purpose of the mutation operator is to prevent the genetic population from converging to a local minimum and to introduce to the population of new possible solutions. The mutation is carried out according to the mutation probability. Mutation rate is usually set to a very low level. However, different references have found that a higher mutation rate is necessary when the GA has converged (Jaramillo et al., 2002). In this paper, crossover in first and second chromosomes is performed using one of these methods: swap, binary and reversion mutations. In third chromosome, the mutation method is only arithmetic mutation. Fig. 2 shows different types of mutation.

4.1.5.1. Binary mutation

In Binary mutation, some bits are selected randomly and their values are changed from zero to one and one to zero. In Fig. 2, in case of binary mutation, colored bits are selected randomly and binary mutated.

4.1.5.2. Swap mutation

In Swap mutation, two bits are selected randomly and are exchanged with each other. In Fig.2, in the case of swap mutation, colored bits are selected randomly and swapped.

4.1.5.3. Reversion mutation

In reversion mutation, two bits are selected randomly and bits between them are reversed. In Fig. 2, in the case of reversion mutation, colored bits are selected randomly and reversed.

4.1.5.4. Arithmetic mutation

The definition of arithmetic mutation is as follows:

$$
\begin{cases}
C_i^{new} = C_i^{old} + \alpha_i & \text{If } C_i^{new} \leq C_i^{old} + \alpha_i \\
u_i & \text{If } C_i^{new} > C_i^{old} + \alpha_i
\end{cases}
\tag{22}
$$

where u_i is upper bound, C^{old} is a parent, C^{new} is an offspring, and α is random matrix (between 0 and u_i). At each iteration, offspring as many as the population size are created. In order to produce new generation, these offspring replace less fit individuals in the existing population. In Fig. 2, in the case of arithmetic mutation, upper bound and random matrix is given ($u_i = 10$). Certain number of bits are selected randomly. According to the Eq. (22) arithmetic mutation is operated.

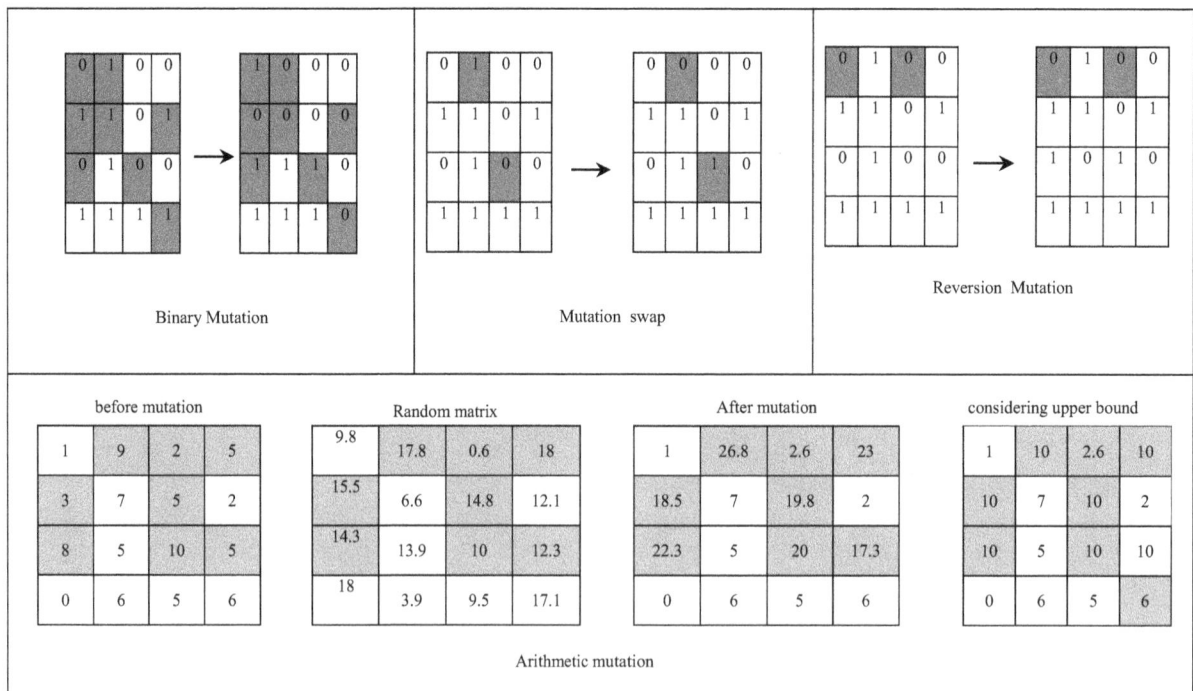

Fig. 2. Mutation

4.1.6. Termination criteria

The algorithm will iterate until the maximum number of iterations is attained.

4.2. Bee algorithm

4.2.1. Review of Bee algorithm

Bee algorithm is one of the swarm-based algorithms which imitate the food foraging behavior of honeybee swarms (Tsai, 2014). The basic version of the algorithm performs a kind of local search combined with random search and it can be used for combinatorial optimization and functional optimization (Pham et al., 2006). The foraging process begins in a colony by scout bees being sent to search for promising flower patches. At first, these bees spread out in flower patches randomly. Each selected food source represents a feasible solution of the problem. The nectar amount in the food source represents the quality of the problems' solution. When the foraging process is finished, based on the nectar amount, food sources are ranked into three categories: elite, good (average) and bad (unselected) sites. Each scout bee performs a dancing known as the waggle dance above a certain quality threshold deposit their nectar. This way, it transfers information of that region (in comparison to the hive), its distance from the hive and its quality rating to other bees. This information helps the colony to send follower bees to flower patches. Most follower bees go to region where are more promising to find nectar. In other words, more follower bees are assigned to elite sites than those of good sites. Assigning follower bees is as generating a neighbor for a solution. In each iteration of the algorithm for each elite site, a certain number of neighbors are generated. In this paper, the method for generating neighbors is similar to mutation methods in GA. The fitness of these neighbors is calculated and for each site the best neighbor is recognized. The quality of the best neighbor is compared to that of current bee. If the quality of the best neighbor bee is better than that of current bee, it will be replaced. Otherwise, current bee stays in its site with no movement. This process is performed on the good site. The difference is that the number of neighbors generated for good site is less than those of elite sites. Bad sites are abandoned and scout bees assigned to bad sites are assigned to other sites randomly. Therefore, the new population is generated. This process continues utile the stop criteria is met. In this algorithm, chromosomes defined in GA are considered as bee and Mutation operator is performed to generate neighbors. Bee algorithm is as follow:

1. The bees of initial population are randomly generated (n: number of initial bees).
2. The fitness are calculated.
3. Certain number of best bees (m) are selected to finding neighborhood. Among selected bees (m), a certain number of them (e) are considered as elite bees. The rest of them (m-e) are good bees.
4. By mutation method, the neighborhood are generated for elite and good bees and their fitness are calculated. The number of neighborhood are generated for each elite bees are more than good bees.
5. Among neighborhoods of each bees, the best bee is selected. These bees are transferred to the next generation.
6. Good bees (n-m) are used to random search and their fitness are calculated. These bees also transmitted to the next generation.
7. If a stop condition is met, the algorithm stops. Otherwise, go to step 3.

4.2.2. Termination criteria

The algorithm will iterate until the maximum number of iterations is attained.

5. Numerical examples

5.1. Test problems

To generate test problems, a similar approach to Revelle et al. (2008) was employed. In this approach, the locations of nodes were randomly generated using a uniform distribution between 0 and 30 for both x and y coordinates. The distances between the nodes are then defined as their Euclidean distance. Populations on the demand nodes for each time period were randomly generated using a uniform distribution between 0 and 100. Fazel Zarandi et al. (2013) by considering 5 periods use this approach for generating dynamic sample problems. This paper by considering time scale uses this method for generating 30 sample problems. The minimum number of new facilities in each period is randomly generated using a uniform distribution between 0 and q in such a way that sum of minimum number of new facilities in the whole planning horizon be less than or equal to q ($\sum_t^T m_t \leq q$). The maximum number of new facilities in each period is generated by random numbers larger than or equal to the minimum number of new facilities in each period ($n_t \geq m_t$). Lingo 8.0 was used to solve these problems and results were compared against those obtained using the GA and bees algorithm.

5.2. Parameter setting

Heuristic and metaheuristic algorithms are sensitive to their parameters; A small change can affect the quality of the solution. So, tuning algorithms are necessary to find a better solutions (Pasandideh et al., 2015). The most widely used method to tune the algorithms is a full factorial design (Chan et al., 2015). This method does not seem effective when the number of parameters significantly increases, since it requires arduous task to conduct the experiment. A family of matrices is used to reduce the number of experiments. In Taguchi method, we utilize the orthogonal arrays to investigate a large number of decision variables with a small number of experiments (Raju et al., 2014).

In this method, factors are classified into two groups: Controllable factors (signal) and uncontrollable factors (noise). Also, objective functions are categorized into three groups: "the smaller the better", "the larger the better" and "the nominal value is expected". The objective functions of the proposed model is "the larger the better". S/N ratio (the large the better) is calculated by Eq. (23).

$$S/N\ ratio\ =\ -10\log\left(\frac{1}{n}\sum_{i=1}^{n}\frac{1}{x_i^2}\right), \tag{23}$$

where x_i = observed response value and n= number of replications. According to Taguchi's design of experiments, for 4 parameters and 3 levels L_9 Taguchi orthogonal array was selected (Table 1 and Table 2). For calibration of each algorithm, 6 sample problems are iterated for 5 times in each scenario. Since sample problems' dimension is not identical, so the differences between their objective functions are large and using raw data cause to wrong analyses. In other words, the dimension of the problem should be excluded form data. So after conversion of raw data to relative deviation index (RDI), the S/N ratio is calculated. Relative deviation index is calculated by Eq. (24):

$$RDI_{ijk} = \begin{cases} \left|\dfrac{OF_{ijk} - u_i}{u_i - l_i}\right| & u_i \neq l_i \\ 0 & u_i = l_i \end{cases} \quad i = 1..6, \quad j = 1..5, k = 1..9 \tag{24}$$

where OF_{ijk} is the objective function value related to iteration j in sample problem i in scenario k. l_i and u_i are minimum and maximum values for ith sample problem respectively. The S/N rate for scenario k can be calculated (using relative deviation index of objectives function) by Eq. (25).

$$S/N_k = \begin{cases} -10log\left(\dfrac{1}{n}\sum_{i=1}^{I}\sum_{j=1}^{J}\dfrac{1}{RDI_{ijk}^2}\right) & RDI_{ijk} \neq 0 \quad k = 1...9 \\ 0 & RDI_{ijk} = 0 \end{cases} \tag{25}$$

In Taguchi method, S/N rate is considered as the first criterion. There could be no meaningful difference between different S/N levels, so, another criterion named RDI_k is introduced for scenario k which is calculated by Eq. (23). $\overline{RDI}_k$ is considered as the smaller the better.

$$\overline{RDI}_k = \frac{1}{n}\sum_{i=1}^{I}\sum_{j=1}^{J}RPI_{ijk} \quad k = 1...9 \tag{26}$$

It is time consuming to set all effective parameters in bee algorithm. Therefore, we set the most important effective parameters and other parameters have been determined by try and error. According to S/N (Fig. 3 and Fig. 5) and RDI (Fig. 4 and Fig. 6) the selected levels are colorful.

Table 1
Parameter and their levels in GA

Parameters	Level 1	Level 2	Level 3
% crossover	0.6	0.65	**0.7**
% mutation	0.2	0.25	**0.3**
Population size	50	**100**	150
Max iteration	30	50	100

Table 2
Parameters and their levels in Bee algorithms

Parameters	Level 1	Level 2	Level 3
% good site	0.6	0.65	**0.7**
% elite site	0.01	0.1	**0.3**
n scout bee	10	50	**80**
Max iteration	20	30	**50**

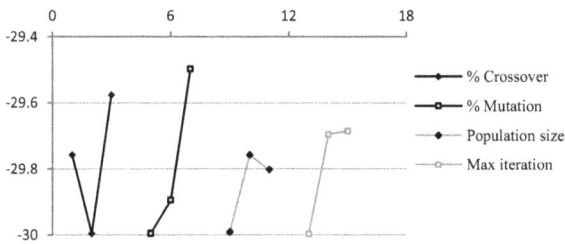

Fig. 3. S/N ratio (GA)

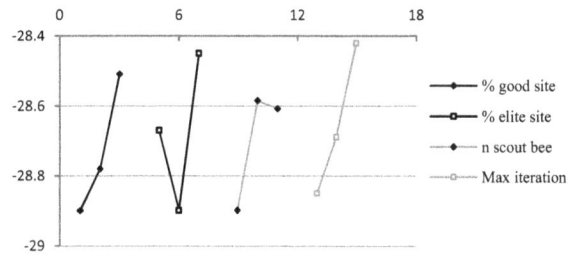

Fig. 5. S/N ratio (bee algorithm)

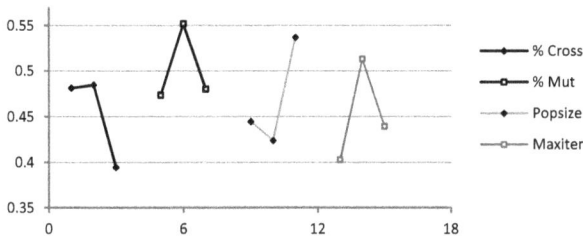

Fig. 4. RDI (GA)

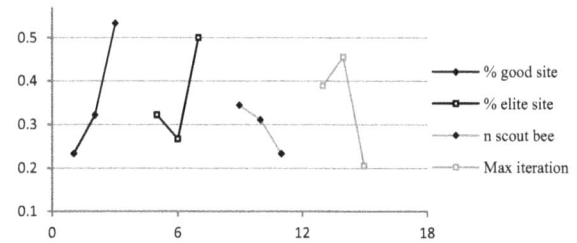

Fig. 6. RDI (bee algorithm)

5.4. Results and discussions

In this paper, Lingo software package has been applied to find the exact solutions for some sample problems. Lingo uses branch and bound method to solve the problems. Objective bound specifies the theoretical bound of objective function. This bound specifies how much a solver can improve the objective function. The best solution cannot exceed the objective bound. Colored rows specify problems in which Lingo cannot find the optimal solution in one hour. In such cases instead of optimal value, the objective bound and the best solution found in one hour is reported (Niroomand, 2008). In such cases, metaheuristic/heuristic algorithms might find better solution than what lingo finds in one hour. In such situation, the gap will be negative (Mehdizadeh & Afrabandpei, 2012). Gap is calculated as follow:

$$Gap = \frac{Exact\ solution - Metaheuristic\ solution}{Exect\ solution} \times 100 \qquad (27)$$

According to computational results of 30 sample problems, it could be concluded that Lingo can find the optimal solution for only one thirds of the problems. In half of the sample problems, GA finds a solution better than or equal to what lingo finds in one hour ($Gap_{GA} < 0$). In other problems except one problem, GA can achieve a gap less than 1.9%. The run time of GA in the largest problems is less than 2.5 minutes. Bee algorithm finds a solution better than or equal to what lingo finds in one hour in less than half of the sample problems ($Gap_{BA} < 0$). In other problems except two problems, Bee algorithm can achieve a gap less than 2.9%. The run time of Bee algorithm is less than 4 minutes in the largest problems. Totally, it can be stated as following:

$$Time_{GA} < Time_{BA}$$
$$Gap_{GA} \leq Gap_{BA}$$

Therefore, GA can find better solution in a shorter time. The average run time and objective function value in 5 iterations is reported in Table 3.

Table 3
Computational results

Sample problems				Lingo		GA			BA		
I	J	T	Z*	Objective bound	Time (s)	Z*	Time (s)	Gap (%)	Z*	Time (s)	Gap (%)
		3	550	-	197	548.8	14.33	0.21	545.2	21.96	0.87
100	50	5	950	-	316	932.6	22.85	1.83	922.2	35.11	2.92
		7	1297	1300	2600	1271.2	35.03	1.98	1240.4	48.38	4.36
		3	550	-	337	547.8	15.09	0.4	547	23.29	0.54
100	100	5	1000	-	2816	991.8	24.17	0.82	986.2	41.32	1.38
		7	1177	1200	3600	1163.8	33.26	1.12	1141.8	50.29	2.99
		3	550	-	319	549	25.18	0.18	546.6	35.55	0.61
150	150	5	999	1000	3600	992.2	40.93	0.68	985.6	56.97	1.34
		7	1119	1200	3600	1144.6	58.09	-2.28	1160.4	79.24	-3.69
		3	550	-	548	549.6	28.68	0.07	548.6	43.31	0.25
200	100	5	996	1000	3600	995.2	46.85	0.08	987.4	70.82	0.86
		7	1099	1200	3600	1150.4	65.10	-4.67	1145	98.96	-4.18
		3	550	-	945	549.2	35.94	0.14	546.6	49.63	0.61
200	200	5	986	1000	3600	992.6	58.79	-0.66	991.2	81.25	-0.52
		7	1108	1200	3600	1141.6	83.86	-3.03	1147.8	112.94	-3.59
		3	550	-	1153	549.8	50.86	0.036	549.4	70.18	0.10
250	250	5	888	1000	3600	996.4	83.47	-12.2	991.2	115.8	-11.62
		7	1096	1200	3600	1169.2	116.99	-6.67	1156.2	162.06	-5.49
		3	550	-	1193	549.2	43.13	0.14	547.6	65.42	0.43
300	100	5	999	1000	3600	995.6	57.34	0.34	982.6	107.89	1.64
		7	1195	1200	3600	1153.4	108.97	4.98	1132	150.01	5.27
		3	549	550	3600	549.2	52.44	-0.03	549	77.69	0
300	200	5	879	1000	3600	989.6	86.79	-12.58	981	128.81	-11.60
		7	1027	1200	3600	1147.8	121.14	-11.76	1143.2	180.71	-11.31
		3	548	550	3600	550	68.03	-0.36	548.8	91.92	-0.14
300	250	5	926	1000	3600	992.4	117.26	-7.17	991.6	151.01	-7.08
		7	993	1200	3600	1166	151.51	-17.42	1161.4	211.24	-16.95
		3	550	-	1760	550	70.71	0	549	104.14	0.18
300	300	5	980	1000	3600	991.8	117.34	-1.20	993	137.22	-1.32
		7	1037	1200	3600	1161	131.00	-11.95	931.8	192.7	10.14

C=10, S=10, q=10, $p_t \in \{10, 20, 25\}$

6. Conclusion and future research areas

In this paper, capacitated MCLP and dynamic MCLP were integrated to each other and dynamic capacity constraint was considered for facilities. Therefore, the MCLP has been extended to the capacitated dynamic MCLP. The developed model was solved by GA and bee algorithm and the results were compared to the exact solutions of Lingo. We have shown that while GA and bee algorithm are superior to the exact method in terms of runtime, there are negligible errors compared to the optimal solutions. GA found better solutions in a shorter amount of time than the bee algorithm. Although GA shows great performance to solve this model, one may assess the performance of other methods in finding solutions to this problem. Another opportunity for research is to add a constraint on the number of vehicles which can be located in each facility. A possible future study could be to integrate this model with gradual covering location problem. Another future research is to consider cost for each vehicle. Cost can be dynamic and changes in each period. Objective function could be maximization covered demands while the cost of vehicles is minimized. Some parameters can be fuzzy such as covering radius. Covering radius can be dynamic, too.

Acknowledgement

The authors would like to thank the anonymous referees for constructive comments on earlier version of this paper.

References

Beasley, J. E., & Chu, P. C. (1996). A genetic algorithm for the set covering problem. *European Journal of Operational Research*, *94*(2), 392–404.

Berlin, G. N., & Liebman, J. C. (1974). Mathematical analysis of emergency ambulance location. *Socio-Economic Planning Sciences*, *8*(6), 323–328.

Boloori Arabani, A., & Zanjirani Farahani, R. (2012). Facility location dynamics: An overview of classifications and applications. *Computers & Industrial Engineering*, *62*(1), 408–420.

Canel, C., Khumawala, B. M., Law, J., & Loh, A. (2001). An algorithm for the capacitated , multi-commodity multi-period facility location problem. *Computer & Operation Research*, *28*(5), 411–427.

Chan, K. Y., Rajakaruna, N., Engelke, U., Murray, I., & Abhayasinghe, N. (2015). Alignment parameter calibration for IMU using the Taguchi method for image deblurring. *Measurement*, 65(Apr 2015), 207-219.

Church, R., & Revelle, C. (1974). The maximal covering location problem. *Papers in Regional Science*, *32*(1), 101–118.

Correia, I., & Captivo, M. E. (2003). A lagrangean heuristic for a modular capacitated location problem. *Annal of Opeation Research*, *122*(1-4), 141–161.

Correia, I., & Captivo, M. E. (2006). Bounds for the single source modular capacitated plant location problem. *Computers & Operations Research*, *33*(10), 2991–3003.

Current, J. R., & Storbeck, J. E. (1988). Capacitated covering models. *Environment and Planning B: Planning and Design*, *15*(2), 153–163.

Current, J., Ratick, S., & Revelle, C. (1998). Dynamic facility location when the total number of facilities is uncertain: A decision analysis approach. *European Journal of Operational Research*, *110*(3), 597–609.

Daskin, M. S., Hopp, W. J., & Medina, B. (1992). Forecast horizons and dynamic facility location planning. *Annals of Operations Research*, *40*(1), 125–151.

Datta, S. (2012). Multi-criteria multi-facility location in Niwai block, Rajasthan. *IIMB Management Review*, *24*(1), 16–27.

Davari, S., Fazel Zarandi, M. H., & Turksen, I. B. (2013). A greedy variable neighborhood search heuristic for the maximal covering location problem with fuzzy coverage radii. *Knowledge-Based Systems*, 41(March 2013), 68–76.

Dell'Olmo, P., Ricciardi, N., & Sgalambro, A. (2014). A multiperiod maximal covering location model for the optimal location of intersection safety cameras on an urban traffi0c network. *Procedia-Social and Behavioral Sciences*, *108*, 106–117.

Fallah, H., Naimi Sadigh, A., & Aslanzadeh, M. (2009). *Covering problem*, in: Zanjirani Farahani, R., & Hekmatfar, M. (Eds.), Facility Location: Concepts, Models, Algorithms and Case studies. Berlin: Springer-Verlag , pp. 145-176.

Fazel Zarandi, M. H., Davari, S., & Haddad Sisakht, S. A. (2013). The large-scale dynamic maximal covering location problem. *Mathematical and Computer Modelling*, *57*(3), 710–719.

Griffin, P. M., Scherrer, C. R., & Swann, J. L. (2008). Optimization of community health center locations and service offerings with statistical need estimation. *IIE Transactions*, *40*(9), 880–892.

Haghani, A. (1996). Capacitated maximum covering location models: Formulations and solution procedures. *Journal of Advanced Transportation*, *30*(3), 101–136.

Hakimi, S. L. (1965). Optimum distribution of switching centers in a communication network and some related graph theoretic problems. *Operations Research*, 13(3), 462-475.

Holland, J. H. (1975). *Adaptation in natural and artificial systems: An introductory analysis with application to biology, control, and artificial intelligence.* Ann Arbor: University of Michigan Press.

Hormozi, A. M., & Khumawala, B. M. (1996). An improved algorithm for solving a multi-period facility location problem. *IIE transactions,* 28(2), 105-114.

Jahantigh, F. F., & Malmir, B. (2016, March). A Hybrid Genetic Algorithm for Solving Facility Location Allocation Problem. In *Proceedings of the 2016 International Conference on Industrial Engineering and Operations Management, Kuala Lumpur, Malaysia.*

Jaramillo, J. H., Bhadury, J., & Batta, R. (2002). On the use of genetic algorithms to solve location problems. *Computers & Operations Research*, 29(6), 761–779.

Köksoy, O., & Yalcinoz, T. (2008). Robust design using pareto type optimization: A genetic algorithm with arithmetic crossover. *Computers & Industrial Engineering*, 55(1), 208–218.

Liao, A., & Approach, D. G. (2008). A clustering-based approach to the capacitated facility location problem. *Transactions in GIS*, 12(3), 323–339.

Matthias, K., Severin, T., & Salzwedel, H. (2013). Variable mutation rate at genetic algorithms: Introduction of chromosome fitness in connection with multi-chromosome representation. *International Journal of Computer Applications*, 72(17), 31–38.

Máximo, V. R., Nascimento, M. C., & Carvalho, A. C. (2017). Intelligent guided adaptive search for the maximum covering location problem. *Computers & Operations Research*, 78(Feb 2017), 129–137.

Mehdizadeh, E., & Afrabandpei, F. (2012). Design of a mathematical model for logistic network in a multi-stage multi-product supply chain network and developing a metaheuristic algorithm. *Journal of Optimization in Industrial Engineering*, 5(10), 35–43.

Miller, T. C., Friesz, T. L., Tobin, R. L., & Kwon, C. (2006). Reaction function based dynamic location modeling in Stackelberg–Nash–Cournot competition. *Networks and Spatial Economics*, 7(1), 77–97.

Murray, T., & Gerrard, R. A. (1998). Capacitated service and regional constraints in location-allocation modeling. *Location Science*, 5(2), 103–118.

Niroomand, I. (2008). *Modeling and analysis of the generalized warehouse location problem with staircase costs* (Doctoral dissertation, Concordia University Montreal, Quebec, Canada).

Pasandideh, S. H. R., Akhavan Niaki, S. T., & Asadi, K. (2015). Bi-objective optimization of a multi-product multi-period three-echelon supply chain problem under uncertain environments: NSGA-II and NRGA. *Information Sciences,* 292(Jan 2015), 57–74.

Pham, D. T., Ghanbarzadeh, A., Koc, E., Otri, S., Rahim, S., & Zaidi, M. (2011, July). The bees algorithm-A novel tool for complex optimisation. In *Intelligent Production Machines and Systems-2nd I* PROMS Virtual International Conference (3-14 July 2006).* sn.

Pirkul, H., & Schilling, D. (1989). The capacitated maximal covering location problem with backup service. *Annal of Opeation Research*, 18(1), 141–154.

Pirkul, H., & Schilling, D. A. (1991). The maximal covering location problem with capacities on total workload. *Management Science*, 37(2), 233–248.

Raju, B. S., Shekar, U. C., Venkateswarlu, K., & Drakashayani, D. N. (2014). Establishment of Process model for rapid prototyping technique (Stereolithography) to enhance the part quality by Taguchi method. *Procedia Technology*, 14(Jan 2014), 380–389.

Revelle, C., Scholsssberg, M., & Williams, J. (2008). Solving the maximal covering location problem with heuristic concentration. *Computers & Operations Research*, 35(2), 427–435.

Salari, M. (2013). An iterated local search for the budget constrained generalized maximal covering location problem. *Journal of Mathematical Modelling and Algorithms in Operations Research*, 13(3), 301–313.

Schilling, D. A. (1980). Dynamic location modeling for public-sector facilities: A multicriteria approach. *Decision Sciences*, 11(4), 714–724.

Schilling, D. A., Jayaraman, V., & Barkhi, R. (1993). A review of covering problem in facility location. *Location Science*, 1(1), 25–55.

Toregas, C., Swain, R., ReVelle, C., & Bergman, L. (1971). The location of emergency service facilities. *Operations Research,* 19(6), 1363-1373.

Tsai, H. (2014). Novel bees algorithm: Stochastic self-adaptive neighborhood. *Applied Mathematics and Computation, 247(Nov 2014)*, 1161–1172.

Vatsa, A. K., & Jayaswal, S. (2016). A new formulation and Benders decomposition for the multi-period maximal covering facility location problem with server uncertainty. *European Journal of Operational Research*, 251(2), 404-418.

Yin, P., & Mu, L. (2012). Modular capacitated maximal covering location problem for the optimal siting of emergency vehicles. *Applied Geography*, 34(May 2012), 247–254.

Zanjirani Farahani, R., & Hekmatfar, M. (2009). *Facility location: Concept, Models, Algorithms and Case Studies.* Heidelberg: Physica-Verlag.

10

Optimum design of a CCHP system based on Economical, energy and environmental considerations using GA and PSO

Masoud Rabbani*, Setare Mohammadi and Mahdi Mobini

Department of Industrial Engineering, University of Tehran, Tehran, Iran

CHRONICLE	ABSTRACT
Keywords: *Combined cooling heating power generation* *Optimised design* *Control strategy* *Particle Swarm Optimisation* *Genetic algorithm*	Optimum design and control of a Combined Cooling, Heating and Power generation (CCHP) system, in addition to the economic benefits, could be profitable in environmental and energy consumption aspects. The aim of this study is to determine the optimal capacity of equipment and define the best control strategy of a CCHP system. Since determination of optimal system control strategy has a huge impact on improving the objective functions, the system's performance under five different strategies (developed based on well-known Following Electrical Load (FEL) and Following Thermal Load (FTL) strategies) is evaluated. In a real case study, a CCHP system is designed for an educational complex located in Mahmoudabad, Mazandaran, Iran. The objective is to minimize capital and operational costs, energy consumption, and CO_2 emissions of the system. Due to the complexities of the model, genetic algorithm (GA) and particle swarm optimisation (PSO) algorithm are used to find the optimal values of the decision variables. The results show that using FEL strategy CO_2 emissions reduces in compression to FTL strategy. Furthermore, using multiple power generation units under FTL strategy eventuates the least cost but increases CO_2 emissions and energy consumption in compression to FEL strategy.

1. Introduction

A large portion of national energy consumption is expended for fulfilling the buildings' heating, cooling, and electricity demand. Consumed energy in the buildings sector, consisting of residential and commercial end users, accounts for 20.1% of the total delivered energy consumption worldwide (International Energy Outlook, 2016). The type and amount of energy consumed by households can vary significantly within and across regions and countries (International Energy Outlook, 2016). In the USA, residential buildings consume 22% of the total final energy use, compared with 26% in the EU. Residential buildings energy consumption is 28% of total energy consumption in the UK, well above Spain at 15%, mainly due to a more severe climate and the building types. In 2030, energy consumption attributed to residential and the non-domestic sectors is predicted to reach to 67% and 33%, respectively (Pérez-Lombard et al., 2008). Lack of efficient construction regulations, in addition to the low energy

* Corresponding author
E-mail: mrabbani@ut.ac.ir (M. Rabbani)

prices in the past have caused careless and inefficient consumption of energy by the Iranian residential sector compared to industrialized countries (Karbassi et al., 2007). Commercial and residential building sector consume about 40% of total energy in Iran. This consists, 11.7% of oil products, 73.13% of natural gas and 13.25% of electricity (Iran Energy Efficiency Organization (IEEO-SABA), 2016). In addition to the economic burdens, this energy consumption trend all around the world is causing severe environmental problems as well as energy security issues (Cai et al., 2009).

Using Combined Cooling, Heating and Power generation system (CCHP) is a proven method for enhancing energy efficiency. Using CCHPs leads to economic savings, while reducing the emissions (Zheng et al., 2014). Also, possible energy sources for CCHP systems include a vast range of fossil fuels, biomass, geothermal and solar power, giving the flexibility desired for installing these systems in different geographical regions. Consequently, CCHP systems are installed in a variety of buildings, such as hotels, offices, hospitals and supermarkets (Ge et al., 2009; Wang et al., 2008).

In this field the goal is to fulfil a building's energy demand while minimizing the costs and environmental consequences (Løken, 2007). In order to do so, structural design and operational planning of the system need to be optimised. Structural design of the system relates to defining the optimum number and capacity of the equipment; and operational planning of the system relates to the determination of the hourly operation of the equipment (Mago & Chamra, 2009). One of the main challenges in energy planning field is access to reliable estimation of energy demand. Additionally, the fluctuations in the building energy demand (in terms of heating, cooling, and electricity) makes the design and operational planning of the system a complex task (Cao, 2009) since reaching the optimum design requires solving an optimisation model at every interval of time. The large scale of the optimisation problem makes the models computationally intractable; therefore, operating strategies are proposed to reduce the complexity of the models. These strategies determine the state of the Power Generation Unit (PGU) and the proportion of the cooling demand fulfilled by the electric chillers (so called "electric cooling to cool load ratio") in each period, which significantly reduce the complexity of the problem. A variety of methods for determining optimum design of CCHP systems are proposed. Initially linear optimisation models were developed to design energy systems. Cao (2009) analysed the influence of energy prices on the system's economic feasibility. The objective function was minimisation of the annual cost, and maximization of the exegetic efficiency. Piacentino and Cardona (2008) presented a Mixed Integer Linear Programing (MILP) model to optimize the economic and environmental performance of a tri-generation system (heating, cooling and electricity production). Nonlinear Programming (NLP) and Mixed Integer Programming (MIP) models were used to find the optimum design of the system in research by Gamou and Yokoyama (1998) and Arcuri et al. (2007), respectively. Reduced gradient method was used by Chen and Hong (1996) to solve the presented mathematical model. In a similar study a matrix approach was employed to model the problem by Geidl and Andresson (2007). They presented the mathematical model of the problem in the matrix form and used Sequential Quadratic Programming to optimize an hourly linear objective function.

Due to their capability in tackling large-scale optimisation problems, artificial intelligence, in the form of heuristic and metaheuristic algorithms are commonly employed to optimize the design and operation of CCHP systems. Metaheuristic algorithms' ability of exploration and exploitation is admissible when evaluation of limited number of feasible solutions is desired (Črepinšek et al., 2013). Genetic Algorithm (GA) and Particle Swarm (PSO) algorithm have been applied to optimize of CCHP design and operational parameters. The PSO algorithm is used by Tichi et al. (2010) for minimizing the cost of operating various CHP and CCHP systems in an industrial dairy unit. Wu (2011) considered the optimisation of operation of a CHP system under uncertainty and used the PSO algorithm to solve the model. Ghaebi et al. (2012) investigated exergoeconomic optimisation of a CCHP system. The presented economic model was based on the Total Revenue Requirement (TRR) and the total cost of the system was defined as the objective function. This model was solved by GA. Designing CCHP systems involves determination of the equipment's capacity as the main goal. Wang et al. (2010) designed a CCHP system with consideration of PGU and storage tank capacity as decision variables. On-off coefficient and "electric cooling to cool load" ratio was considered as decision variables too. This research was extended by an investigation a biomass gasification CCHP system (Wang et al., 2014). In this research the capacity of the gasification reactor, PGU, absorption chiller, electric chiller, and heat exchanger were considered as decision variables. In another study by Sanaye et al. (2015) a CCHP system was designed with equipment's capacity, partial load of PGU in each month, and electric cooling to cool load ratio as decision variables. This study considered a more comprehensive design compared with previous studies; in addition to the capacity of PGU, the number of them was considered as the decision variables. When a high-capacity PGU is installed, due to fluctuations in electric demand, the optimum solution dictates that the PGU is in off state a number of courses. This will lead to purchase of the whole electricity demand from the grid and supplement of heating demand by auxiliary boiler. Therefore, energy consumption and pollution increase during these periods; also, the cost of buying electricity will increase significantly. Consequently, it might be more beneficial to consider a number of smaller-capacity PGU instead of a large-scale one. This is why the number of PGUs is considered as a decision variable.

In the previous research the number and the capacity of the equipment, the on-off coefficient, and "electric cooling to cool load ratio" under different strategies were not simultaneously considered as decision variables. Therefore, the interconnections between these variables have not been taken into account, which is investigated in this study. In this research various strategies, for simultaneous utilization of several power generation units and adaptation of the operation status of chillers throughout the year, are explored so that the performance of the system under different circumstances is evaluated. Commonly employed strategies such as Following Electrical Load (FEL) and Following Thermal Load (FTL) are implemented for an actual set of buildings to optimize the performance of the CCHP system. Moreover, different strategies are implemented for a real case and results are analysed. As mentioned before, main goal of using the CCHP systems is to lower the economic costs and the environmental consequences. In this study, it is endeavoured to reflect the influence of optimum design and operation planning of the system in reduction of economic costs, environmental footprint measured in terms of CO_2 emissions, and energy consumption. As a summary, the followings are the contribution of this paper:

- Three commonly employed strategies in addition to two novel strategies for operational planning of CCHP systems are explored.
- GA and PSO algorithms are employed to obtain the optimum values of design parameters and their performance in solving this optimisation problem are compared.
- Eight design parameters (decision variables), including the capacity of gas turbine as the prime mover, their number and operational strategy, the capacity of the backup boiler and storage tank, the capacity of electrical and absorption chillers, the electric cooling ratio, and the on–off coefficient of PGUs are considered and the results under various strategies are compared.
- The developed strategies and algorithms are applied to a real case study.

2. Problem Description

Conventionally, in Separate Production (SP) systems, electric chillers are used to fulfil cooling demand, while heating demand of the buildings is supplied with a boiler (commonly a gas boiler), and the electricity is purchased off the grid. CCHP systems, however, consist of several separated segments that perform in an integrated fashion to fulfil the electricity, cooling, and heating demands. Fig. 1 shows general structure of a CCHP system. PGU generates electricity power by consuming the fuel; the heat exchanger retrieves heat generated during generation of electricity; depending on the implemented strategy, the recovered heat is either used to fulfil the heating demand or is directed to the absorption chiller to fulfil the cooling demand; the electric chiller is used to complement the absorption chillers and fulfil the cooling demand when needed; the auxiliary boiler and energy storage tank reduce the risk of system failure and increase system reliability.

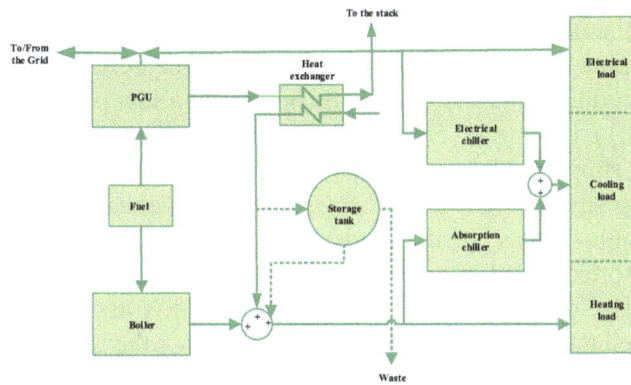

Fig. 1. Schematic of a CCHP system (Sanaye & Hajabdollahi, 2015)

Operational planning of CCHP systems is usually conducted based on two strategies: FEL and FTL. The core of FEL strategy is fulfilment of the electrical demand. If the electrical demand of the buildings exceeds the capacity of PGU, it works in full load; otherwise it works in partial load to provide the required amount of electrical power. The cooling load provided by the electric chiller is determined in each period (an hour) based on the electrical power production. When systems operate based on FEL, overproduction of thermal energy would be wasted. When the energy generated by the PGU is insufficient, lack of electricity is purchased from the grid. Also, thermal storage tanks are used to enhance thermal efficiency of the CCHP systems. Recovered heat from the PGU will be used by the absorption chiller for cooling, or by the heat exchanger to supply the heat demand of the buildings.

On the contrary to the FEL strategy, in the FTL strategy, the purpose is to fulfil the thermal demand of the building, so there is no excess of thermal energy and in case of shortage, thermal energy would be supplied by an auxiliary boiler. When CCHP system operates based on the FTL, surplus or shortage of electrical power is possible. In case of surplus, if selling the excess electricity to the grid is not possible, surplus electricity would be wasted; and in case of shortage, unmet electricity demand is fulfilled by purchasing from the grid.

Another commonly adopted strategy for operational planning of CCHP systems is similar to FEL but with an additional decision variable (x) which defines the ratio of electric cooling to cool load. In other words, in this strategy the proportion of the cooling demand supplied by the electric chiller is a decision variable, compared to the FEL strategy where the priority is always given to the electric chiller and the capacity of the PGU defines the amount of the cooling demand supplied by the electric chiller. Using FEL with no restriction on the electric chiller utilization might lead to significant heat waste which could have been used by the absorption chiller to supply the cooling demand. In order to prevent increasing the complexity of the model, the value of x is commonly considered to be fixed throughout the year (Sanaye & Khakpaay, 2014; J. Wang et al., 2010) (one decision variable is added instead of 8760 variables).

As mentioned before, using several smaller PGUs instead of a single large-capacity PGU could be beneficial in some cases. At the first glance, deployment of several PGUs will dramatically increase the capital cost of the system; however, the operational costs could be reduced to the extent that compensate the extra capital cost. Therefore, considering the multi-PGU case provides the opportunity to evaluate the trade-off between the higher capital cost and the reduced operational costs which is missed when a single PGU is considered. Specifically, under the FEL strategy, it is anticipated that multi-PGU approach offers more favourable results when the minimum and maximum electrical load of the system throughout the year are widely different. If the electric load fluctuations in the system is significant and we have a high capacity PGU, it will be turned off in many periods when the partial load would fall below its economical operational threshold (as explained in section 3.1) leading to higher purchased amount of the electricity from the grid and utilizing the auxiliary boiler to fulfil the heat demand. On the contrary when several PGUs are available their status (on/off) could be adjusted respective to the partial load of the system, hence, reducing the operational costs. Similarly, when using a single high-capacity PGU under the FTL strategy, in a number of periods during spring and fall the PGU is turned off because the heating demand of the system is reduced and the PGU must operate at a low partial load that is not economical (as explained in section 3.1). Having several PGU with smaller capacity could reduce purchasing power from the grid and reduce supplying heating demand by the auxiliary boiler, hence, improving system's performance. Considering several PGU will change how the operational strategies are applied. When using FEL strategy in a multi-PGU system, in order to determine the status (on/off) of each PGU at each time step, the PGUs are sorted based on their capacity and the smallest unit with the lowest capacity is placed in active status. The reminder of the electrical demand is assigned to the next PGU and it is activated. This is continued until the demand is fully responded or the generation of the electricity by the next PGU is not economically viable. This approach is expected to increase the efficiency of the PGUs by increasing their utilization rate. When using the FTL strategy, in order to determine the status (on/off) of each PGU, the PGUs are sorted based on their capacity and the smallest unit with the lowest capacity is placed in active status. The reminder of heating demand is assigned to the next PGU and it is activated. This is continued until the heating demand is fully responded or the activation of the next PGU is not economically viable.

3. Model formulation

In the followings, the parameters of the model are introduced and the objective functions and constraints are presented. Table 1 shows a list of the parameters and symbols used in the model.

Table 1
Parameters and symbols of the developed model

Parameter		Subscripts	
η_r	Heat recovery factor	e	Electricity
η_{grid}	Transmission efficiency	r	Recovery
η_{plant}	Plant efficiency	f	Fuel
η_s	Heat storage efficiency	h	Hour
i	Interest rate	$temp$	Temporary
n	Lifetime of the equipment (year)	nom	Nominal
$C_{e,buy}$	Purchase price of the electricity ($/kWh)	p	Pumps and other equipment
$C_{e,sale}$	Selling price of the electricity ($/kWh)	hea	Heating
$C_{f,b}$	Buying price of the fuel for boiler ($/kWh)	j	Number of PGU (counter)
$C_{f,pgu}$	Buying price of the fuel for PGU ($/kWh)	k	Equipment number (counter)
μ_{co2_e}	CO$_2$ emission of electricity (gr/kWh)	req	Required
μ_{co2_f}	CO$_2$ emission of fuel (gr/kWh)	pgu	Power Generation Unit
El	Electrical demand of the buildings (kW)	max	Maximum capacity
Q_c	Cooling demand of the buildings (kW)	min	Minimum capacity
Q_{hea}	Thermal demand of the building (kW)	$total$	Total amount
COP_{ec}	Coefficient of performance of electric chiller	$best$	The best amount
COP_{ac}	Coefficient of performance of absorption chiller	symbols	
Ta_c	Carbon tax ($/kg)	PL	Partial load

Subscripts		K	Total number of PGUs
out	Exit	N	Total number of equipment
in	Entrance	C	Capital cost per kW ($/kW)
E	Electricity Generated/Required	*Cap*	Capital cost of equipment($)
ac	Absorption chiller	*fr*	Instantaneous fraction
ec	Electric chiller	x	Electric cooling to cool load ratio
c	Cooling	Q	Heating load (kW)
b	Boiler	F	Fuel consumption (kWh)
S	Storage tank	*Sl*	Salvage value ($)
grid	Local grid		

3.1. Under FEL strategy

When FEL is the adopted strategy, the ratio of cooling load supplemented by the electric chiller per hour is calculated by (1). The amount of cooling load supplied by the electric chiller is calculated by (2).

$$x_h = \begin{cases} 1 & if \quad E_{pgu,max} \geq El_h + E_{p,h} + Q_{c,h}/COP_{ec} \\ ((E_{pgu,max} - El_h - E_{p,h}) \times COP_{ec})/Q_{c,h} & if \quad El_h + E_{p,h} \leq E_{pgu,max} < El_h + E_{p,h} + Q_{c,h}/COP_{ec} \\ 0 & if \quad E_{pgu,max} < El_h + E_{p,h} \end{cases} \tag{1}$$

$$Q_{ec,h} = x_h . Q_{c,h} \tag{2}$$

The electricity consumption of the electric chiller is calculated as shown in (3). The capacity constraint of the electric chiller is shown in (4). The total electricity requirement of buildings is shown in (5). The capacity utilization of the PGU is denoted by $fr_{pgu,h}$, called the instantaneous fraction of the PGU, and is calculated using (6).

$$E_{ec,h} = Q_{ec,h}/COP_{ec} \tag{3}$$

$$Q_{ec,min} \leq Q_{ec} \leq Q_{ec,max} \tag{4}$$

$$E_{req,h} = (El_h + E_p) + E_{ec,h} \tag{5}$$

$$fr_{pgu,h} = \frac{E_{req,h}}{E_{pgu,max}} \tag{6}$$

The efficiency of the PGU is highly dependent on its capacity utilization (Wang et al., 2011) . Below a certain threshold it is not rational to use the PGU since most of the consumed fuel is wasted on heat generation rather than electricity generation. In (7), α is considered as the lower bound on the instantaneous fraction to ascertain the PGU is either turned off or is working above the predetermined level. α is a decision variable in the model and its upper limit is equal to 1.

$$E_{pgu,h} = \begin{cases} 0 & if \quad fr_{pgu,h} < \alpha \\ E_{req,h} & if \quad \alpha \leq fr_{pgu,h} < 1 \\ E_{pgu,max} & if \quad fr_{pgu,h} \geq 1 \end{cases} \tag{7}$$

Partial load of PGU is calculated by (8. The efficiency of the PGU is determined by (9). Increasing the partial load of PGU increases its efficiency (Sanaye & Hajabdollahi, 2013; Sanaye, Meybodi, & Shokrollahi, 2008) as shown in the (9). The electricity purchased from the grid or sold to the grid is calculated using (10).

$$PL_{pgu,h} = \frac{E_{pgu,h}}{E_{pgu,\max}} \tag{8}$$

$$\frac{\eta_{pgu,h}}{\eta_{pgu,nom}} = -0.0001591(PL_{pgu,h})^2 + 0.024(PL_{pgu,h}) + 0.1904 \tag{9}$$

$$E_{grid,h} = E_h + E_p + E_{ec,h} - E_{pgu,h} \tag{10}$$

(11) declares that power exchange with the grid was purchase off the grid or sold to the grid. The fuel consumed by PGU is estimated by the (12). Based on the fuel consumption and the efficiency of the PGU, the generated heat is calculated by (13).

$$-u_{1,h}.E_{sale,h} + u_{2,h}E_{buy,h} = E_{grid,h}$$
$$u_1 + u_2 \leq 1 \qquad u_1, u_2 \in \{0,1\} \tag{11}$$

$$F_{pgu,h} = \frac{E_{pgu,h}}{\eta_{pgu,h}} \tag{12}$$

$$Q_{pgu,h} = F_{pgu,h}(1 - \eta_{pgu,h}) \tag{13}$$

The amount of the recovered heat is calculated by (14). Fuel consumption related to the electricity purchased from the grid is calculated by (15). The supplied cooling via absorption chiller is determined by (16). Absorption chiller's cooling range is between minimum and the maximum capacity of absorption chiller, represented in (17).

$$Q_{r,h} = Q_{pgu,h}.\eta_r \tag{14}$$

$$F_{e,h} = \frac{E_{buy,h}}{\eta_{grid} \times \eta_{plant}} \tag{15}$$

$$Q_{ac,h} = (1 - x_h).Q_{c,h} \tag{16}$$

$$Q_{ac,\min} \leq Q_{ac,h} \leq Q_{ac,\max} \tag{17}$$

The heat consumed by the absorption chiller is calculated by its coefficient of performance in (18). The total amount of heat requirement of the system is calculated as (19).

$$Q_{ac,in,h} = (1 - x_h) \times Q_{c,h} / COP_{ac} \tag{18}$$

$$Q_{req,h} = Q_{ac,in,h} + Q_{hea,h} \tag{19}$$

Part of the heat requirement is supplied by recovered heat from the PGU and the rest is provided by the backup boiler. The generated heat from the auxiliary boiler can be calculated by (20). Capacity constraint of the boiler is represented in (21). Partial load of the boiler is determined by (22) and its efficiency can be calculated by (23) (Sanaye & Hajabdollahi, 2013; Sanaye et al., 2008).

$$Q_{b,h} = \begin{cases} 0 & \text{if } Q_{r,h} + Q_{s,h} - Q_{req,h} \geq 0 \\ Q_{req,h} - Q_{r,h} - Q_{s,h} & \text{if } Q_{req,h} - Q_{r,h} - Q_{s,h} > 0 \end{cases} \tag{20}$$

$$Q_{b,\min} \leq Q_{b,h} \leq Q_{b,\max} \tag{21}$$

$$PL_{b,h} = \frac{Q_{b,h}}{Q_{b,\max}} \tag{22}$$

$$\frac{\eta_{b,h}}{\eta_{b,nom}} = 0.0951 + 1.525(PL_{b,h}) - 0.6249(PL_{b,h})^2 \tag{23}$$

Fuel consumption of the auxiliary boiler is calculated by (24) and the total fuel consumption in the CCHP system is determined in (25). The amount of heat charged in or discharged from storage tank can be determined by (26). The initial investment cost per kW capacity of equipment ($/kW) is determined using (27) and (28) (Sanaye & Hajabdollahi, 2013).

$$F_{b,h} = \frac{Q_{b,h}}{\eta_{b,h}} \tag{24}$$

$$F_h = F_{b,h} + F_{pgu,h} \tag{25}$$

$$Q_{h+1,s} = \eta_s . Q_{h,s} + U_{h,s,in} . Q_{h,s,in} - U_{h,s,out} . Q_{h,s,out}$$

$$U_{h,s,in} \& U_{h,s,out} \in \{0,1\} \tag{26}$$

$$U_{h,s,in} + U_{h,s,out} \leq 1$$

$$C_1 = -0.014 \times 10^{-6} \times E_{pgu,max} + 600 \tag{27}$$

$$Cap_1 = C_1 \times E_{pgu,max} \tag{28}$$

The gas turbine maintenance cost ($/kW), expressed based on its capacity, is assumed equal to $0.0055 kWh (Smith et al., 2010). The initial cost per kW of the boiler ($/kW) is estimated by (29 and its total capital cost is calculated by (30. Operating cost of the boiler is assumed $0.0027 kWh (Sanaye & Hajabdollahi, 2013). Initial investment per kW ($/kW) of absorption and electric chiller are estimated as shown in (31 and (32, respectively (Sanaye et al., 2008). Their total capital cost is calculated by (33 and (34. Operating cost for both chillers is assumed to be 0.003 $/kWh (Sanaye & Hajabdollahi, 2013). The initial cost of a storage tank per kW is considered 33 $/kW (Sanaye & Hajabdollahi, 2013) so the total capital cost of storage tank is calculated by (35.

$$C_2 = 205 \left(Q_{b,max} \times 10^{-6} \right)^{-0.13} \tag{29}$$

$$Cap_2 = C_2 \times Q_{b,max} \tag{30}$$

$$C_3 = 540 (Q_{ab,max} \times 10^{-6})^{-0.128} \tag{31}$$

$$C_4 = 482 (Q_{ec,max} \times 10^{-6})^{-0.07273} - 159.7 \tag{32}$$

$$Cap_3 = C_3 \times Q_{ab,max} \tag{33}$$

$$Cap_4 = C_4 \times Q_{ec,max} \tag{34}$$

$$Cap_5 = 33 \times Q_{s,max} \tag{35}$$

3.2. Under FTL strategy

For the FTL strategy mathematical relationships are similar to FEL except for the chiller's operation parameters that are determined in a different way. At first the efficiency of PGU at full load ($\eta_{pgu,max}$) is calculated using (9) and the heat generated by the PGU is calculated by (36). Using the heat recovery factor, recovered heat from the CCHP system at the highest capacity is calculated in (37) and the ratio of electric cooling load to cool load is determined as shown in (38).

$$Q_{pgu,max} = F_{pgu,h}(1 - \eta_{pgu,h}) = \frac{E_{pgu,max}}{\eta_{pgu,max}}(1 - \eta_{pgu,max}) \tag{36}$$

$$Q_{r,max} = Q_{pgu,max} . \eta_r \tag{37}$$

$$x_h = \begin{cases} 0 & if \quad Q_{r,max} \geq Q_{hea,h} + Q_{c,h} / COP_{ac} \\ 1 - ((Q_{r,max} - Q_{hea,h}) \times COP_{ac}) / Q_{c,h} & if \quad Q_{hea,h} < Q_{r,max} < Q_{hea,h} + Q_{c,h} / COP_{ac} \\ 1 & if \quad Q_{r,max} \leq Q_{hea,h} \end{cases} \tag{38}$$

By determination of x_h, the required heat can be calculated according to (19). If the required heat is less than $Q_{r,max}$, recovered heat would be equal to its required heat. By using (39), $E_{pgu,h,temp}$ is calculated and using (40 temporary partial load ($PL_{pgu,h,temp}$) is calculated. As shown in (41), d_h is a binary variable which equals to 1 when PGU is on and equals to zero when PGU is off. Using (42), heat recovered from the PGU is calculated based on heat requirement of the system. Based on the recovered heat of PGU the amount of electrical generation ($E_{pgu,h}$) is estimated and the partial load of PGU is calculated by (43). In other words, at first $PL_{pgu,h,temp}$ is calculated to determine if PGU should be in on or off mode. After determination of PGU's state, the amount of $PL_{pgu,h}$ is calculated.

$$E_{pgu,h,temp} = \frac{Q_{req,h} \cdot \eta_{pgu,h}}{(1-\eta_{pgu,h})\eta_r} \tag{39}$$

$$PL_{pgu,h,temp} = \frac{E_{pgu,h,temp}}{E_{pgu,max}} \tag{40}$$

$$d_h = \min\left\{ \left\lceil \frac{PL_{pgu,h,temp}}{\alpha} \right\rceil, 1 \right\} \tag{41}$$

$$Q_{rec,h} = \begin{cases} Q_{rec,max} & if\ Q_{req,h} \geq Q_{rec,max} \\ d_h \times Q_{req,h} & otherwise \end{cases} \tag{42}$$

$$PL_{pgu,j,h} = \frac{E_{pgu,j,h}}{E_{pgu,j,max}} \tag{43}$$

3.3. Under FEL strategy with fixed ratio of electric cooling to cool load

As previously mentioned, in the third strategy, the ratio of electric cooling to cool load is considered as a decision variable. PGU operates based on FEL but x is determined on the basis of both electrical and thermal load and Eq. (1) is omitted from the optimisation model.

3.4. Multiple PGUs, Under FEL strategy with fixed ratio of electric cooling to cool load

The power generation for each of the PGUs in each period of time is calculated in (7) until condition $f_{pgu,h} < \alpha$ is met. By this condition the other units will become inactive. PGUs are sorted in descending order of $E_{pgu,max}$ as shown in (44). In this stage, based on FEL strategy the total electrical requirement of the system is determined. Total electricity requirement for each unit can then be calculated by (45) and (46). Initially the instantaneous load factor of the active unit is calculated in (47); then, the amount of electricity generated for each active unit is determined using (48).

$$E_{pgu,j,max} \leq E_{pgu,j+1,max} \quad j \in \{1,...,K-1\} \tag{44}$$

$$E_{req,1,h} = E_{req,h} \tag{45}$$

$$E_{req,j+1,h} = E_{req,j,h} - E_{pgu,j,h} \quad j \in \{1,...,K-1\} \tag{46}$$

$$fr_{pgu,j,h} = \frac{E_{req,j,h}}{E_{pgu,j,max}} \tag{47}$$

$$E_{pgu,j,h} = \begin{cases} 0 & if\ \ fr_{pgu,j,h} < \alpha \\ E_{req,j,h} & if\ \ \alpha \leq fr_{pgu,j,h} < 1 \\ E_{pgu,j,max} & if\ \ fr_{pgu,j,h} \geq 1 \end{cases} \tag{48}$$

At this stage, the electricity generated by the CCHP system in each period is determined. If unit number j is placed on inactive status all next units will be in inactive status. This restriction is shown in (49); then, the total electrical load production is calculated by (50). (49) shows that PGUs are activated respectively and (50) represents total electricity generated by PGUs. Purchased electricity from the grid can be calculated by (51). Calculation of total heat retrieved from the CCHP system in each period is based on (52). The rest of the equations are similar to FEL strategy.

$$\gamma_{j+1} \leq \gamma_j \quad j \in \{1,2,...,K-1\} \quad \gamma_j \in \{0,1\} \tag{49}$$

$$E_{total,pgu,h} = \sum_{j=1}^{K} \gamma_j \times E_{pgu,j,h} \tag{50}$$

$$E_{grid,h} = E_{req,K,h} - E_{pgu,K,h} \tag{51}$$

$$Q_{rec,h} = \sum_{j=1}^{K} Q_{rec,j,h} \tag{52}$$

3.5. Multiple PGUs, under FTL strategy with fixed ratio of electric cooling to cool load

(49 is used to sort the PGUs. First, the entire heating system requirement is determined by (19). For the first activated unit the total heating demand of the system is considered as the required heat amount which is shown in (53). By (54) required heat of other units can be determined. After the last PGU, if still there is unmet demand, the auxiliary boiler is used for which the amount of heating demand is calculated using (55).

$$Q_{req,1,h} = Q_{req,h} \tag{53}$$

$$Q_{req,j+1,h} = Q_{req,j,h} - Q_{rec,j,h} \quad j \in \{1,2,...,K-1\} \tag{54}$$

$$Q_{boiler,h} = Q_{req,K,h} - Q_{rec,K,h} \tag{55}$$

3.6. Evaluation Criteria

Capital and variable costs, the amount of CO_2 emissions, and amount of energy consumption are considered as the criteria in the objective function in the optimisation model. The first criterion evaluates the economic costs of the system. It consists of the capital cost of equipment, cost of fuel consumed by the boiler and the PGU, operation cost of equipment, and cost/profit from transfer of electricity from/to the grid. Salvage value of equipment is considered 10% of their capital cost (Gibson et al., 2015).

$$R = \frac{i(1+i)^n}{(1+i)^n - 1} \tag{56}$$

$$A = \frac{i}{((1+i)^n - 1)} \tag{57}$$

$$Z_1 = ATC = R \times [\sum_{k=1}^{N} C_k \times Cap_k] - A \times Sl + \sum_{h=1}^{8760} (E_{buy,h}.C_{e,buy} + F_{b,h}C_{f,b} + F_{pgu,h}.C_{f,pgu} - E_{sale,h}C_{e,sale}$$
$$+ 0.0055 \times E_{pgu,h} + 0.0027 \times Q_{b,h} + 0.003 \times Q_{c,h}) - Ta_c \times 10^{-3} \times CDE \tag{58}$$

$$Z_2 = CDE = \sum_{h=1}^{8760} \mu_{co2_f}.F_h + \mu_{co2_e}.E_{buy,h} \tag{59}$$

R is the capital recovery factor and calculated as shown in (56 and A is the uniform series sinking fund and its value is calculated using (57), n represents the service life of the equipment and i is the interest rate. Similar to Bahrami and Farahbakhsh (2013), it is assumed that the values of i and n are equal for all equipment. The Annual Total Cost (ATC) is calculated by (58. The second criterion evaluate the amount of CO_2 Emissions (CDE) to reflect environmental concerns. CO_2 emissions are released by fuel consumption of auxiliary boiler and the PGU. Also, when electricity is purchased from the grid, the related CO_2 emissions are accounted for as shown in (59).

The third criterion represents the Primary Energy Consumption (PEC) of the system which is composed of two parts. The first part is amount of fuel consumed by the boiler and the PGU, and the second part is the fuel related to the electrical energy purchased from the grid. Total energy consumption of the system is shown in (60). In this study we use a weighted sum of these three criterion to form a single objective function as shown in (61).

$$Z_3 = PEC = \sum_{h=1}^{8760} F_h + F_{e,h} \qquad (60)$$

$$\min Z = \omega_1 \frac{Z_1}{Z_{1,best}} + \omega_2 \frac{Z_2}{Z_{2,best}} + \omega_3 \frac{Z_3}{Z_{3,best}} \qquad (61)$$

To calculate the optimal value of the multi objective function, first each of the single-objective functions are optimised to obtain the optimum value for each of them, denoted as $Z_{1,best}$, $Z_{2,best}$, and $Z_{3,best}$ in (61. Each single-objective function is normalized and then sum of them is calculated. By changing the weights different results can be achieved. Since the economic cost of fuel consumption (fuel consumption of the PGU, boiler and electricity purchased from the grid) in CCHP system in addition to CO_2 tax are considered in ATC function, the objective functions are largely aligned with each other which makes using the normalized weighted sum a proper method in handling the multi-objective optimisation model.

4. Evolutionary Algorithms

GA and PSO are both population-based algorithms commonly employed in the field of energy planning. Genetic algorithm was introduced by John Holland (Mitchell, 1998) as an evolutionary algorithm inspired by biology concepts such as inheritance, mutation, selection and crossover. Small random changes are determined by mutation which is determinative of GA diversity. Crossover operator determines how the algorithm combines two selected parents, to generate children for the next generation. Candidate solutions are assessed by the evaluation function (also known as fitness function).

PSO was invented by Kennedy and Eberhart in the mid1990s (1995), inspired by the movement of the particles. The PSO algorithm includes three phases, namely, generating particles' positions and velocities, updating the velocity of particles, and updating the position of particles. The three values that affect the new direction of a particle are its current motion, the best position in its memory, and swarm influence. (62) shows how the movement speed of the particle is updated and (63) shows how the new position of the particle is determined where the inertia coefficient for particle i in its next movement is represented by $v^i[t + 1]$, new position of the particle i is represented $x^i[t + 1]$, and w is the current motion factor, c_1 is particle own memory factor, and c_2 is the swarm influence factor.

$$v^i[t+1] = wv^i[t] + c_1 r_1 \left(x^{i,best}[t] - x^i[t] \right) + c_2 r_2 \left(x^{gbest}[t] - x^i[t] \right) \qquad (62)$$

$$x^i[t+1] = x^i[t] + v^i[t+1] \qquad (63)$$

4.1. Solution Representation

Solution representation scheme adopted in this research are similar for the both algorithms, with minor differences respective to the implemented strategy. In GA, a chromosome is a set of parameters which define a solution for the problem. An example chromosome for each strategy is shown Fig. 2. In the FEL strategy a chromosome with six bits is used. Chromosomes encompass normalized variables, i.e., all the bits filled with continuous variables between zero and one. Bits represent the capacity of PGU, auxiliary boiler, absorption chillers, electric chiller, heating storage tank and on/off coefficient. In the FTL strategy chromosome consist of 5 bits. In this strategy chromosome structure is similar to the FEL strategy just the bit corresponding to heat storage tank capacity is removed, because in this strategy excess heat is not produced at any time and thus the storage tank is removed. In FEL with fixed ratio of electric cooling load to cool load the chromosome has one bit more than in the FEL strategy which is related to the electric cooling load to cool load ratio. In the multi-PGUs strategy an upper limit for the number of PGUs is considered and binary bits are used to indicate the instalment of the PGUs; corresponding to each binary bit there is one bit related to that PGU's capacity. Therefore, assuming n as the upper limit on the number of the PGUs, 2*n bits are considered for PGUs to determine their instalment and capacity.

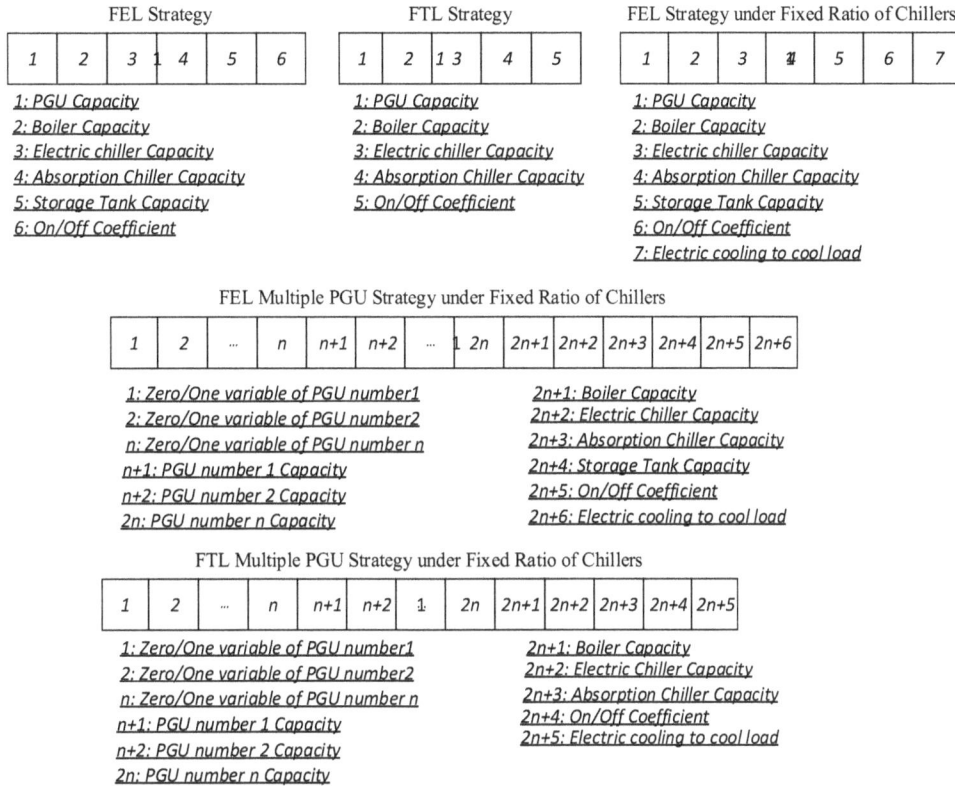

FEL Strategy

1	2	3	4	5	6

1: PGU Capacity
2: Boiler Capacity
3: Electric chiller Capacity
4: Absorption Chiller Capacity
5: Storage Tank Capacity
6: On/Off Coefficient

FTL Strategy

1	2	3	4	5

1: PGU Capacity
2: Boiler Capacity
3: Electric chiller Capacity
4: Absorption Chiller Capacity
5: On/Off Coefficient

FEL Strategy under Fixed Ratio of Chillers

1	2	3	4	5	6	7

1: PGU Capacity
2: Boiler Capacity
3: Electric chiller Capacity
4: Absorption Chiller Capacity
5: Storage Tank Capacity
6: On/Off Coefficient
7: Electric cooling to cool load

FEL Multiple PGU Strategy under Fixed Ratio of Chillers

1	2	...	n	n+1	n+2	...	2n	2n+1	2n+2	2n+3	2n+4	2n+5	2n+6

1: Zero/One variable of PGU number1
2: Zero/One variable of PGU number2
n: Zero/One variable of PGU number n
n+1: PGU number 1 Capacity
n+2: PGU number 2 Capacity
2n: PGU number n Capacity

2n+1: Boiler Capacity
2n+2: Electric Chiller Capacity
2n+3: Absorption Chiller Capacity
2n+4: Storage Tank Capacity
2n+5: On/Off Coefficient
2n+6: Electric cooling to cool load

FTL Multiple PGU Strategy under Fixed Ratio of Chillers

1	2	...	n	n+1	n+2	...	2n	2n+1	2n+2	2n+3	2n+4	2n+5

1: Zero/One variable of PGU number1
2: Zero/One variable of PGU number2
n: Zero/One variable of PGU number n
n+1: PGU number 1 Capacity
n+2: PGU number 2 Capacity
2n: PGU number n Capacity

2n+1: Boiler Capacity
2n+2: Electric Chiller Capacity
2n+3: Absorption Chiller Capacity
2n+4: On/Off Coefficient
2n+5: Electric cooling to cool load

Fig. 2. Chromosomes in different strategies

4.2. Operators

Mutation and crossover are the two essential operators in GA. In crossover operator, at first the parent chromosomes are selected, then by selection of genes from the parents, new off-springs are created. Parent chromosomes are selected according to their fitness; chromosomes which have better fitness function value have higher chance of being selected. After crossover, mutation takes place which prevents algorithm from premature convergence to a local optimum by inserting randomly created solutions based on the existing ones. In Fig. 3 examples of crossover and mutation operators for FEL strategy are shown. In this study a single point crossover is implemented. The crossover point is randomly selected, and two new solutions are created by swapping the two sides of the point in the parents to form a new solution. Applying the mutation operator, two points of a given chromosome are randomly selected and swapped to form a new solution. An example of crossover and mutation operators for Muti-PGUs strategy is shown in Fig. 4. These chromosomes possess binary and continues variables limited to [0,1]. Crossover operator is exactly the same for the multi-PGU strategy.

To apply mutation two random points are selected. If this selected point is binary, the bit will change to its contrary form. The capacity related to this binary variable will change based on this bit's new value. For example, if it becomes zero the capacity related to this bit will change to zero; otherwise the capacity related to this bit is updated to a new random value. If selected bit is continuous, it will become updated to a new continuous random variable by mutation operator.

In the implemented PSO algorithm the particles structure is similar to the chromosome structure in GA. Mutation is the only operator in PSO which is same as GA.

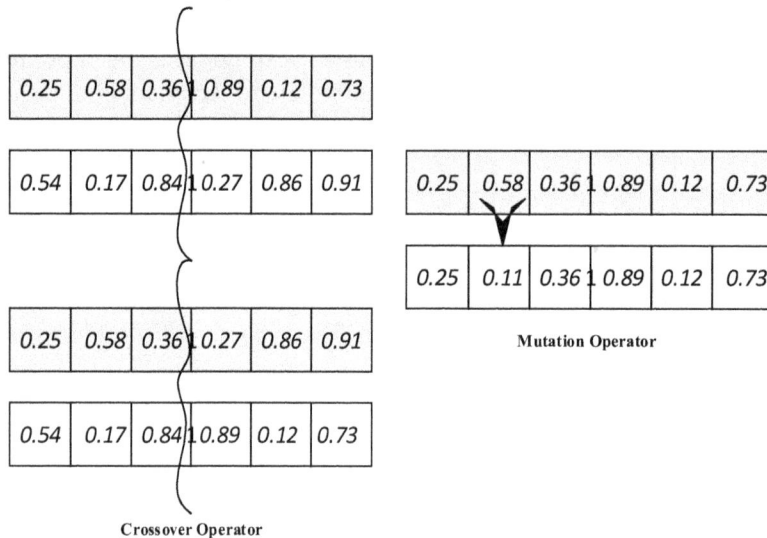

Fig. 3. Mutation and crossover example of FEL

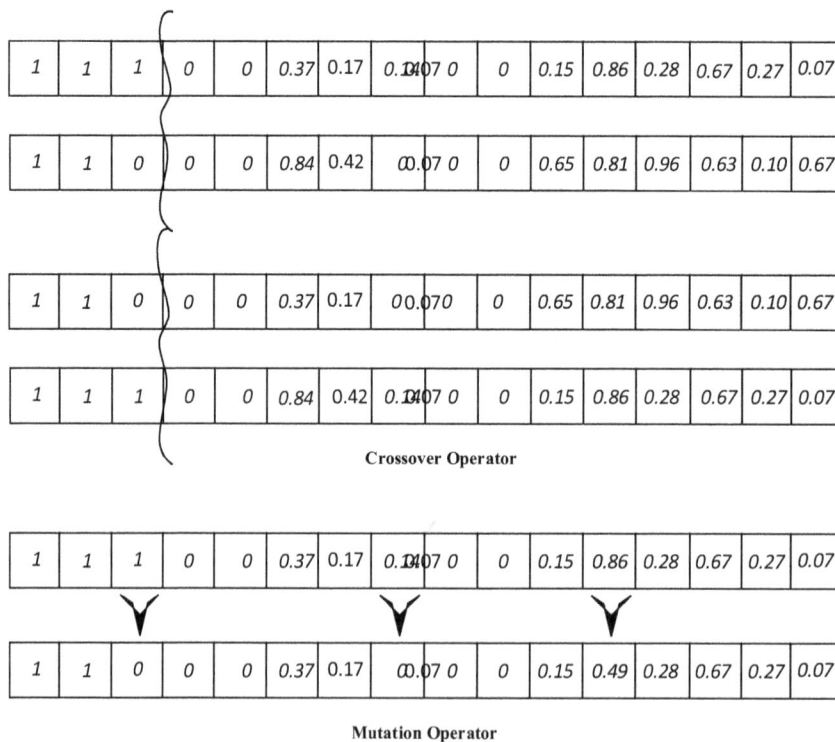

Fig. 4. Crossover and mutation example of Multi-PGU

5. Case study

The proposed CCHP optimisation model and the solution procedures are applied to a real case study, an educational complex located in Mazandaran, Iran. The CCHP system provides electricity, heating (for both space heating and domestic hot water), and cooling. List of existing buildings, the total area of outer wall, windows, doors and usable area are shown in Table 2.

Table 2

List of existing buildings and their characteristics

Type of building	Buildings number	Outer wall area	Windows area	Doors area	Usable area
Common Villa	40	6,102	644	31	8,817
VIP villa	28	576.5	89	34	1,150.5
Guest house	77	299	25	63	315
Restaurant	3	882	127	32	1,100
Residential building	8	245.5	46.5	7	371
Market	1	334	104	66	504
Mechanic Laboratory	1	560	111.7	10.25	1,233
Electric Laboratory	1	568	169.7	7	1,225
Educational sites	2	822	140	33.6	2,951
Sport hall	1	1,454	118.2	15.5	1,609
Central Kitchen	1	644	142	17.7	1,356

The technical parameters of CCHP system and the specifications of the components are listed in Table 3. Design parameters (decision variables) and the acceptable range of their variations are listed in Table 4. The range of the decision variables is determined by load demand of the buildings; the range of the boiler and storage tank capacity are determined according to heating demand; and chillers capacity range is determined according to cooling demand of the buildings.

Table 3

Technical parameters and specifications of components

Parameter	Explanation	Value	Reference
η_r	Heat recovery factor	0.8	(Wang & Fang, 2011)
η_{grid}	Transmission Efficiency	0.9	(Iranian Electricity management, 2016)
η_{plant}	Plant Efficiency	0.37	(Iranian Electricity management, 2016)
η_s	Heat Storage Efficiency	0.9	(Iranian Fuel Conservation Organization, 2016)
$\eta_{b,nom}$	Max Efficiency of boiler	0.8	(Liu et al., 2013)
$\eta_{pgu,nom}$	Max Efficiency of PGU	0.4	(Iranian Electricity management, 2016)
i	Interest rate	0.12	(Ghaebi et al., 2012)
n	Lifetime of the equipment	15	(Q. Wang & Fang, 2011)
Sl	Salvage value	0.1	(Gibson et al., 2015)
$C_{e,buy}(\$/kWh)$	Purchase price of electricity	0.12	(Iranian Electricity management, 2016)
$C_{e,sale}(\$/kWh)$	Selling price of electricity	0.09	(Iranian Electricity management, 2016)
$C_{f,b}(\$/kWh)$	Buying price of gas for boiler	0.04	(Iranian Fuel Conservation Organization, 2016)
$C_{f,pgu}(\$/kWh)$	Buying price of gas for PGU	0.03	(Iranian Fuel Conservation Organization, 2016)
$\mu_{co2_e}(gr/kWh)$	Pollution emission of electricity	968	(Liu et al., 2013)
$\mu_{co2_f}(gr/kWh)$	Pollution emission of fuel	220	(Liu et al., 2013)
$Ta_c(\$/kg)$	The carbon tax	0.03	(Liu et al., 2013)
COP_{ec}	Coefficient of performance of electric chiller	3	(Liu et al., 2013)
COP_{ac}	Coefficient of performance of absorption chiller	0.7	(Liu et al., 2013)

Table 4

Range of variations of the decision variables.

number of prime movers	1-5
Nominal power range of prime movers	500-5000
the on–off coefficient of PGU	0.2-1
Boiler heating capacity	100-6000
Electrical chiller cooling capacity	100-4000
Absorption chiller cooling capacity	100-4000
Electric cooling ratio	0-1
Storage tank heating capacity	0-5000

The electricity, cooling, and heating load curves are estimated using Energy Plus ("Energy Plus ", 2014) shown in Fig. 5.

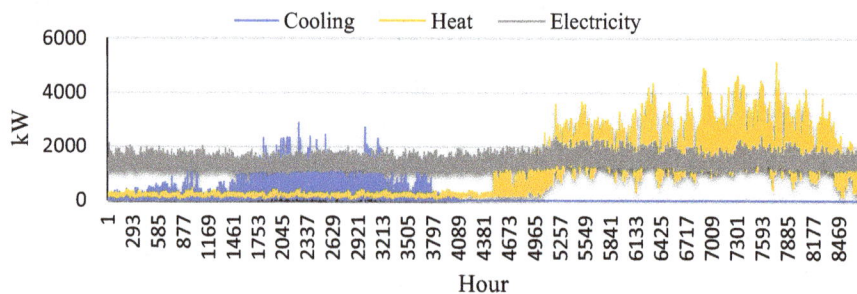

Fig. 5. Electricity, Cooling, and Heating load curves during a year

6.   Results and discussions

In this section the results of the algorithms under each strategy is presented and the comparison is drawn between them. In this research GA and PSO are coded in MATLAB R *2013b* ("MATLAB," 2013) Stopping criteria and parameters for GA and PSO are represented in Table 5.

Table 5

Parameters for GA and PSO

Variable	Value
GA	
Population size	100
Maximum iteration number	200
Crossover probability	0.6
Mutation probability	0.4
PSO	
Population size	100
Maximum iteration number	200
Inertia weight w	0.9
Particle own memory factor C_1	2
Swarm influence factor C_2	2

6.1. Results for different strategies for different weights

The results obtained when different weights for each objective function is used are shown in Fig. 6; 19 different weights results are represented. It is started by the (1,0,0) vector and ended to (0,0,1) vector for different scenarios. Extreme points which are related to single objectives are represented in Table 6; since minimizing PEC is equivalent to minimizing CDE, only one row (under each strategy) is considered for these two objectives. As ATC decreases, PEC and CDE increase because investment decreases and less electricity is generated by the CCHP system. As a result, buying electricity from the grid increases rising PED and CED. PEC and CDE are aligned because consumption of more fuel consequences more emission. The range of the results for the three objectives is less than 1% because of the relationship between the objective functions, reaffirming the suitability of the using the weighted-sum method to handle the three objectives simultaneously. The equal weights method is implemented in many decision-making problems; this method's results are most of the time close to the optimal weighting methods as discussed in (Wang et al., 2009, 2010). Therefore, in the following all of weights are assumed equal and $w_1 = w_2 = w_3 = \frac{1}{3}$.

Table 6
Results of different strategies in thresholds

FEL			FTL			Weights
TC	CDE	PEC	TC	CDE	PEC	
2,617,692	9,119,994,819	41,210,138	3,180,603	15,542,546,552	59,642,749	[1,0,0]
2,617,714	9,115,345,992	40,954,988	3,181,048	15,533,221,024	59,571,263	[0,0,1] [0,1,0]
FEL. Fixed ratio of cooling load			Multiple PGU.FTL			Weights
TC	CDE	PEC	TC	CDE	PEC	
2,614,252	9,021,807,496	40,145,361	1,826,494	10,913,736,182	41,285,055	[1,0,0]
2,614,514	9,007,103,652	40,020,287	1,826,585	10,910,462,062	41,158,490	[0,0,1] [0,1,0]

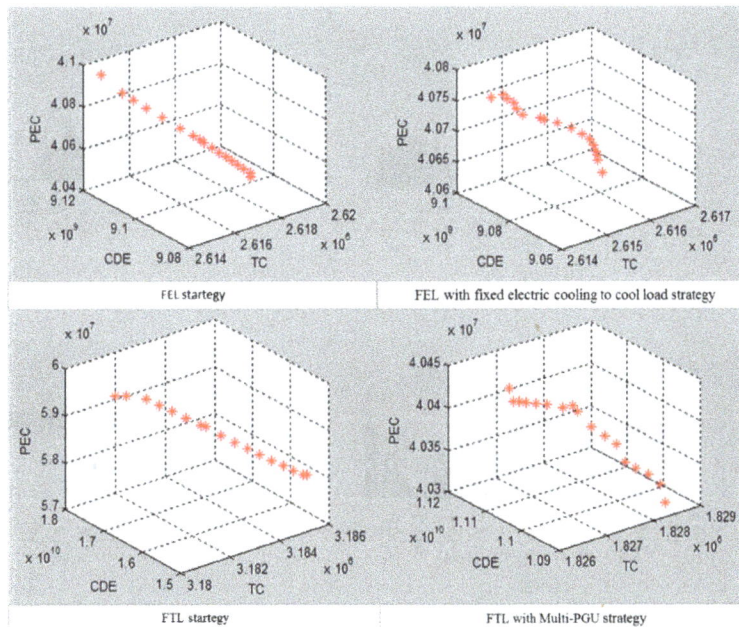

Fig. 6. Objectives amount for different strategies in different weights

6.2. Results for FEL strategy

The best design of the system in this strategy is presented in the Table 7 for both the GA and PSO algorithm. PSO outperforms GA in this settings as reflected in the last three rows of Table 7. Numerical results in various segments of the system are shown in Fig. 7. Following this strategy, PGU is operating in its maximum capacity in 5870 periods and in partial load larger than 0.8 in 8684 periods (see Fig. 7 d. E PGU). The results regarding the heat storage tank (Fig. 7 c. Q Storage) indicate that during two periods of time the storage tank discharges completely. First in summer, to fulfil the heat requirement of the absorption chiller; and second in the winter, to meet the heat demand of the buildings. As shown in Fig. 7 a. Q Boiler, the thermal load provided by the boiler is maximized during winter, because recovered heat from the PGU is not sufficient to fulfil the heat demand of the system and the boiler supplies the remaining heat demand. In summer, the boiler is operational in 21 periods because recovered heat is not sufficient to fulfil absorption chiller's heat demand. As shown in Fig. 7 b. Q Electric Chiller, despite the priority given to the use of electric chiller under this strategy, only 34% of the cooling demand is supplied with electric chiller and the remaining cooling demand is fulfilled by the absorption chiller.

Table 7

Best design of the system at FEL strategy

Decision Variable	PSO Results	GA Results
Nominal power of prime movers (kW)	1,345	1,489
the on–off coefficient of PGU	0.73	0.65
Boiler heating capacity (kW)	3,883	3,617
Electrical chiller cooling capacity (kW)	1,046	1,480
Absorption chiller cooling capacity (kW)	2,801	2,368
Storage tank heating capacity (kW)	2,263	2,126
Annual Total Cost ($)	2,615,183	2,631,256
Carbon Dioxide Emission (gr)	9,089,898,936	9,125,387,007
Primary Energy Consumption (kWh)	40,707,123	40,787,638

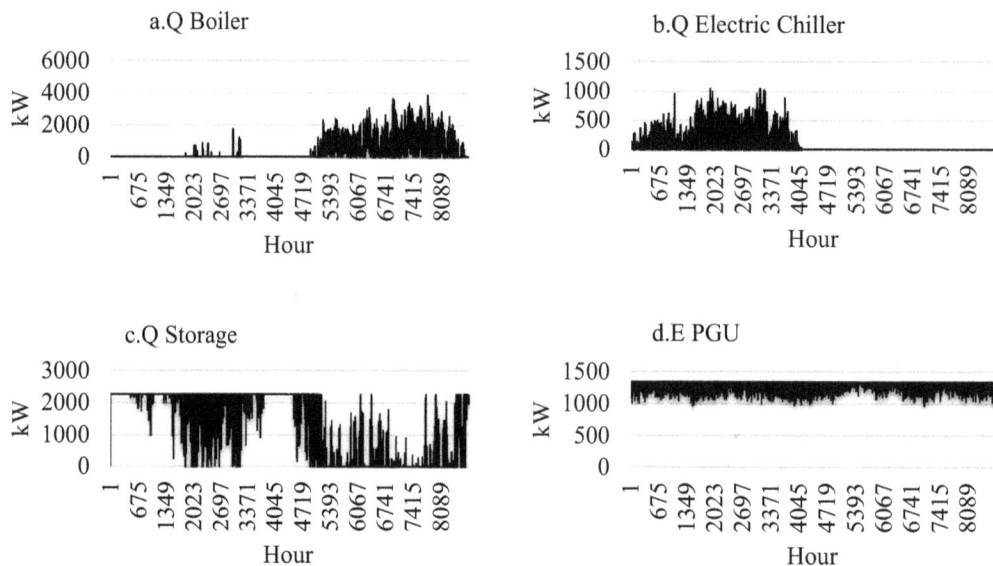

Fig. 7. Performance of system all year long under FEL

6.3. Results for FTL strategy

The obtained results for this strategy are listed in Table 8. The on-off coefficient factor (α) is valued at the lowest limit (0.2). When the PGU is turned off the heat requirement of the buildings has to be provided by the boiler as shown in Fig. 8 d. Q Boiler. The priority in this strategy is with the absorption chiller, thus, cooling load provided by the absorption chiller throughout the year is more than FEL because, 98 % of cooling load is provided by the absorption chiller (c. Q Electric Chiller & b. Q Absorption chiller). The electric chiller is used in 45 periods throughout the year. The obtained results from the FTL strategy are dominated by the results from the FEL strategy.

Table 8

Best design of the system at FTL strategy

Decision Variable	PSO Results	GA Results
Nominal power of prime movers (kW)	2,140	2,665
the on–off coefficient of PGU	0.2	0.2
Boiler heating capacity (kW)	3,840	3,207
Electrical chiller cooling capacity (kW)	1258	817
Absorption chiller cooling capacity (kW)	1,736	2,177
Annual Total Cost ($)	3,183,595	3,198,811
Carbon Dioxide Emission (gr)	15,897,884,200	16,012,517,593
Primary Energy Consumption (kWh)	58,687,631	59,159,430

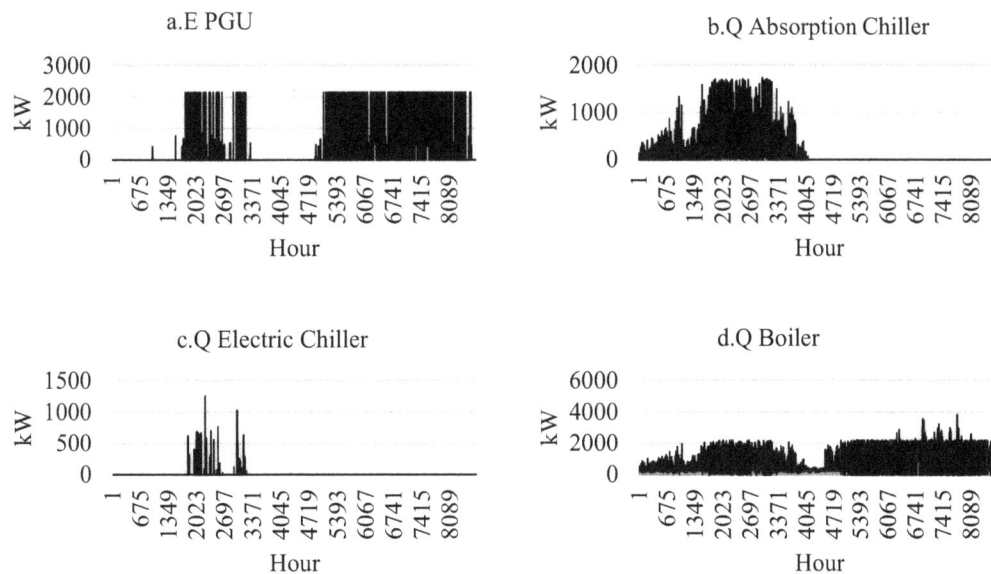

Fig. 8. Performance of system all year long (FTL)

In addition to higher economic costs, FTL strategy cause higher energy consumption and environmental pollution. In 6648 periods PGU is turned off as shown in Fig. 8 a. E PGU and power requirement of the system is supplied through the network which in addition to the cost of buying electricity imposes higher emissions and energy consumption to the system.

6.4. Results for FEL, fixed electric cooling ratio

The results of this strategy, shown in Table 9, indicate that best performance of the system is obtained when electric chiller supplies 24% of cooling demand of the system; the electric cooling to cool load ratio is set to 0.24 in the PSO as shown in Fig. 9 (d. Q Electric Chiller) and to 0.26 in the GA. The results show that this approach incurs less ATC, PEC and CDE than the two previous strategies. 96% of the power requirement of the system is provided via PGU. The total heat requirement is 11,586,092 kWh which is less than FTL (11,924,618 kWh) and more than FEL (11,436,859 kWh).

Table 9

Best design of the system at FEL, fixed electric cooling ratio strategy

Decision Variable	PSO Results	GA Results
Nominal power of prime movers (kW)	1,488	1,507
the on–off coefficient of PGU	0.65	0.65
Boiler heating capacity (kW)	3,325	3,291
Electrical chiller cooling capacity (kW)	685	747
Absorption chiller cooling capacity (kW)	2,214	2,151
Storage tank heating capacity (kW)	1,743	1,735
Electric cooling to cool load ratio	0.24	0.26
Annual Total Cost ($)	2,607,587	2,611,042
Carbon Dioxide Emission (gr)	9,015,246,427	9,017,618,247
Primary Energy Consumption (kWh)	40,111,759	40,136,568

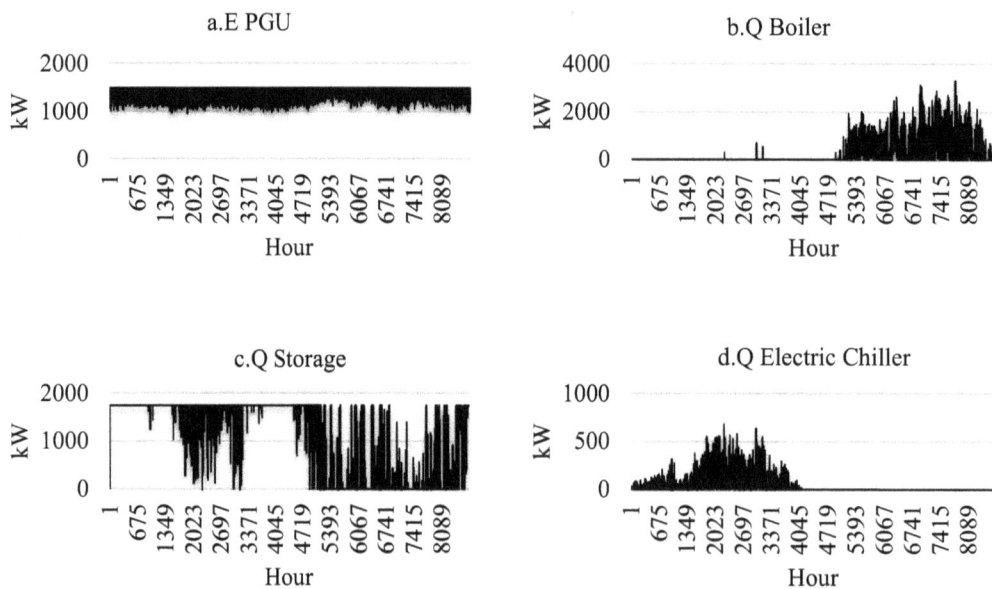

Fig. 9. Performance of system all year long (Fixed ratio of electric chiller cool load to cooling demand)

Heat required in this strategy is between FEL and FTL strategy because under the FEL strategy, the priority is with the electric chiller and heat requirement of the system will decrease. Under FTL and FEL strategies, the priority of the chillers is given so the performance of the chillers is partially predefined while under FEL with fixed electric cooling ratio, the model determines the best ratio of cooling load supplied by chillers; therefore, the results show a better performance. The capacity of the storage tank in this strategy is lower than under the FEL strategy (c. Q Storage). Following this strategy PGU operates in maximum capacity (a. E PGU) in 2974 periods and in 7761 periods operates in partial load above 80% of its capacity.

6.5. Results for FEL multi PGU based on fixed electric cooling ratio

The obtained results show that the selection of several PGUs in this strategy is not efficient and only increases the system costs. The reason is that PGU under the FEL strategy operates in partial load higher than 80% in 7761 periods indicating an efficient use of its capacity; as a result, selection of multiple PGUs does not improve the performance and only increases the capital cost of the system. So if the system is to operate based on FEL strategy, it is suggested to use a single PGU in this case study.

6.6. Results for FTL, Multi PGU based on fixed electric cooling ratio

As shown in Table 10 under FTL strategy using three PGUs is recommended. As shown Fig. 10 b. Q Boiler, a boiler with lower capacity compared to the previous strategies is chosen. This is because the recovered heat from the PGUs in most periods can fulfill the majority of the heat demand. Most of cooling demand is supplied by the absorption chiller which is more efficient and electric cooling to cool load ratio is 0.96. In the peak of the heat demand, during winter and summer, all three units are in active status and operate in their maximum capacity as shown in Fig. 10 (a. E PGU). The costs of FTL strategy with one PGU is more than the FTL strategy with multiple PGUs. It is because at least one PGU is in the active mode with high partial load in 4805 periods, therefore, provided thermal and electrical load covers the demand.

Table 10

The best design of the system under the FTL strategy.

Decision Variable	PSO Results	GA Results
Nominal power of prime movers1 (kW)	634	625
Nominal power of prime movers2 (kW)	634	678
Nominal power of prime movers2 (kW)	1152	1189
the on–off coefficient of PGU	0.71	0.70
Boiler heating capacity (kW)	1,140	1,065
Electrical chiller cooling capacity (kW)	172	156
Absorption chiller cooling capacity (kW)	2,774	2,783
Electric cooling ratio	0.059	0.063
Annual Total Cost ($)	1,828,168	1,829,785
Carbon Dioxide Emission (gr)	11,055,104,613	11,061,070,542
Primary Energy Consumption (kWh)	41,219,129	41,609,366

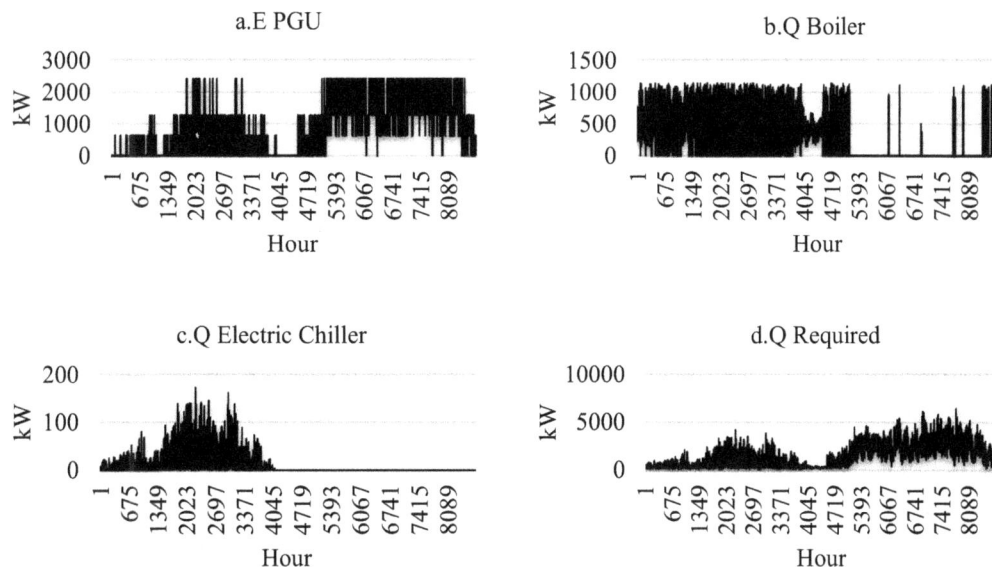

Fig. 10. Performance of system all year long (FTL, Multi PGU)

As a result, in addition to purchasing less electricity from the grid, less thermal load is supplied via auxiliary boiler in compression to the FTL strategy.

6.7.Electricity exchange with the grid

Electricity exchanged with the grid under different strategies is shown in Fig. 11.

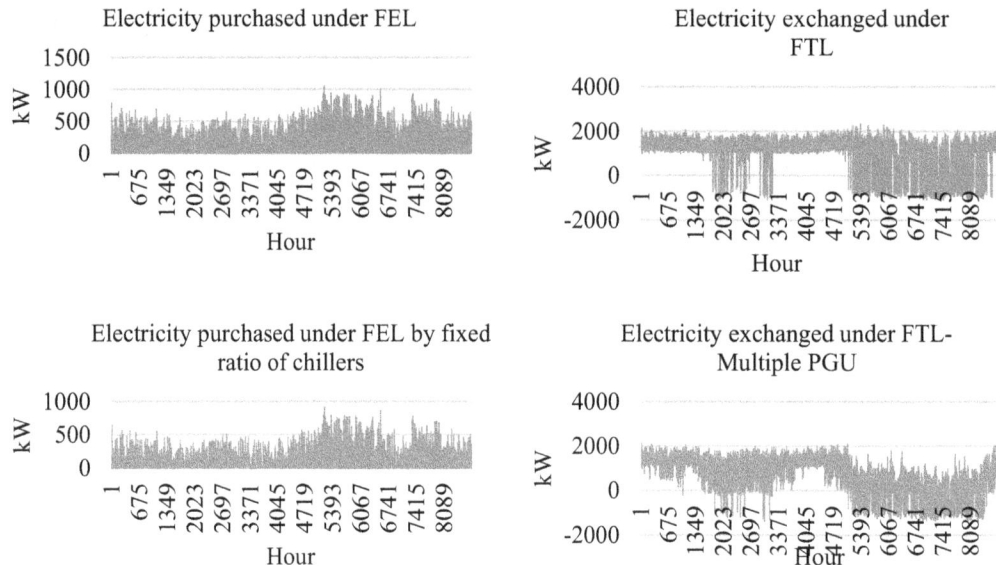

Fig. 11. Electricity exchanged with the grid

The total electricity purchased from the grid in FEL, FTL, FEL with fixed ratio of electric cooling to cool load, and FTL with multiple-PGU are respectively 1,070,688 kWh; 9,717,615 kWh; 516,235 kWh; and 7,081,472 kWh. Electricity sold to the grid in FTL and FTL with Multiple PGU are respectively 978,028 kWh, and 1,026,760. Total electricity purchased form the grid under the FEL strategy with fixed ratio of electric cooling to cool load is less than the other strategies; so if increases in the price of electricity is predicted, this strategy's chance of being chosen by the decision makers increases. If the selling price of the electricity rises, multiple-PGU under FTL results less ATC which can alter the decision makers' decision.

7. Conclusion

A combined cooling, heating and power generation (CCHP) system was optimally designed. The decision variables were the number of prime movers (PGUs), their capacity and operational strategy, backup boiler, storage tank heating, absorption chiller and electric chiller capacity, electric cooling ratio, and the on-off coefficient of the PGU. This combination of the decision variables, along with the various strategies considered in this study, provide a more comprehensive view of the real system and, to the best of authors' knowledge, is presented for the first time. Due to the complexity of the developed model, PSO and GA were used to solve it. PSO showed a better performance in solving the optimisation problem although the difference between PSO and GA was maximum 0.7%.

The results obtained for the case study show that under the FEL strategy, CDE and PEC are significantly reduced in compression to the FTL strategy. Under FTL strategy with one PGU, the PGU is put in the inactive mode in a large number of time. Under FTL strategy with multiple PGU significant reduction in costs was observed but, CDE and PEC were still higher in compression to the FEL strategy. If the approach of decision makers is to reduce the economic costs, it is better to work based on multi-PGU under FTL strategy; and if they rather to reduce CDE and PEC or are seeking to buy the lowest amount of electricity from the grid, system should work based on FEL strategy with fixed ratio of electric cooling to cool load. If selling price of electricity were rising, the system should operate based on multi-PGU under the FTL strategy; and if buying price of electricity were rising, system should operate based on the FEL strategy with fixed ratio of electric cooling to cool load.

As a direction for future research, analysing the potential usage of municipality waste in the presented case study, and generally biomass in other jurisdictions, as the primary source of energy in the CCHP systems is suggested. Another avenue of the future research could be the consideration of the chill storage tanks in the CCHP system and analysing its effect on system's performance. Also, estimating the input parameters such as electricity, cooling and heating demand of the CCHP by design of a system dynamic model or time series prediction is an interesting topic. These methods can be used to predict input data regarding energy consumption. When only the energy consumption data of last years and physical futures of the buildings are used to estimate energy demand of the system, the possible trends in the energy consumption of the system are neglected.

Acknowledgement

We thank Iranian Fuel Conservation Company (IFCO) for assistance with the case study, and Mr Mohammad Babagolzadeh for constructive discussions.

References

Arcuri, P., Florio, G., & Fragiacomo, P. (2007). A mixed integer programming model for optimal design of trigeneration in a hospital complex. *Energy, 32*(8), 1430-1447.

Bahrami, S., & Safe, F. (2013). A financial approach to evaluate an optimized combined cooling, heat and power system. *Energy and Power Engineering, 5*(05), 352.

Cai, W., Wu, Y., Zhong, Y., & Ren, H. (2009). China building energy consumption: situation, challenges and corresponding measures. *Energy Policy, 37*(6), 2054-2059.

Cao, J. (2009). Evaluation of retrofitting gas-fired cooling and heating systems into BCHP using design optimization. *Energy Policy, 37*(6), 2368-2374.

Chen, B.-K., & Hong, C.-C. (1996). Optimum operation for a back-pressure cogeneration system under time-of-use rates. *Power Systems, IEEE Transactions on, 11*(2), 1074-1082.

Črepinšek, M., Liu, S.-H., & Mernik, M. (2013). Exploration and exploitation in evolutionary algorithms: A survey. *ACM Computing Surveys (CSUR), 45*(3), 35.

Eberhart, R. C., & Kennedy, J. (1995). *A new optimizer using particle swarm theory.* Paper presented at the Proceedings of the sixth international symposium on micro machine and human science.

Energy Plus (2014). 8.5.0. . 2016, from www.energyplus.net

Ge, Y., Tassou, S., Chaer, I., & Suguartha, N. (2009). Performance evaluation of a tri-generation system with simulation and experiment. *Applied Energy, 86*(11), 2317-2326.

Geidl, M., & Andersson, G. (2007). Optimal power flow of multiple energy carriers. *Power Systems, IEEE Transactions on, 22*(1), 145-155.

Ghaebi, H., Saidi, M., & Ahmadi, P. (2012). Exergoeconomic optimization of a trigeneration system for heating, cooling and power production purpose based on TRR method and using evolutionary algorithm. *Applied thermal engineering, 36*, 113-125.

Gibson, C. A., Meybodi, M. A., & Behnia, M. (2015). How Carbon Pricing Impacts the Selection and Optimization of a Gas Turbine Combined Heat and Power System: An Australian Perspective. *Journal of Clean Energy Technologies, 3*(1).

International Energy Outlook. (2016). Retrieved 31 August, 2016, from http://www.eia.gov

Iran Energy Efficiency Organization (IEEO-SABA). (2016). Retrieved 31 August, 2016, from http://en.saba.org.ir

Iranian Electricity management. (2016). Retrieved 31 August, 2016, from www.tavanir.org.ir

Iranian Fuel Conservation Organization. (2016). Retrieved 31 August, 2016, from www.ifco.org

Ito, K., Gamou, S., & Yokoyama, R. (1998). Optimal unit sizing of fuel cell cogeneration systems in consideration of performance degradation. *International Journal of Energy Research, 22*(12), 1075-1089.

Karbassi, A., Abduli, M., & Abdollahzadeh, E. M. (2007). Sustainability of energy production and use in Iran. *Energy Policy, 35*(10), 5171-5180.

Liu, M., Shi, Y., & Fang, F. (2013). Optimal power flow and PGU capacity of CCHP systems using a matrix modeling approach. *Applied Energy, 102*, 794-802.

Løken, E. (2007). Use of multicriteria decision analysis methods for energy planning problems. *Renewable and Sustainable Energy Reviews, 11*(7), 1584-1595.

Mago, P., & Chamra, L. (2009). Analysis and optimization of CCHP systems based on energy, economical, and environmental considerations. *Energy and Buildings, 41*(10), 1099-1106.

MATLAB. (2013). 2013. from http://www.mathworks.com/products/matlab/

Mitchell, M. (1998). *An introduction to genetic algorithms*: MIT press.

Pérez-Lombard, L., Ortiz, J., & Pout, C. (2008). A review on buildings energy consumption information. *Energy and Buildings, 40*(3), 394-398.

Piacentino, A., & Cardona, F. (2008). EABOT–energetic analysis as a basis for robust optimization of trigeneration systems by linear programming. *Energy Conversion and Management, 49*(11), 3006-3016.

Sanaye, S., & Hajabdollahi, H. (2013). 4E analysis and Multi-objective optimization of CCHP using MOPSOA. *Proceedings of the Institution of Mechanical Engineers, Part E: Journal of Process Mechanical Engineering*, 0954408912471001.

Sanaye, S., & Hajabdollahi, H. (2015). Thermo-economic optimization of solar CCHP using both genetic and particle swarm algorithms. *Journal of Solar Energy Engineering, 137*(1), 011001.

Sanaye, S., & Khakpaay, N. (2014). Simultaneous use of MRM (maximum rectangle method) and optimization methods in determining nominal capacity of gas engines in CCHP (combined cooling, heating and power) systems. *Energy, 72*, 145-158.

Sanaye, S., Meybodi, M. A., & Shokrollahi, S. (2008). Selecting the prime movers and nominal powers in combined heat and power systems. *Applied thermal engineering, 28*(10), 1177-1188.

Smith, A., Luck, R., & Mago, P. J. (2010). Analysis of a combined cooling, heating, and power system model under different operating strategies with input and model data uncertainty. *Energy and Buildings, 42*(11), 2231-2240.

Tichi, S., Ardehali, M., & Nazari, M. (2010). Examination of energy price policies in Iran for optimal configuration of CHP and CCHP systems based on particle swarm optimization algorithm. *Energy Policy, 38*(10), 6240-6250.

Wang, J.-J., Jing, Y.-Y., Zhang, C.-F., & Zhao, J.-H. (2009). Review on multi-criteria decision analysis aid in sustainable energy decision-making. *Renewable and Sustainable Energy Reviews, 13*(9), 2263-2278.

Wang, J.-J., Xu, Z.-L., Jin, H.-G., Shi, G.-h., Fu, C., & Yang, K. (2014). Design optimization and analysis of a biomass gasification based BCHP system: A case study in Harbin, China. *Renewable Energy, 71*, 572-583.

Wang, J., Jing, Y., Zhang, C., & Zhang, B. (2008). *Distributed combined cooling heating and power system and its development situation in China.* Paper presented at the ASME 2008 2nd International Conference on Energy Sustainability collocated with the Heat Transfer, Fluids Engineering, and 3rd Energy Nanotechnology Conferences.

Wang, J., Zhai, Z. J., Jing, Y., & Zhang, C. (2010). Particle swarm optimization for redundant building cooling heating and power system. *Applied Energy, 87*(12), 3668-3679.

Wang, J., Zhai, Z. J., Jing, Y., Zhang, X., & Zhang, C. (2011). Sensitivity analysis of optimal model on building cooling heating and power system. *Applied Energy, 88*(12), 5143-5152.

Wang, Q., & Fang, F. (2011). *Optimal configuration of CCHP system based on energy, economical, and environmental considerations.* Paper presented at the Intelligent Control and Information Processing (ICICIP), 2011 2nd International Conference on.

Wu, Z., Gu, W., Wang, R., Yuan, X., & Liu, W. (2011). *Economic optimal schedule of CHP microgrid system using chance constrained programming and particle swarm optimization.* Paper presented at the 2011 IEEE Power and Energy Society General Meeting.

Permissions

All chapters in this book were first published in IJIEC, by Growing Science; hereby published with permission under the Creative Commons Attribution License or equivalent. Every chapter published in this book has been scrutinized by our experts. Their significance has been extensively debated. The topics covered herein carry significant findings which will fuel the growth of the discipline. They may even be implemented as practical applications or may be referred to as a beginning point for another development.

The contributors of this book come from diverse backgrounds, making this book a truly international effort. This book will bring forth new frontiers with its revolutionizing research information and detailed analysis of the nascent developments around the world.

We would like to thank all the contributing authors for lending their expertise to make the book truly unique. They have played a crucial role in the development of this book. Without their invaluable contributions this book wouldn't have been possible. They have made vital efforts to compile up to date information on the varied aspects of this subject to make this book a valuable addition to the collection of many professionals and students.

This book was conceptualized with the vision of imparting up-to-date information and advanced data in this field. To ensure the same, a matchless editorial board was set up. Every individual on the board went through rigorous rounds of assessment to prove their worth. After which they invested a large part of their time researching and compiling the most relevant data for our readers.

The editorial board has been involved in producing this book since its inception. They have spent rigorous hours researching and exploring the diverse topics which have resulted in the successful publishing of this book. They have passed on their knowledge of decades through this book. To expedite this challenging task, the publisher supported the team at every step. A small team of assistant editors was also appointed to further simplify the editing procedure and attain best results for the readers.

Apart from the editorial board, the designing team has also invested a significant amount of their time in understanding the subject and creating the most relevant covers. They scrutinized every image to scout for the most suitable representation of the subject and create an appropriate cover for the book.

The publishing team has been an ardent support to the editorial, designing and production team. Their endless efforts to recruit the best for this project, has resulted in the accomplishment of this book. They are a veteran in the field of academics and their pool of knowledge is as vast as their experience in printing. Their expertise and guidance has proved useful at every step. Their uncompromising quality standards have made this book an exceptional effort. Their encouragement from time to time has been an inspiration for everyone.

The publisher and the editorial board hope that this book will prove to be a valuable piece of knowledge for researchers, students, practitioners and scholars across the globe.

List of Contributors

Suresh Nipanikar
Research Scholar, Department of Mechanical Engineering, Dr. Babasaheb Ambedkar Technological University, Lonere-402103, Maharashtra, India

Vikas Sargade
Professor, Department of Mechanical Engineering, Dr. Babasaheb Ambedkar Technological University, Lonere-402103, Maharashtra, India

Ramesh Guttedar
PG Student, Department of Mechanical Engineering, Dr. Babasaheb Ambedkar Technological University, Lonere-402103, Maharashtra, India

Yeison Díaz-Mateus, Bibiana Forero and Gabriel Zambrano-Rey
Industrial Engineering Department, Pontificia Universidad Javeriana, Cra 7 #40-62, Ed. Jose Gabriel Maldonado P.3, Bogotá, Colombia

Héctor López-Ospina
Department of Industrial Engineering, Universidad del Norte, Km. 5 Vía Puerto Colombia, Barranquilla, Colombia

Oussama Zerti, Mohamed Athmane Yallese, Abderrahmen Zerti and Salim Belhadi
Mechanics and Structures Research Laboratory (LMS), Mechanical Engineering Dept., May 8th 1945 University, Guelma 24000, Algeria

Francois Girardin
Laboratoire Vibrations Acoustique, INSA-Lyon, 25 bis avenue Jean Capelle, F-69621 Villeurbanne Cedex, France

Eduyn López-Santana, William Camilo Rodríguez-Vásquez and Germán Méndez-Giraldo
Sistemas Expertos y Simulación (SES), Facultad de Ingeniería, Universidad Distrital Francisco José de Caldas, Bogotá, Colombia

Lakhdar Bouzid
Mechanical Engineering Department, Mechanics and Structures Research Laboratory (LMS), May 8th 1945 University, P.O. Box 401, Guelma 24000, Algeria
University of Larbi Ben M'Hidi, Oum el Bouaghi, Algeria

Sofiane Berkani and Mohamed Athmane Yallese
Mechanical Engineering Department, Mechanics and Structures Research Laboratory (LMS), May 8th 1945 University, P.O. Box 401, Guelma 24000, Algeria

Fronçois Girardin
Laboratoire Vibrations Acoustique, INSA-Lyon, 25 bis avenue Jean Capelle, F-69621 Villeurbanne Cedex, France

Tarek Mabrouki
University of Tunis El Manar, ENIT, Tunis, Tunisia

Pedram Beldar
Department of Industrial and Mechanical Engineering, Qazvin Branch, Islamic Azad University, Qazvin, Iran

Antonio Costa
University of Catania, DICAR, Viale Andrea Doria 6, 95125 Catania, Italy

Robert Matindi, Phil Hobson and Geoff Kent
School of Chemistry, Physics and Mechanical Engineering Science and Engineering Faculty, Brisbane Qld 4001 Australia

Mahmoud Masoud
School of Mathematical Sciences, Queensland University of Technology, 2 George St, Brisbane Qld 4001 Australia

Shi Qiang Liu
School of Economics and Management, Fuzhou University, Fuzhou, 350108, China

Rajiv Kumar
Reseach Scholar, Department of Mechanical Engineering, National Institute of Technology, Kurukshetra, Haryana, India -136119

P.C. Tewari and Dinesh Khanduja
Professor, Department of Mechanical Engineering, National Institute of Technology, Kurukshetra, Haryana, India -136119

Jafar Bagherinejad
Associate Professor, Department of Industrial Engineering, Alzahra University, Tehran, Iran

Mahnaz Shoeib
MSc student of Industrial Engineering, Alzahra University, Tehran, Iran

Masoud Rabbani, Setare Mohammadi and Mahdi Mobini
Department of Industrial Engineering, University of Tehran, Tehran, Iran

Index